Noncompact Problems at the Intersection of Geometry, Analysis, and Topology

Haïm Brezis

Felix Browder

Felix Browder and Haïm Brezis

CONTEMPORARY MATHEMATICS

350

Noncompact Problems at the Intersection of Geometry, Analysis, and Topology

Proceedings of the Brezis-Browder Conference
Noncompact Variational Problems
and General Relativity
October 14–18, 2001
Rutgers, The State University of New Jersey,
New Brunswick, NJ

Abbas Bahri
Sergiu Klainerman
Michael Vogelius
Editors

American Mathematical Society
Providence, Rhode Island

2000 *Mathematics Subject Classification.* Primary 58-06, 53-06.

Library of Congress Cataloging-in-Publication Data

Brezis-Browder Conference, Noncompact Variational Problems and General Relativity (2001: New Brunswick, N.J.)

Noncompact problems at the intersection of geometry, analysis, and topology: proceedings of the Brezis-Browder Conference, Noncompact Variational Problems and General Relativity, October 14–18, 2001, Rutgers, the State University of New Jersey, New Brunswick, NJ / Abbas Bahri, Sergiu Klainerman, Michael Vogelius, editors.

     p. cm.– (Contemporary mathematics, ISSN 0271-4132; 350)

Includes bibliographical references.

ISBN 0-8218-3635-8 (alk. paper)

1. Differential geometry–Congresses.   2. Functional analysis–Congresses.   3. Calculus of variations–Congresses.   I. Brezis, H. (Haim)   II. Browder, Felix E.   III. Bahri, Abbas.   IV. Klainerman, Sergiu, 1950–   V. Vogelius, M. (Michael)   VI. Title.   VII. Contemporary mathematics (American Mathematical Society); v. 350.

QA614.9.B74   2001

516.3′6–dc22

                                             2004047646

Contents

Preface

It is a great pleasure and an honour to write this introduction to the Proceedings of the Brezis-Browder Conference which was held at Rutgers, The State University of New Jersey from October 14th to October 18th 2001.

The Conference had two purposes: it was conceived and meant to honour two great living mathematicians, Haim Brezis and Felix Browder who have had and continue to have, each in his own way and through their intense collaboration, a profound impact on the fields of Partial Differential Equations, Functional Analysis, and Geometry.

This conference was also conceived as a gathering (with appropriate timing and momentum) of mathematicians with interests in non compact variational problems, pseudo-holomorphic curves, singular and smooth solutions to problems admitting a conformal (or some group) invariance, Sobolev spaces on manifolds, and Configuration spaces. In addition S. Klainerman organized a day around Einstein equations and related topics.

The speakers and participants came from all around the world: the U.S., France, Italy, Germany, China, Israel, and from the Third World (Tunisia, Mauritania).

I would like to point out that Haim Brezis and Felix Browder, besides their outstanding contribution to mathematics, have also contributed in an exceptional way to the thriving of the mathematical community. The students of H. Brezis have multiplied throughout countries and continents, in an almost biblical way (for a mathematician). The contribution of Felix Browder to the mathematical community is underlined by the number of conferences and symposia, seminars etc. which he has organized and also by his role as president of the AMS.

Both of them also have the very special distinction that they have engaged, directly and indirectly, in quite successful efforts to promote mathematics in areas of the world where it used to have little impact (e.g. North and Black Africa).

The Conference, with respect to its original design, was a success.

However, it was overshadowed by the tragic events of September 11th 2001, which filled the hearts of all speakers, participants, organizers with a deep sense of sadness and delusion.

After September 11th, all speakers and participants thought of this conference also in a new way, as a collective act of protest, of friendship, of knowledge. It was rededicated to the memory of the victims of September 11th, 2001.

One year and a half later, I think that we have achieved our purpose, which in the meantime had become at least threefold: the one which we have chosen early on, that is to honour Haim and Felix and to discuss noncompact phenomena, Einstein equations etc; the one which we came to choose: to honour the memory of the victims of September 11th, and then also the simpler one of going on with continuous questioning in Science, in a friendly and positive atmosphere, through our talks, papers, seminars, and conferences.

For the organizing Committee,
Abbas Bahri

Contemporary Mathematics
Volume **350**, 2004

Conformal Deformations of Riemannian Metrics via "Critical Point Theory at Infinity" : the Conformally Flat Case with Umbilic Boundary

Mohameden Ould Ahmedou

Dedicated to Haim Brezis and Felix Browder, and to the memory of the victims of **9/11**

ABSTRACT. In this paper we prove that every Riemannian metric on a locally conformally flat manifold with umbilic boundary can be conformally deformed to a scalar flat metric having constant mean curvature. This result can be seen as a generalization to higher dimensions of the well known Riemann mapping Theorem in the plane.

1. Introduction

In [**16**], José F. Escobar raised the following question: Given a compact Riemannian manifold with boundary, when it is conformally equivalent to one that has zero scalar curvature and whose boundary has a constant mean curvature ? This problem can be seen as a "generalization" to higher dimensions of the well known Riemannian mapping Theorem. The later states that an open, simply connected proper subset of the plane is conformally diffeomorphic to the disk. In higher dimensions few regions are conformally diffeomorphic to the ball. However one can still ask whether a domain is conformal to a manifold that resembles the ball into ways : namely, it has zero scalar curvature and its boundary has constant mean curvature. In the above the term "generalization" has to be understood in that sens. The above problem is equivalent to finding a smooth positive solution to the following nonlinear boundary value problem on a Riemannian manifold with boundary (M^n, g), $n \geq 3$:

$$(P) \qquad \begin{cases} -\Delta_g u + \frac{(n-2)}{4(n-1)} R_g u = 0, & u > 0 \quad \text{in } \overset{\circ}{M}; \\ \partial_\nu u + \frac{n-2}{2} h_g u = Q(M, \partial M) u^{\frac{n}{n-2}}, & \text{on } \partial M. \end{cases}$$

where R is the scalar curvature of M, h is the mean curvature of ∂M, ν is the outer normal vector with respect to g and $Q(M, \partial M)$ is a constant whose sign is uniquely determined by the conformal structure. Indeed if $\overline{g} = u^{\frac{4}{n-2}} g$, then the metric $\overline{g}$ has

---

Key words and phrases. Critical trace Sobolev exponent, curvature, conformal invariance, lack of compactness, critical point at infinity .

zero scalar curvature and the boundary has constant mean curvature with respect to $\bar{g}$.

Solutions of equation (P) correspond , up to a multiple constant, to critical points of the following functional J defined on $H^1(M) \setminus \{0\}$

$$(1) \qquad J(u) = \frac{\left(\int_M \left(|\nabla_g u|^2 + \frac{n-2}{4(n-1)} R_g\, u^2 \right) dV_g + \frac{n-2}{2} \int_{\partial M} h_g\, u^2 d\sigma_g \right)^{\frac{n-1}{n-2}}}{\int_{\partial M} |u|^{2\frac{n-1}{n-2}} d\sigma_g}.$$

where dV_g and $d\sigma_g$ denote the Riemannian measure on M and ∂M induced by the metric g.

The regularity of the H^1 solutions of (P) was established by P. Cherrier [13]; and related problems regarding conformal deformations of metrics on manifold with boundary were studied in [1] ,[8] [11] , [14], [19] , [20][21], [22] , [23], [26] , [29] and the references therein.

The exponent $\frac{2(n-1)}{n-2}$ is critical for the Sobolev trace embedding $H^1(M) \to L^q(\partial M)$. This embedding being not compact , the functional J does not satisfy the Palais Smale condition. For this reason standard variational methods cannot be applied to find critical points of J.

Following the original arguments introduced by T. Aubin [2], [3] and R. Schoen [31] to prove Yamabe conjecture on closed manifolds, Escobar proved the existence of a smooth positive solution u of (P) on $(M^n, g), n \geq 3$ for many cases. To state his results we need some preliminaries:

Let H denote the second fondamental form of ∂M in (M, g) with respect to the inner normal. Let us denote the traceless part of the second fundamental form by U that is $U(X, Y) = H(X, Y) - h_g\, g(X, Y)$

DEFINITION 1.1. *A point $q \in \partial M$ is called an* **umbilic point** *if $U = 0$ at q. ∂M is called umbilic if every point of ∂M is umbilic.*

Regarding the above problem Escobar proved the following Theorem [16, 18]:

THEOREM 1.1. *Let (M^n, g) be a compact Riemannian manifold with boundary, $n \geq 3$. Assume that M^n satisfies one of the following conditions:*

: *(i) $n \geq 6$ and M has a nonumbilic point on ∂M*
: *(ii) $n \geq 6$ and M is conformally locally flat with umbilic boundary*
: *(iii) $n = 4, 5$ and ∂M is umbilic*
: *(vi) $n = 3$*

then there exists a smooth metric $u^{\frac{4}{n-2}}\, g$, $u > 0$ on M of zero scalar curvature and constant mean curvature on ∂M.

In his proof Escobar uses strongly an extension of the positive mass Theorem of R. Schoen and S.T. Yau [33], [32] to some type of manifolds with boundary. Such an extension was proved by Escobar in [17]. Besides the proof of T.Aubin and R.Schoen of the Yamabe conjecture, another proof by A. Bahri [5] and A.Bahri and H. Brezis [6] of the same conjecture is available by techniques related to the Theory of critical point at Infinity of A. Bahri [4].

We plan to give a complete positive answer to the above problem based on the topological argument of Bahri-Coron [7], as Bahri and Brezis did for the Yamabe conjecture. In this first part we study the case where the manifold is locally conformally flat with umbilic boundary. Namely we prove the following Theorem

THEOREM 1.2. *Suppose that (M^n, g), $n \geq 3$ is a compact locally conformally flat manifold with umbilic boundary, then equation (P) has a solution.*

Let us observe that while the solution obtained by Escobar is a minimum of J, our solution is in general, a critical point of J of higher Morse index, more precisely we have the following characterization of solutions obtained by Bahri-Coron existence scheme(see [**12**]) :

THEOREM 1.3. *The solution u obtained in Theorem 1.2 satisfies, for some integer p_0:*

- : *(i) $p_0^{\frac{1}{n-2}} S \leq J(u) \leq (p_0 + 1)^{\frac{1}{n-2}} S$*
- : *(ii) $ind(J, u) \leq (p_0 + 1)(n - 1) + p_0$, $ind(J, u) + dimker d^2 J(u) \geq p_0(n - 1) + p_0$*
- : *(iii) u induces some difference of topology at the level $p_0^{\frac{1}{n-2}} S$.*

where $ind(J, u)$ is the Morse index of J at u, and $S = 2^{1-n}\omega_{n-1}$, where ω_{n-1} is the volume of the $n - 1$ dimensional unit sphere. Moreover, if (P) has only nondegenerate solutions, $ind(J, u) = (n - 1)p_0 + p_0$.

The remainder of the paper is organized as follows: in section 2 we construct some "almost solutions" which are solutions of the "problem at Infinity". In section 3 we collect some standard results regarding the description of the lack of compactness and some local deformation Lemma. In section 4 we perform an expansion of J at 'Infinity' and we give the proof of Theorem 1.2 in section 5. Lastly we devote the appendix to establish some technical Lemma and to recall some well known results.

Acknowledgements

This paper is dedicated to Prof. H. Brezis and F. Browder and to the memory of the Victims of **9/11**. The author is indebted to Pr. Abbas Bahri for teaching him his Theory of critical point at Infinity and he is gratefull to Pr. Antonio Ambrosetti for his interest in his work and his constant support.

2. Construction of "almost solutions"

In this paper we assume that (M^n, g) is a compact Riemannian manifold with boundary and dimension $n \geq 3$. Let R_{pq} and $R = g^{pq} R_{pq}$ be the Ricci curvature and the scalar curvature, respectively; let h_{ij} and $h = \frac{1}{n-1} g^{ij} h_{ij}$ be the second fundamental form of the boundary of M, ∂M and the mean curvature , respectively. Let $\tilde{g} = u^{\frac{4}{n-2}} g$ be a metric conformally related to g. We denote by a tilde all quantities computed with respect to the metric $\tilde{g}$. The transformation law for the scalar curvature is

$$(2) \qquad \tilde{R} = \frac{4(n - 1)}{n - 2} \frac{Lu}{u^{\frac{n+2}{n-2}}}$$

where L is the conformal Laplacian $L = \Delta - \frac{n-2}{4(n-1)} R$ on M; while the transformation law for the mean curvature is

$$(3) \qquad \tilde{h} = \frac{2}{n - 2} \frac{Bu}{u^{\frac{n}{n-2}}}$$

where B is the boundary operator $B = \frac{\partial}{\partial \nu} + \frac{n-2}{2} h$ on ∂M.

Consider now the following eigenvalue problem on (M,g) :

$$(E) \qquad \begin{cases} L_g u = \lambda\, u & \text{on } \mathring{M} \\ B_g u = 0 & \text{on } \partial M \end{cases}$$

Let λ_1 the first eigenvalue of (E).

DEFINITION 2.1. *We say that a manifold M is of positive(negative, zero) type if $\lambda_1 > 0(< 0, = 0)$.*

As it is well known the existence problem is easy when the manifold is of negative or zero type, so we treat only in this paper the case of manifold of positive type.

Now we construct some almost solutions of (P), which will play a central role in the description of the lack of compactness.

Let f_1 denote a positive eigenfunction corresponding to the first eigenvalue of (E) , and consider $g_1 = (f_1)^{\frac{4}{n-2}} g$ then , according to (2) and (3) we have: $R_{g_1} > 0$ and $h_{g_1} = 0$ on ∂M. We can work with g_1 instead of g, but for simplicity we still denote it by g. Let $a \in \partial M$; since M is a compact locally conformally flat manifold one can find a neighborhood of a, $\mathcal{U}(a) \supset B_\rho^M(a)$, $\rho > 0$ uniform and a conformal diffeomorphism φ which maps $B_\rho^M(a)$ into $\mathbb{R}^n$ with $\varphi(0) = a$. Therefore, denoting g_0 the flat metric on $\mathbb{R}^n$, there exist a positive function u_a such that $\varphi^*(g_0) = u_a^{\frac{4}{n-2}} g$. Since the boundary is umbilic, $\varphi(\partial M \cap B_\rho^M(a))$ has to be a piece of sphere or a piece of a hyperplane (See [**34**]) and since spheres and hyperplanes are locally conformal to each other, we can assume without loss of generality that $\partial B_2^+(0) \cap \partial \mathbb{R}^n_+ \subset \varphi(\partial M \cap B_\rho^M(a))$ and $\varphi(\mathring{M} \cap B_\rho^M(a)) \subset \mathbb{R}^n_+$. Since $\partial B_2^+(0) \cap \partial \mathbb{R}^n_+$ has zero mean curvature in $\overline{B_2^+}$, we deduce from (3) that $\frac{\partial u_a}{\partial \nu} = 0$ on $\partial M \cap B_\rho^M(a)$. We extend u_a to be a smooth positive function on M such that $\frac{\partial u_a}{\partial \nu} = 0$ on ∂M and $u_a = 0$ on $M \setminus B_{2\rho}^M(a)$. Consider now the conformal metric $\overline{g_0} = u^{\frac{4}{n-2}} g$, then $\overline{g_0}$ has the property that $h_{\overline{g_0}} = 0$ and it is Euclidean in $B_\rho^M(a)$. Moreover this metric can be chosen to depend smoothly on a (see [**5**]).

For $a \in \partial M$, define the function:

$$\delta_{a,\lambda}(y) = \bar{c} \, \frac{\lambda^{\frac{n-2}{2}}}{\left((1 + \lambda x^n)^2 + \lambda^2 |x'|^2\right)^{\frac{n-2}{2}}}$$

where $(x', x^n) = \varphi(y)$, and $\bar{c}$ is chosen such that $\delta_{a,\lambda}$ satisfies the following equation

$$\begin{cases} -\Delta_{\overline{g}_0} u = 0, & \text{in } B_\rho^M \cap \mathring{M}; \\ \partial_\nu u = \delta_{a,\lambda}^{\frac{n}{n-2}}, & \text{on } B_\rho^M \cap \partial M \end{cases}$$

Set $\hat{\delta}_{a,\lambda} = \omega_a\, u_a\, \delta_{a,\lambda}$ where ω_a is a cutoff function $\omega_a = 1$ on $B_\rho^M(a)$ and $\omega_a = 0$ on $M \setminus B_{2\rho}^M$.

We define now a familly of almost solutions $\varphi_{a,\lambda}$ to be the unique solution of

$$\begin{cases} -L_g u = 0, & \text{in } \mathring{M}; \\ B_g u = \hat{\delta}_{a,\lambda}^{\frac{n}{n-2}}, & \text{on } \partial M \end{cases}$$

Let us recall that the operators L_g and B_g are conformally invariant under the conformal change of metrics, namely we have:

LEMMA 2.1. [**16**]
Let $\psi \in C^2(B_\rho(a))$, we have

$$L_g(u_a\psi) = u_a^{\frac{n+2}{n-2}} L_{\overline{g}_0}(\psi)$$

and

$$B_g(u_a\psi) = u_a^{\frac{n}{n-2}} B_{\overline{g}_0}(\psi)$$

In the remainder of this section we establish some properties of our almost solutions $\varphi_{a,\lambda}$.

LEMMA 2.2. *There are two positive constants C and B , such that for all $a \in \partial M$ and $\lambda \geq B$, we have*

$$\left|\varphi_{a,\lambda} - \hat{\delta}_{a,\lambda}\right|_\infty \leq \frac{C}{\lambda^{\frac{n-2}{2}}}$$

PROOF.
Let $H_{a,\lambda} = \lambda^{\frac{n-2}{2}}(\varphi_{a,\lambda} - \hat{\delta}_{a,\lambda})$, we have

$$L_g H_{a,\lambda} = \lambda^{\frac{n-2}{2}} \; L_g\left(\omega_a u_a \delta_{a,\lambda}\right) = \lambda^{\frac{n-2}{2}} \; u_a^{\frac{n+2}{n-2}} \; L_g\left(\omega_a \delta_{a,\lambda}\right)$$

Since on B_ρ , $\omega_a = 0$, we deduce that on B_ρ we have $L_g H_{a,\lambda} = 0$, whereas on $M \setminus B_\rho$ there holds $L_g H_{a,\lambda} \leq C$.

From another part

$$B_g H_{a,\lambda} = \lambda^{\frac{n-2}{2}} \; \left[B_g\,\varphi_{a,\lambda} - B_g\left(\omega_a u_a \delta_{a,\lambda}\right)\right] = \lambda^{\frac{n-2}{2}} \left[\hat{\delta}_{a,\lambda} \; - u_a^{\frac{n}{n-2}} \; B_g\left(\omega_a \delta_{a,\lambda}\right)\right]$$

on $B_\rho(a) \cap \partial M$, $\omega_a = 1$, therefore $B_g H_{a,\lambda} = 0$, while on $M \setminus B_\rho$ there holds $B_g H_{a,\lambda} \leq C$. Thus our Lemma follows from Lemma 6.3 quoted in the appendix.

∎

LEMMA 2.3. *There are two positive constants C and B , such that for all $a \in \partial M$ and $\lambda \geq B$, we have*

$$\varphi_{a,\lambda} \geq \frac{C}{\lambda^{\frac{n-2}{2}}}$$

PROOF.
Using Lemma 2.2, we know that if $\rho_1 < \rho$ is chosen small enough, independent of λ, the following inequality holds on $B(a, \rho_1)$

$$\varphi_{a,\lambda} \geq \hat{\delta}_{a,\lambda} - \frac{C}{\lambda^{\frac{n-2}{2}}} \geq \frac{C}{\lambda^{\frac{n-2}{2}}}$$

Let $\Sigma_1 = \partial B(a, \rho) \cap \overset{\circ}{M}$ and $\Sigma_2 = \partial M \setminus \Sigma_1$.
Then we have

$$(4) \qquad \begin{cases} L_g(\varphi_{a,\lambda} - \frac{C}{\lambda^{\frac{n-2}{2}}}) & \leq 0 \quad \text{in } \overset{\circ}{M} \\ \varphi_{a,\lambda} - \frac{C}{\lambda^{\frac{n-2}{2}}} & \geq 0, \qquad \text{on } \Sigma_1 \\ \frac{\partial}{\partial \nu}(\varphi_{a,\lambda} - \frac{C}{\lambda^{\frac{n-2}{2}}}) & \geq 0 \quad \text{on } \Sigma_2 \end{cases}$$

Then by the hopf maximum principle, we deduce from (4) that

$$\varphi_{a,\lambda} \geq \frac{C}{\lambda^{\frac{n-2}{2}}} \quad \text{for } x \in M$$

∎

LEMMA 2.4. *Let $\theta > 0$ be given. There are positive constants C and B, such that the following estimates hold, provided $\lambda \geq B$*

: (i)

$$\left| \int_{\partial M} B_g\, \varphi_{a,\lambda}\, \varphi_{a,\lambda} d\sigma_g \;-\; \bar{c}^{\frac{2(n-1)}{n-2}} \int_{\mathbb{R}^{n-1}} \frac{dx}{(1+|x|^2)^{n-1}} \right| \;\leq\; \frac{C}{\lambda^{n-2}} \quad \textit{for } a \in \partial M$$

: (ii)

$$\left| \int_{\partial M} \varphi_{a,\lambda}^{\frac{2(n-1)}{n-2}} d\sigma_g \;-\; \bar{c}^{\frac{2(n-1)}{n-2}} \int_{\mathbb{R}^{n-1}} \frac{dx}{(1+|x|^2)^{n-1}} \right| \;\leq\; \frac{C}{\lambda^{n-2}} \quad \textit{for } a \in \partial M$$

: (iii)

$$\int_{\partial M} \varphi_{a_1,\lambda}^{\frac{n}{n-2}}\, \varphi_{a_2,\lambda} d\sigma_g \;\geq\; \frac{C}{\lambda^{n-2}} \quad \textit{for } a_1, a_2 \in \partial M$$

: (vi)

$$\int_{\partial M} B_g\, \varphi_{a,\lambda}\, \varphi_{a,\lambda} d\sigma_g \;\leq\; (1+\theta) \int_{\partial M} \varphi_{a_1,\lambda}^{\frac{n}{n-2}}\, \varphi_{a_2,\lambda} d\sigma_g$$

PROOF.

Proof of (i)

From the definition of $\varphi_{a,\lambda}$, we derive:

$$\int_{\partial M} B_g\, \varphi_{a,\lambda}\, \varphi_{a,\lambda} d\sigma_g = \int_{\partial M} (\omega_a\, \delta_{a,\lambda} u_a)^{\frac{n}{n-2}}\, \varphi_{a,\lambda} d\sigma_g$$

Using Lemma 2.2, we deduce:

$$(5) \quad \int_{\partial M} B_g\, \varphi_{a,\lambda}\, \varphi_{a,\lambda} d\sigma_g$$

$$= \int_{\partial M \cap B_{2\rho}} (\omega_a\, \delta_{a,\lambda} u_a)^{\frac{2(n-1)}{n-2}} d\sigma_g + O(\frac{1}{\lambda^{n-2}}) \int_{\partial M \cap B_{2\rho}} (\omega_a\, \delta_{a,\lambda} u_a)^{\frac{n}{n-2}} d\sigma_g$$

$$= \int_{\partial M \cap B_\rho} (\delta_{a,\lambda})^{\frac{2(n-1)}{n-2}} dv + O(\frac{1}{\lambda^{n-2}}) \int_{\partial M \cap B_{2\rho}} (\omega_a\, \delta_{a,\lambda} u_a)^{\frac{n}{n-2}} dv_{g_0} + O(\lambda^{n-1})$$

$$= \bar{c}^{\frac{2(n-1)}{n-2}} \int_{\mathbb{R}^{n-1}} \frac{dx}{(1+|x|^2)^{n-1}} + O(\frac{1}{\lambda^{n-2}}).$$

The proof of (ii) is essentially reduced, up to minor differences to the same computations involved in the proof of (ii).

Proof of (iii)

From Lemma 2.3 we deduce

$$\int_{\partial M} \varphi_{a_1,\lambda}^{\frac{n}{n-2}}\, \varphi_{a_2,\lambda} d\sigma_g \geq \frac{1}{C\lambda^{n-2}} \int_{\partial M} \varphi_{a_1,\lambda}^{\frac{n}{n-2}} d\sigma_g$$

Then from Lemma 2.2 and Lemma 2.3 we derive

$$\int_{\partial M} \varphi_{a_1,\lambda}^{\frac{n}{n-2}}\, \varphi_{a_2,\lambda} d\sigma_g \;\geq\; \frac{1}{C\lambda^{n-2}} \int_{B_\rho(a) \cap \partial M} \hat{\delta}_{a_1,\lambda}^{\frac{n}{n-2}} d\sigma_g$$

$$\geq\; \frac{1}{C\lambda^{n-2}} \int_{B_\rho(a) \cap \partial M} \delta_{a_1,\lambda}^{\frac{n}{n-2}} d\sigma_{g_0} = O(\frac{1}{\lambda^{n-2}})$$

Proof of (vi)

$$\int_{\partial M} B_g \varphi_{a_1,\lambda}\, \varphi_{a_2,\lambda} dv_g$$

$$= \int_{\partial M} \hat{\delta}_{a_1,\lambda}^{\frac{n}{n-2}}\, \varphi_{a_2,\lambda} dv_g$$

$$= \int_{\partial M \cap B_\rho} \varphi_{a_1,\lambda}^{\frac{n}{n-2}} \varphi_{a_2,\lambda} + O(\frac{1}{\lambda^{\frac{n}{2}}}) \int_{\partial M \backslash B_\rho} \varphi_{a_2,\lambda}$$

$$= \int_{\partial M} \varphi_{a_1,\lambda}^{\frac{n}{n-2}}\, \varphi_{a_2,\lambda} d\sigma_g + O(\frac{1}{\lambda^{\frac{n-2}{2}}}) \int_{\partial M} \hat{\delta}_{a_1,\lambda}^{\frac{2}{n-2}}\, \varphi_{a_2,\lambda} d\sigma_g + O(\frac{1}{\lambda^{n-1}})$$

$$= \int_{\partial M} \varphi_{a_1,\lambda}^{\frac{n}{n-2}}\, \varphi_{a_2,\lambda} dv_g\, \varphi_{a_2,\lambda} + O(\lambda^{\frac{n-2}{2}}) \int_{\partial M} \delta_{a_1,\lambda}^{\frac{2}{n-2}}\, \varphi_{a_2,\lambda} d\sigma_{\overline{g}_0}$$

$$= \int_{\partial M} \varphi_{a_1,\lambda}^{\frac{n}{n-2}}\, \varphi_{a_2,\lambda} dv_g + O(\frac{1}{\lambda^{\frac{n-2}{2}}}) \int_{\partial M} \delta_{a_1,\lambda}^{\frac{2}{n-2}}\, \delta_{a_2,\lambda} d\sigma_{\overline{g}_0} + O(\frac{1}{\lambda^{n-1}})$$

Now from Lemma 6.1 in the Appendix we deduce :

$$O(\frac{1}{\lambda^{\frac{n-2}{2}}}) \int_{\partial M \cap B_\rho(a_1)} \delta_{a_1,\lambda}^{\frac{n}{n-2}}\, \delta_{a_2,\lambda} d\sigma_{\overline{g}_0} = o\left(\int_{\partial M} \delta_{a_1,\lambda}^{\frac{n}{n-2}}\, \delta_{a_2,\lambda} d\sigma_{\overline{g}_0}\right)$$

Therefore using (iii) we have

$$\int_{\partial M} B_g \varphi_{a_1,\lambda}\, \varphi_{a_2,\lambda} dv_g = \int_{\partial M} \varphi_{a_1,\lambda}^{\frac{n}{n-2}}\, \varphi_{a_2,\lambda} dv_g (1 + o(1))$$

The proof of (vi) and the proof of Lemma 2.4 are thereby completed. ∎

3. Some standard facts

We recall that solutions of Problem (P) arises , up to a constant, as critical points of the functional J is defined by

$$J(u) = \left(\int_M -L_g u\, u\, dv_g + \int_{\partial M} B_g u\, u\, d\sigma_g\right)^{\frac{n-1}{n-2}} \left(\int_{\partial M} u^{\frac{2(n-1)}{n-2}}\right)^{-1}$$

where u belongs to Σ^+ defined as follows:

$$\Sigma^+ = \{u \in H^1(M), u \geq 0, \|u\| = 1\}$$

Let us observe that Σ^+ is invariant by the flow of $-\partial J$.

The functional J is known to not satisfy Palais Smale condition(PS for short) , which leads to the failure of classical existence mecanism. In order to describe this failure we need some notation.

For $\varepsilon > 0$ and $p \geq 1$, let

$$V(p, \varepsilon) =$$

$$\left\{ u \in \Sigma^+ \text{such that } \exists (a_1, \cdots, a_p) \in (\partial M)^p \text{ and } \exists (\lambda_1, \cdots, \lambda_p) \in (\mathbb{R}_+^*)^p \text{such that} \right.$$

$$\left. \left\| u - \frac{\sum_{i=1}^p \varphi_{a_i,\lambda_i}}{\|\sum_{i=1}^p \varphi_{a_i,\lambda_i}\|} \right\| < \varepsilon, \text{ with } \lambda_i \geq \frac{1}{\varepsilon} \text{ and } \varepsilon_{ij} < \varepsilon \right\}$$

where $\varepsilon_{ij} = \left(\dfrac{1}{\frac{\lambda_i}{\lambda_j} + \frac{\lambda_j}{\lambda_i} + \lambda_i \lambda_j\, d(a_i,a_j)^2}\right)^{\frac{n-2}{2}}$ and d denotes the geosedic distance.

If u is a function in $V(p,\varepsilon)$, one can find an optimal representation, arguing as in Proposition 7 of [**7**] , namely we have:

LEMMA 3.1. *For every $p \geq 1$, there exists $\varepsilon > 0$ such that $\forall u \in V(p,\varepsilon)$ the minimization problem*

$$\inf_{\alpha_i,b_i,\mu_i} \left\| u - \sum_{i=1}^{p} \alpha_i \varphi_{b_i,\mu_i} \right\|$$

has a unique solution, up to permutation on the set of indices $\{1,\cdots,p\}$

At this point we introduce the following notations:

Let S be defined as $S = \left(\int_{\mathbb{R}^{n-1}} \frac{1}{(1+|x|^2)^{n-1}} \right)^{\frac{1}{n-2}}$, $b_p = p^{\frac{1}{n-2}} S$ and

$$W_p = \{ u \in \Sigma^+, \text{such that } J(u) < b_{p+1} \}$$

We are ready now to state the characterization of the Palais Smale sequences failing the P.S condition.

PROPOSITION 3.1. *Under the assumption that (P) has no solution, let $u_k \subset \Sigma^+$ be a sequence satisfying $J(u_k) \to c$, a positive number and $\partial J(u_k) \to 0$. There exist an integer $p \geq 1$ and a sequence $(\varepsilon_k)_k$ such that $u_k \in V(p,\varepsilon)$. Conversely, let $p \in \mathbb{N}^+$, let ε_k be a positive sequence with $\lim_{k \to +\infty} \varepsilon_k = 0$ and let $u_k \in V(p,\varepsilon)$ then $\partial J(u_k) \to 0$ and $J(u_k) \to b_p$.*

PROOF.   The proof of this Proposition is by now standard, taking into account the uniqueness result of Li-Zhu [**28**], see also [**15**] and using the Liouville Theorem to rule out the possibility of interior blow up. ∎.

We also have the following local deformation Lemma , whose proof is similar to the proof of Lemma 17 in [**6**] :

LEMMA 3.2. *Under the asumption that (P) has no solution, for $\varepsilon > 0$, the pair (W_p, W_{p-1}) retracts by deformation onto the pair $(W_{p-1} \cup A_p, W_{p-1})$ where $A_p \subset V(p,\varepsilon)$.*

4. Expansion of the functional near its potential critical point at infinity

This section is devoted to an asymptotic expansion of J in the neighborhood of its potential critical points at infinity, that is in some $V(p,\varepsilon)$. This expansion displays the fact that when the number of the bubbles are large enough, their interaction increases to force their energy to be under their critical level.Such a fact is a key point in the topological argument.

LEMMA 4.1. *There holds:*

: *(i) For every $p \in \mathbb{N}^*$ and every $\varepsilon_1 > 0$, there exists $\lambda_p = \lambda(p,\varepsilon_1)$ such that for any $(\alpha_1,\cdots,\alpha_p)$ satisfying $\alpha_i \geq 0, \sum_{i=1}^{p} \alpha_i = 1$, for any $(x_1,\cdots,x_p) \in (\partial M)^p$ for any $\lambda \geq \lambda_p$, we have:*

$$\text{If } \sum_{i \neq j} \int_{\partial M} \varphi_{a_i,\lambda}^{\frac{n}{n-2}} \varphi_{a_j,\lambda} \geq \varepsilon_1 \quad \text{then} \quad J(\sum_{i=1}^{p} \alpha_i \varphi_{a_i,\lambda}) \leq p^{\frac{1}{n-2}} S.$$

: *(ii) There exist $0 < \theta_0 < 1$, $C > 0$, and $\overline{\varepsilon_1} \geq \varepsilon_1$, such that for any $p \in \mathbb{N}^*$, for any $(\alpha_1,\cdots,\alpha_p)$ satisfying $\alpha_i \geq 0$, $\sum_{i=1}^{p} \alpha_i = 1$, $\frac{\alpha_i}{\alpha_j} \geq \theta_0$, for* *any $(a_1, \cdots, a_p) \in (\partial M)^p$, for any $\lambda \geq 1$, the following inequality holds*

$$J(\sum_{i=1}^{p} \alpha_i \varphi_{\alpha_i,\lambda}) \leq p^{\frac{1}{n-2}} S \left(1 + O(\frac{1}{\lambda^{n-2}}) + \frac{(p+1)C}{\lambda^{n-2}} \right)$$

If

$$\sum_{i \neq j} \int_{\partial M} \varphi_{a_i,\lambda}^{\frac{n}{n-2}} \varphi_{a_j,\lambda} \leq \bar{\varepsilon}_1$$

$\vdots$ *(iii) If we drop in (ii) the condition $\frac{\alpha_i}{\alpha_j} \geq \theta_0$, then the following weaker inequality still holds:*

$$J(\sum_{i=1}^{p} \alpha_i \varphi_{\alpha_i,\lambda}) \leq p^{\frac{1}{n-2}} S(1 + \frac{1}{C\lambda^{n-2}} + \frac{1}{C} \sum_{i \neq j} \int_{\partial M} \varphi_{a_i,\lambda}^{\frac{n}{n-2}} \varphi_{a_j,\lambda}$$

PROOF.

Let

$$\begin{cases} b_i = \dfrac{\alpha_i \hat{\delta}_{a_i,\lambda}}{\sum_{j=1}^{p} \alpha_j \hat{\delta}_{a_j,\lambda}} & \text{if } \hat{\delta}_{a_i,\lambda} \neq 0 \\ b_i = 1 & \text{if } \hat{\delta}_{a_i,\lambda} = 0 \end{cases}$$

then Lemma B2 of [**7**], which extends to our functional, implies that

$$J(\sum_{i=1}^{p} \varphi_{a_i,\lambda}) \leq \left(\frac{\int_{\partial M} (\sum_{i=1}^{p} \hat{\delta}_{a_i,\lambda})^{\frac{2(n-1)}{n-2}}}{\int_{\partial M} (\sum_{i=1}^{p} \varphi_{a_i,\lambda})^{\frac{2(n-1)}{n-2}}} \right)^{\frac{1}{2}} \left(\sum_{i=1}^{p} b_i \int_{\partial M} \hat{\delta}_{a_i,\lambda}^{\frac{2(n-1)}{n-2}} \right)^{\frac{1}{n-2}}$$

Using Lemma 2.2, and $\int_{\partial M} \hat{\delta}_i^{\frac{n}{n-2}} dv_g = O(\frac{1}{\lambda^{\frac{n-2}{2}}})$, we derive easily

$$J(\sum_{i=1}^{p} \varphi_{a_i,\lambda}) \leq (1 + O(\frac{1}{\lambda^{n-2}})) \left(\sum_{i=1}^{p} b_i \int_{\partial M} \hat{\delta}_{a_i,\lambda}^{\frac{2(n-1)}{n-2}} \right)^{\frac{1}{n-2}}$$

Thus we obtain

$$J(\sum_{i=1}^{p} \varphi_{a_i,\lambda}) \leq (1 + O(\frac{1}{\lambda^{n-2}})) \left((p-1)S^{n-2} + \int_{\partial M} \frac{\alpha_1 \hat{\delta}_{a_1,\lambda}}{\alpha_1 \hat{\delta}_{a_1,\lambda} + \alpha_2 \hat{\delta}_{a_2,\lambda}} \hat{\delta}_{a_1,\lambda}^{\frac{2(n-1)}{n-2}} \right)^{\frac{1}{n-2}}$$

We may assume without loss of generality that

$\vdots$ (i) $\frac{\alpha_1}{\alpha_2} \leq 1$

$\vdots$ (ii)

$$\int_{\partial M} (\varphi_{a_1,\lambda}^{\frac{n}{n-2}} \varphi_{a_2,\lambda} + \varphi_{a_1,\lambda} \varphi_{a_2,\lambda}^{\frac{n}{n-2}}) = \sup_{i \neq j} \int_{\partial M} \varphi_{a_i,\lambda}^{\frac{n}{n-2}} \varphi_{a_j,\lambda} + \varphi_{a_i,\lambda} \varphi_{a_j,\lambda}^{\frac{n}{n-2}}$$

We then have using that $\int_{\partial M \backslash B_{\frac{\varrho}{2}}(a_1)} \hat{\delta}_{a_1,\lambda}^{\frac{2(n-1)}{n-2}} dv_g = O(\frac{1}{\lambda^{n-2}})$

$$J(\sum_{i=1}^{p} \varphi_{a_i,\lambda}) \leq (1 + O(\frac{1}{\lambda^{n-2}})) \left((p-1)S^{n-2} + \int_{\partial M} \frac{\alpha_1 \hat{\delta}_{a_1,\lambda}}{\alpha_1 \hat{\delta}_{a_1,\lambda} + \alpha_2 \hat{\delta}_{a_2,\lambda}} \hat{\delta}_{a_1,\lambda}^{\frac{2(n-1)}{n-2}} \right)^{\frac{1}{n-2}}$$

The continuity of u_y with respect to y implies the existence of $\eta > 0$ such that

$$\frac{1}{2} \leq \frac{u_{a_1}}{u_{a_2}} \leq 2 \quad \text{if} \quad x_2 \in B_\eta(a_1)$$

Thus if $d(a_2, a_2) \leq \eta$, we have

$$
(6) \quad J(\sum_{i=1}^{p} \varphi_{a_i,\lambda})
$$

$$
\leq \; (1 + O(\frac{1}{\lambda^{n-2}})) \left((p-1)S^{n-2} + \int_{\partial M} \frac{\alpha_1 \hat{\delta}_{a_1,\lambda}}{\alpha_1 \hat{\delta}_{a_1,\lambda} + \alpha_2 \hat{\delta}_{a_2,\lambda}} \hat{\delta}_{a_1,\lambda}^{\frac{2(n-1)}{n-2}} \right)^{\frac{1}{n-2}}
$$

$$
\leq \; (1 + O(\frac{1}{\lambda^{n-2}})) \left((p-1)S^{n-2} - \int_{\partial M \cap B_{a_1}(\rho)} \frac{\alpha_1 \delta_{a_1,\lambda}}{\alpha_1 \hat{\delta}_{a_1,\lambda} + \delta_{a_2,\lambda}} \hat{\delta}_{a_1,\lambda}^{\frac{2(n-1)}{n-2}} \right)^{\frac{1}{n-2}} .
$$

At this point we state the following Claim which proof is postponed until the end of this section.

Claim : There exists ε_0 small enough, and a positive constant $\overline{C}'$ such that

$$
\int_{\partial M} \frac{\delta_{a_2,\lambda}}{2\delta_{a_2,\lambda} + \delta_{a_2,\lambda}} \delta_{a_1,\lambda}^{\frac{2(n-1)}{n-2}} \, dv_{\overline{g}_0} \geq \overline{C}' \varepsilon_0 \int_{\partial M} \varphi_{a_1,\lambda}^{\frac{n}{n-2}} \varphi_{a_2,\lambda} + \varphi_{a_1,\lambda} \varphi_{a_2,\lambda}^{\frac{n}{n-2}} \, dv_{\overline{g}_0}
$$

From another part , by assumption, we know that

$$
\sum_{i \neq j} \int_{\partial M} \left(\varphi_{a_i,\lambda}^{\frac{n}{n-2}} \varphi_{a_j,\lambda} dv_{\overline{g}} \right) \leq p^2 \int_{\partial M} \left(\varphi_{a_1,\lambda}^{\frac{n}{n-2}} \varphi_{a_2,\lambda} + \varphi_{a_2,\lambda}^{\frac{n}{n-2}} \varphi_{a_1,\lambda} \right) dv_{\overline{g}}
$$

Thus we derive from the above claim and (6) , that:

$$
(7) \quad J(\sum_{i=1}^{p} \alpha_i \varphi_{a_i,\lambda})
$$

$$
\leq \; (1 + O(\frac{1}{\lambda^{n-2}})) \left(pS^{n-2} - \frac{\overline{C}\varepsilon_0}{p^2} \sum_{i \neq j} \int_{\partial M} \varphi_{a_i,\lambda}^{\frac{n}{n-2}} \varphi_{a_j,\lambda} dv_{\overline{g}} \right)^{\frac{1}{n-2}}
$$

$$
(8) \quad \leq \; p^{\frac{1}{n-2}} S(1 + O(\frac{1}{\lambda^{n-2}}) - \frac{\overline{C_1}\varepsilon_0\varepsilon_1}{p^3})
$$

clearly implies (i), which is therefore proven if $d(a_1, a_2) \leq \eta$. Now we rule out the case where $d(a_1, a_2) \geq \eta$ as follows:

If $d(a_1, a_2) \geq \eta$ then

$$
\int_{\partial M} \left(\varphi_{a_1,\lambda}^{\frac{n}{n-2}} \varphi_{a_2,\lambda} + \varphi_{a_2,\lambda}^{\frac{n}{n-2}} \varphi_{a_1,\lambda} \right) = O(\frac{1}{\lambda^{n-2}})
$$

Therefore

$$
\sum_{i \neq j} \varphi_{a_i,\lambda}^{\frac{n}{n-2}} \varphi_{a_j,\lambda} \leq p^2 . O(\frac{1}{\lambda^{n-2}}) = O(\frac{1}{\lambda^{n-2}})
$$

Taking λ very large we have $\sum_{i \neq j} \varphi_{a_i,\lambda}^{\frac{n}{n-2}} \varphi_{a_j,\lambda} < \varepsilon_1$ therefore the proof of (i) is reduced to the case $d(a_1, a_2) \leq \eta$. The proof of (i) is thereby established.

The proofs of (ii) and (iii) will be completed together since they rest on the same expansion of the functional J.

Using Lemma 6.2 we derive the following inequality

$$
(9) \qquad J(\sum_{i-1}^{p} \alpha_i \, \varphi_{a_i,\lambda}) \leq \frac{\left(\sum_{i=1}^{p} \alpha_i^2 \int_{\partial M} \hat{\delta}^{\frac{n}{n-2}} \varphi_{a_i,\lambda} + \sum_{i\neq j}^{p} \alpha_i \alpha_j \int_{\partial M} \hat{\delta}_{a_i,\lambda}^{\frac{n}{n-2}} \varphi_{a_j,\lambda} \right)^{\frac{n-1}{n-2}}}{\sum_{i=2}^{p} \alpha_i^{\frac{2(n-1)}{n-2}} \int_{\partial M} \varphi_{a_i,\lambda}^{\frac{2(n-1)}{n-2}} + \sum_{i\neq j}^{p} \frac{\gamma(n-1)}{n-2} \alpha_i^{\frac{n}{n-2}} \alpha_j \int_{\partial M} \varphi_{a_i,\lambda}^{\frac{n}{n-2}} \varphi_{a_j,\lambda}}
$$

Then (9) implies using Lemma 2.2 and Lemma 2.4

$$
(10) \qquad J(\sum_{i-1}^{p} \alpha_i \varphi_{a_i,\lambda})
$$

$$
\leq \frac{\left(\sum_{i=1}^{p} \alpha_i^2 S^{n-2} + + (1+\theta) \sum_{i\neq j}^{p} \alpha_i \alpha_j \int_{\partial M} \varphi_{a_i,\lambda}^{\frac{n}{n-2}} \varphi_{a_j,\lambda} + \sum_{i=1}^{p} \alpha_i^2 \frac{\overline{C}}{\lambda^{n-2}} \right)^{\frac{n-1}{n-2}}}{\sum_{i=2}^{p} \alpha_i^{\frac{2(n-1)}{n-2}} S^{n-2} + \sum_{i\neq j}^{p} \frac{\gamma(n-1)}{n-2} \alpha_i^{\frac{n}{n-2}} \alpha_j \int_{\partial M} \varphi_{a_i,\lambda}^{\frac{n}{n-2}} \varphi_{a_j,\lambda} - \sum_{i=1}^{p} \alpha_i^{\frac{2(n-1)}{n-2}} \frac{\overline{C}}{\lambda^{n-2}}}
$$

where $\overline{C}$ and $\overline{C}'$ are positive constants independant of $(\alpha_1, \cdots, \alpha_p)$, λ and p. Let us assume that

$$
\sum_{i\neq j} \int_{\partial M} \varphi_{a_i,\lambda}^{\frac{n}{n-2}} \varphi_{a_j,\lambda} < \varepsilon_1.
$$

If ε_1 is chosen small enough, then for λ large enough , we have

$$
(11) \qquad J(\sum_{i-1}^{p} \alpha_i \, \varphi_{a_i,\lambda})
$$

$$
\leq \frac{\left(\sum_{i=1}^{p} \alpha_i^2 S^{n-2} + + (1+\theta) \sum_{i\neq j}^{p} \alpha_i \alpha_j \int_{\partial M} \varphi_{a_i,\lambda}^{\frac{n}{n-2}} \varphi_{a_j,\lambda} + \sum_{i=1}^{p} \alpha_i^2 \frac{\overline{C}}{\lambda^{n-2}} \right)^{\frac{n-1}{n-2}}}{\sum_{i=2}^{p} \alpha_i^{\frac{2(n-1)}{n-2}} S^{n-2} + \sum_{i\neq j}^{p} \frac{\gamma(n-1)}{n-2} \alpha_i^{\frac{n}{n-2}} \alpha_j \int_{\partial M} \varphi_{a_i,\lambda}^{\frac{n}{n-2}} \varphi_{a_j,\lambda} - \sum_{i=1}^{p} \alpha_i^{\frac{2(n-1)}{n-2}} \frac{\overline{C}}{\lambda^{n-2}}}.
$$

Making an expansion we have

$$
(12) \qquad J(\sum_{i-1}^{p} \alpha_i \, \varphi_{a_i,\lambda})
$$

$$
\leq \frac{\left(\sum_{i=1}^{p} \alpha_i^2 \, S^{n-2} + + (1+\theta) \sum_{i\neq j}^{p} \alpha_i \alpha_j \int_{\partial M} \varphi_{a_i,\lambda}^{\frac{n}{n-2}} \varphi_{a_j,\lambda} + \sum_{i=1}^{p} \alpha_i^2 \frac{\overline{C}}{\lambda^{n-2}} \right)^{\frac{n-1}{n-2}}}{\sum_{i=2}^{p} \alpha_i^{\frac{2(n-1)}{n-2}} S^{n-2} + \sum_{i\neq j}^{p} \frac{\gamma(n-1)}{n-2} \alpha_i^{\frac{n}{n-2}} \alpha_j \int_{\partial M} \varphi_{a_i,\lambda}^{\frac{n}{n-2}} \varphi_{a_j,\lambda} - \sum_{i=1}^{p} \alpha_i^{\frac{2(n-1)}{n-2}} \frac{\overline{C}}{\lambda^{n-2}}}
$$

Finally we obtain:

$$
(13) \qquad J(\sum_{i-1}^{p} \alpha_i \, \varphi_{a_i,\lambda})
$$

$$
\leq \frac{\left(\sum_{i=1}^{p} \alpha_i^2 \right)^{\frac{n-1}{n-2}}}{\sum_{i=1}^{p} \alpha_i^{\frac{2(n-1)}{n-2}}} S \left(1 + \frac{n-1}{n-2} \sum_{i\neq j} \frac{\alpha_i \alpha_j}{\sum_i \alpha_i^2 S^{n-2}} \int_{\partial M} \varphi_{a_i,\lambda}^{\frac{n}{n-2}} \varphi_{a_j,\lambda} \right.
$$

$$
\left. - \frac{n-1}{n-2} \sum_{i\neq j} \frac{\alpha_i \alpha_j}{\sum_i \alpha_i^2 S^{n-2}} \int_{\partial M} \varphi_{a_i,\lambda}^{\frac{n}{n-2}} \varphi_{a_j,\lambda} + \frac{\overline{C}''}{\lambda^{n-2}} \right).
$$

Let us observe that (iii) follows from (13), so it remains only to prove (ii).

Proof of (ii)

Let us now assume that there exists θ_0 , $0 < \theta_0 < 1$ and $\frac{\alpha_i}{\alpha_j} \geq \theta_0$ for any (i,j), then

$$\frac{\alpha_i \alpha_i}{\sum_{r=1}^{p} \alpha_r^2} \leq \frac{1}{p\theta_0^2} \quad \text{and} \quad \frac{\alpha_i^{\frac{n}{n-2}} \alpha_j}{\sum_{r=1}^{p} \alpha_r^{\frac{2(n-1)}{n-2}}} \geq \frac{\theta_0^{\frac{2(n-1)}{n-2}}}{p}$$

Now we choose $0 < \theta_0 < 1$ and $\theta > 0$ such that $(1+\theta)\frac{1}{\theta_0^2} - \gamma\theta_0^{\frac{2(n-1)}{n-2}} < 0$.

Let $\delta = S^{n-2}\left(\frac{\gamma(n-1)}{n-2}\theta_0^{\frac{2(n-1)}{n-2}} - \frac{n-1}{n-2}\frac{1+\theta}{\theta_0^2} \right)$

We then derive

(14)
$$J(\sum_{i-1}^{p} \alpha_i \varphi_{a_i,\lambda}) \leq \frac{(\sum_{i=1}^{p} \alpha_i^2)^{\frac{n-1}{n-2}}}{\sum_{i=1}^{p} \alpha_i^{\frac{2(n-1)}{n-2}}} S \left(1 - \frac{\delta}{p} \sum_{i\neq j} \frac{\alpha_i \alpha_j}{\sum_i \alpha_i^2 S^{n-2}} \int_{\partial M} \varphi_{a_i,\lambda}^{\frac{n}{n-2}} \varphi_{a_j,\lambda} + \frac{\overline{C''}}{\lambda^{n-2}} \right).$$

Then using (iii) of Lemma 2.4 , we derive (ii) from (14). The proof of Lemma 4.1 is thereby complete.$\blacksquare$

Proof of the claim

Let $\varepsilon_0 > 0$ be given and let

$$E_0 = \{x \in \partial M \quad \text{such that } \delta_{a_1,\lambda}(x) \geq \varepsilon_0 \left(2\delta_{a_1,\lambda} + \delta_{a_2,\lambda}\right)(x)\}$$

We have:

(15)
$$\int_{\partial M} \frac{\delta_{a_2,\lambda}}{2\delta_{a_1,\lambda} + \delta_{a_2,\lambda}}$$
$$\geq \varepsilon_0 \int_{E_0} \delta_{a_1,\lambda}^{\frac{n}{n-2}} \delta_{a_2,\lambda} \, dv_{\overline{g}_0}$$
$$\geq \varepsilon_0 \left(\int_{\partial M} \delta_{a_1,\lambda}^{\frac{n}{n-2}} \delta_{a_2,\lambda} \, dv_{\overline{g}_0} - (\frac{\varepsilon_0}{1 - 2\varepsilon_0})^{\frac{1}{n-2}} \int_{\partial M} \delta_{a_1,\lambda}^{\frac{n-1}{n-2}} \delta_{a_2,\lambda}^{\frac{n-1}{n-2}} \right)$$

Let us obseve that

$$\int_{\partial M} \delta_{a_1,\lambda}^{\frac{n}{n-2}} \delta_{a_2,\lambda} = \int_{\partial M} \delta_{a_2,\lambda}^{\frac{n}{n-2}} \delta_{a_1,\lambda}$$

and

$$\int_{\partial M} \delta_{a_1,\lambda}^{\frac{n-1}{n-2}} \delta_{a_2,\lambda}^{\frac{n-1}{n-2}} \leq \frac{1}{2} \left(\int_{\partial M} \delta_{a_1,\lambda}^{\frac{n}{n-2}} \delta_{a_2,\lambda} + \delta_{a_1,\lambda}\delta_{a_2,\lambda}^{\frac{n}{n-2}} \right)$$

Then (15) becomes

(16) $$\int_{\partial M} \frac{\delta_{a_2,\lambda}}{2\delta_{a_1,\lambda} + \delta_{a_2,\lambda}} \geq \frac{\varepsilon_0}{2} \left(1 - (\frac{\varepsilon_0}{1 - \varepsilon_0})^{\frac{1}{n-2}}\right) \int_{\partial M} \delta_{a_1,\lambda}^{\frac{n}{n-2}} \delta_{a_2,\lambda} + \delta_{a_1,\lambda}\delta_{a_2,\lambda}^{\frac{n}{n-2}}$$

Thus for ε_0 small enough , we have:

$$\int_{\partial M} \frac{\delta_{a_2,\lambda}}{2\delta_{a_1,\lambda} + \delta_{a_2,\lambda}} \;\geq\; C\varepsilon_0 \int_{\partial M} \delta_{a_1,\lambda}^{\frac{n}{n-2}} \delta_{a_2,\lambda} + \delta_{a_1,\lambda} \delta_{a_2,\lambda}^{\frac{n}{n-2}},$$

$$\geq\; C'\varepsilon_0 \int_{\partial M \cap B_\rho(a_1)} \hat{\delta}_{a_1,\lambda}^{\frac{n}{n-2}} \overline{\delta}_{a_2,\lambda} + \overline{\delta}_{a_1,\lambda} \hat{\delta}_{a_2,\lambda}^{\frac{n}{n-2}}$$

$$\geq\; C''\varepsilon_0 \int_{\partial M \cap B_\rho(a_1)} \varphi_{a_1,\lambda}^{\frac{n}{n-2}} \varphi_{a_2,\lambda} + \varphi_{a_1,\lambda} \varphi_{a_2,\lambda}^{\frac{n}{n-2}}$$

$$(17) \qquad\qquad \geq\; C''\varepsilon_0 \int_{\partial M} \varphi_{a_1,\lambda}^{\frac{n}{n-2}} \varphi_{a_2,\lambda} + \varphi_{a_1,\lambda} \varphi_{a_2,\lambda}^{\frac{n}{n-2}}.$$

Hence our claim is proved.

Lemma 4.1 implies the following proposition:

PROPOSITION 4.1. *There exists an integer p_0 and a positive real number $\lambda_0 > 0$ such that for any $(\alpha_1, \cdots, \alpha_p)$ satisfying $\alpha_i \geq 0$, $\sum_{i=1}^p \alpha_i = 1$, for any $(a_1, \cdots, a_p) \in \partial M$, for any $\lambda \geq \lambda_0$, we have*

$$J\left(\sum_{i=1}^p \alpha_i \varphi_{\alpha_i,\lambda}\right) \leq p_0^{\frac{1}{n-2}} S$$

PROOF. The proof of Proposition 4.1 follows from (i), (ii) and (iii) of Lemma 4.1. We first choose $0 < \varepsilon_1 < \overline{\varepsilon_1}$ and λ_0 so that :

$$(18) \qquad \frac{\left(\sum_{i=1}^p \alpha_i^2\right)^{\frac{n-1}{n-2}}}{\sum_{i=1}^p \alpha_i^{\frac{2(n-1)}{n-2}}} \left(1 + O\left(\frac{1}{\lambda^{n-2}}\right) + \frac{\varepsilon_1}{C}\right) < p^{\frac{1}{n-2}}.$$

Considering $(\alpha_1, \cdots, \alpha_p)$, $(a_1, \cdots, a_p)$ and $\lambda \geq \lambda_0$, we study various cases:

1st case: : There exists (i_0, j_0) such that $\frac{\alpha_{i_0}}{\alpha_{j_0}} \leq \theta_0$, then taking, $\lambda \geq \lambda_{p_0} = \sup(\lambda(p_0, \varepsilon), \lambda_0)$ where $\lambda(p_0, \varepsilon_1)$ is given by (i) of Lemma 4.1 , we derive :

$$J\left(\sum_{i=1}^p \alpha_i \varphi_{a_i,\lambda}\right) \leq p^{\frac{1}{n-2}} S \quad \text{if} \quad \sum_{i \neq j}^p \int_{\partial M} \varphi_{a_i,\lambda} \varphi_{a_j,\lambda} \, d\sigma_g \geq \varepsilon_1$$

If on the contrary $\sum_{i \neq j}^p \int_{\partial M} \varphi_{a_i,\lambda} \varphi_{a_j,\lambda} \, d\sigma_g \leq \varepsilon_1$ we apply (iii). Since we have choose ε_1 such that :

$$(19) \qquad \frac{\left(\sum_{i=1}^p \alpha_i^2\right)^{\frac{n-1}{n-2}}}{\sum_{i=1}^p \alpha_i^{\frac{2(n-1)}{n-2}}} \left(1 + O\left(\frac{1}{\lambda^{n-2}}\right) + \frac{\varepsilon_1}{C}\right) < p^{\frac{1}{n-2}}$$

we derive that $J(\sum_{i=1}^p \alpha_i \varphi_{a_i,\lambda}) \leq p^{\frac{1}{n-2}} S$ and the proof of Proposition 4.1 is established in this case.

2nd case: : Let us assume that $\frac{\alpha_i}{\alpha_j} > \theta_0$ for any (i, j) , then either (i) or (ii) of Lemma 4.1 holds. If (i) holds then Proposition 4.1 holds, so let assume that (ii) holds and then choose p_0 such that : $(p_0 + 1)c^2 > 1$ then

$$J\left(\sum_{i=1}^p \alpha_i \varphi_{a_i,\lambda}\right) \leq p_0^{\frac{1}{n-2}} S$$

The proof of Proposition 4.1 is thereby complete. ∎

5. Proof of Theorem 1.2

For the proof of the Theorem 1.2 , we introduce the following notations:
For any $p \geq 1$ and $\lambda > 0$, let

$$B_p = B_p(\partial M) = \{\sum_{i=1}^{p} \alpha_i \delta_{a_i} , \alpha_i \geq 0, \sum_{i=1}^{p} \alpha_i = 1, \, a_i \in \partial M \,\}$$

and $B_0 = B_0(\partial M) = \emptyset$

Set also $f_p(\lambda)$ to denote the map from $B_p(\partial M)$ to Σ^+ defined by

$$f_p(\lambda)(\sum_{i=1}^{p} \alpha_i \, \delta_{a_i}) = \frac{\sum_{i=1}^{p} \varphi_{a_i,\lambda_i}}{\|\sum_{i=1}^{p} \varphi_{a_i,\lambda_i}\|}$$

Clearly we have $B_{p-1} \subset B_p$ and $W_{p-1} \subset W_p$.
Moreover $f_p(\lambda)$ enjoys the following properties:

PROPOSITION 5.1. *The function $f_p(\lambda)$ has the following properties:*

: *(i) For any integer $p \geq 1$, there exists a real number $\lambda_p > 0$ such that*

$$f_p(\lambda) : \, B_p(\partial M) \to W_p \text{ for any } \lambda \geq \lambda_p$$

: *(ii) There exists an integer $p_0 > 1$, such that for any integer $p \geq p_0$, and for any $\lambda \geq \lambda_{p_0}$, the map of pairs $f_p(\lambda) : (B_p, B_{p-1}) \to (W_p, W_{p-1})$ satisfies $(f_p)_*(\lambda) \equiv 0$ where*

$$(f_p(\lambda))_* : \, H_*(B_p, B_{p-1}) \to H_*(W_p, W_{p-1})$$

and H_* is the $*$th homology group with $\mathbb{Z}_2$ coefficients.

PROOF.
(i) is a direct consequence of the inequalities (i), (ii) and (iii) of Lemma 4.1 , indeed :

$$J(f_p(\lambda)(\sum_{i=1}^{p} \alpha_i \, \delta_{a_i})) = J(\sum_{i=1}^{p} \alpha_i \, \varphi_{a_i,\lambda_i})$$

(ii) follows from Propostion 4.1. ∎

For the sequel we need the following notations: Let $\Delta_{p-1} = \{(\alpha_1, \cdots, \alpha_p), \alpha_i \geq 0, \sum_{i=1}^{p} \alpha_i = 1\}$ and $F_p = \{(a_1, \cdots, a_p) \in (\partial M)^p$ such that $\exists i \neq j$ with $a_i = a_j\}$. Let σ_p be the symmetric group of order p, which acts on F_p, and let T_p be a σ_p- equivariant tubular neighborhood of F_p, in $(\partial M)^p$ (The existence of a such neighborhood is derived in the book of G. Bredon [9])

From another part, considering the topological pair (B_p, B_{p-1}) we observe that $(B_p \setminus B_{p-1})$ can be described as $((\partial M)^p)^* \times_{\sigma_p} (\Delta_p \setminus \partial \Delta_{p-1})$ where $((\partial M)^p)^* = \{(a_1, \cdots, a_p) \in (\partial M)^p$ such that $a_i \neq a_j, \forall i \neq j\}$ We notice that $((\partial M)^p)^* \times_{\sigma_p} (\Delta_p \setminus \partial \Delta_{p-1})$ is a noncompact manifold of dimension $(n-1)p + p - 1$. Let for $0 < \theta < 1$, $\mathcal{M}_p = V_p \times_{\sigma_p} \Delta_{p-1}^\theta$, where $V_p = \overline{M^p \setminus T_p}$ and $\Delta_{p-1}^\theta = \{(\alpha_1, \cdots, \alpha_p) \in \Delta_{p-1}$ such that $\frac{\alpha_i}{\alpha_j} \in [1 - \theta, 1 + \theta], \forall i, \forall j\}$. $\mathcal{M}_p$ is a manifold which can be seen as a subst of B_p, and the topological pair $(B_p, \mathcal{M}^c)$ retracts by deformation onto (B_p, B_{p-1}), we thus have

$$H_*(B_p, B_{p-1}) = H_*(B_p, \mathcal{M}_p^c)$$

Thus by excision we have

$$H_*(B_p, B_{p-1}) = H_*(\mathcal{M}, \partial \mathcal{M}_p)$$

Since any manifold is orientable modulo its boundary with $\mathbb{Z}_2$ coefficients, we have a nonzero orientation class in $H_{(n-1)p+p-1}(B_p, B_{p-1})$ which we denote by ω_p.

In contrast with Proposition 5.1, we have the following Proposition:

PROPOSITION 5.2. *Under the assumption that (P) has no solution , we have ,*

$$\text{for every } p \in \mathbb{N}^* \quad (f_p(\lambda))_*(\omega_p) \neq 0.$$

PROOF.

An abstract topological argument displayed in [7] , pp 260-265 , see also [6], which extends virtually to our framework shows that:

$$\text{If } (f_1(\lambda))_* \not\equiv 0 \quad \text{then } (f_p(\lambda))_* \not\equiv 0 \text{ for every } p \geq 2.$$

Since $J_{S+\varepsilon}$, for $\varepsilon > 0$ small enough satisfies $J_{S+\varepsilon} \subset V(1,\delta)$, where $\delta \to 0$ if . $\varepsilon \to 0$, one can define using Lemma 3.1 a continuous map $s : J_{S+\varepsilon} \to \partial M$ which associates to $u = \overline{\alpha}\varphi_{\overline{a},\overline{\lambda}} + v \in J_{S+\varepsilon} \to a \in \partial M$. Here $(\overline{\alpha}, \overline{a}, \overline{\lambda})$ are the unique solution of the minimization : $\min\{\|u - \alpha\varphi_{a,\lambda}\|, \alpha \geq 0, \lambda > 0, a \in \partial M\}$

So if $r : W_1 \to J_{S+\varepsilon}$ denotes the retraction by deformation of W_1 onto $J_{S+\varepsilon}$, the existence of a such retraction by deformation follows from the assumption that (P) has no solution from one part and from Proposition 3.1 from another part. Let us observe that $s \circ r \circ f_1(\lambda) = id_{\partial M}$ hence $(f_1(\lambda))_*(\omega_1) \neq 0$, where ω_1 is the orientation class of ∂M. Therefore the proof of Proposition 5.2 is reduced to the abstract topological argument of Bahri-Coron [7]. ∎

PROOF OF THEOREM 1.2 COMPLETED

Proposition 5.2 is in contradiction with Proposition 5.1. Therefore (P) has a solution and Theorem 1.2 is thereby established. ∎

6. Appendix

LEMMA 6.1. *There holds*

$$\frac{1}{\lambda^{\frac{n-2}{2}}} \int_{\partial M \cap B_\rho(a_1)} \delta_{a_1,\lambda}^{\frac{2}{n-2}} \delta_{a_2,\lambda} dv_{g_0} = o\left(\int_{\partial M} \delta_{a_1,\lambda}^{\frac{n}{n-2}} \delta_{a_2,\lambda} dv_{g_0}\right)$$

PROOF. For $\varepsilon > 0$ a fixed number, let

$$A_\varepsilon = \left\{x \in B_\rho(a_1) \cap \partial M; \delta_{a_1,\lambda} \geq \frac{1}{\varepsilon\lambda^{\frac{n-2}{2}}}\right\}$$

Then

$$\frac{1}{\lambda^{\frac{n-2}{2}}} \int_{\partial M \cap B_\rho(a_1)} \delta_{a_1,\lambda}^{\frac{2}{n-2}} \delta_{a_2,\lambda} dv_{g_0}$$

$$\leq \varepsilon \int_{A_\varepsilon} \delta_{a_1,\lambda}^{\frac{n}{n-2}} \delta_{a_2,\lambda} dv_{g_0} + \frac{1}{\lambda^{\frac{n-2}{2}}} \int_{\partial M \cap B_\rho(a_1) \setminus A_\varepsilon} \frac{1}{\lambda\varepsilon^{\frac{2}{n-2}}} \delta_{a_2,\lambda} dv_{g_0}$$

$$\leq \varepsilon \int_{A_\varepsilon} \delta_{a_1,\lambda}^{\frac{n}{n-2}} \delta_{a_2,\lambda} dv_{g_0} + \frac{1}{\lambda^{\frac{n-2}{2}}} \int_{\partial M \cap B_\rho(a_1) \setminus A_\varepsilon} \frac{1}{\lambda\varepsilon^{\frac{2}{n-2}}} \delta_{a_2,\lambda} dv_{g_0}$$

$$\leq \varepsilon \int_{\partial M} \delta_{a_1,\lambda}^{\frac{n}{n-2}} \delta_{a_2,\lambda} dv_{g_0} + \frac{1}{\varepsilon^{\frac{2}{n-2}}} O\left(\frac{1}{\lambda^{\frac{n}{2}+\frac{(n-2)^2}{2n}}}\right)$$

Since $\frac{n}{2} + \frac{(n-2)^2}{2n} > n-2$ and since

$$\int_{\partial M} \delta_{a_1,\lambda}^{\frac{n}{n-2}} \delta_{a_2,\lambda} dv_{g_0} \geq \frac{C}{\lambda^{n-2}}$$

then our Lemma follows. $\blacksquare$

LEMMA 6.2. [5] *Let $q > 2$ be given. There exists $\gamma > 1$ such that for any* $(a_1, \cdots, a_p), a_i > 0$, *we have*

$$\left(\sum_{i=1}^{p} \alpha_i \right)^q \geq \sum_{i=1}^{p} \alpha_i^p + \frac{\gamma q}{2} \sum_{i \neq j} a_i^{q-1} a_j$$

LEMMA 6.3. [18] *[Maximum Principle]*
Under the assumption that $R_g \geq 0$, $h_g \geq 0$ and $R_g > 0$, or $h_g > 0$, let $u \in C^2(\overset{\circ}{M}) \cap C^1(\partial M)$ *satisfying*

$$L_g u \geq 0 \quad on \ \overset{\circ}{M} \quad and \ B_g u \leq 0 \quad on \ \partial M$$

Then $u \leq 0 \quad on \ M$

References

[1] A. Ambrosetti,Y.Y. Li, , A. Malchiodi, *Yamabe and scalar curvature problems under boundary conditions* , Math. Ann., **322** (2002), 667-699.

[2] T. Aubin , *Equations differentiélles non linéaires et Problème de Yamabe concernant la courbure scalaire*, J. Math. Pures et Appl. **55** (1976), 269-296.

[3] T. Aubin, Some Nonlinear Problems in Differential Geometry, Springer, (1998).

[4] A. Bahri, , Critical points at infinity in some variational problems, Pitman Research Notes in Mathematics Series no 182, Longman, London (1988)

[5] A. Bahri , *Proof of the Yamabe conjecture, without the positive mass theorem, for locally conformally flat manifolds.* In Einstein metric and Yang-Mills connections (Mabuchi T., Mukai S. ed.), Marcel Dekker, New York, (1993).

[6] A. Bahri , H. Brezis , Elliptic Differential Equations involving the Sobolev critical exponent on Manifolds, PNLDE 20, Birkäuser.

[7] A. Bahri , J. M. Coron , *On a nonlinear elliptic equation involving critical Sobolev exponent: The effect of the topology of the domain* , Comm.P.A.M **31** (1988) 253-294.

[8] M. Ben Ayed, K. El Mehdi, M. Ould Ahmedou, Prescribing the scalar curvature under minimal boundary conditions on the half sphere, Advances in Nonlin. Studies, **2** (2002), 93 - 116.

[9] Bredon, G. Introduction to compact transformations Groups, Academic Press, New York(1972).

[10] S-Y. Chang, M. Gursky,, P. Yang, *The scalar curvature equation on 2- and 3- spheres*, Calc. Var. **1** (1993), 205-229.

[11] Chang, S. A., Yang, P., *A perturbation result in prescribing scalar curvature on on S^n*, Duke Math. J. **64** (1991), 27-69.

[12] Y. Chen, , *On a nonlinear elliptic equation involving critical Sobolev exponent* , Nonlinear Anal., T.M.A., **33** (1998), 41-49.

[13] P. Cherrier, *Problemes de Neumann nonlinéaires sur les variétés Riemanniennes*, J. Funct. Anal. **57** (1984), 154-207.

[14] Z. Djadli, A. Malchiodi, M. Ould Ahmedou, *Prescribing scalar and mean curvature on the three dimensional half sphere*, J. Geom. Anal. **13** (2003), 233-267.

[15] J. Escobar , *Sharp constant in a Sobolev trace Inequality*, Indiana Univ.Math.J., **37** (1988), 687-698.

[16] J. Escobar, *Conformal deformation of a Riemannian metric to a scalar fla metric with constant mean curvature on the boundary*, Ann. Math. **136** (1992), 1-50.

[17] J. F. Escobar, *The Yamabe problem on manifolds with boundary*, J. Diff. Geom., **35** (1992), no. 1, 21-84.

[18] J. Escobar, *Conformal metrics with prescribed mean curvature on the boundary*, Cal. Var. and PDE's **4** (1996), 559-592.

[19] V. Felli, M. Ould Ahmedou, *Compactness result in conformal deformations on manifolds with boundaries*, Math. Zeit. **244** (2003), 175-210.

[20] V. Felli, M. Ould Ahmedou, *On a geometric equation with critical nonlinearity on the boundary* , to appear in Pacific Journal of Mathematics.

[21] H. Hamza, *Sur les transformations conformes des variétes Riemanniennes a bord*, J.Funct.Anal. 92(1990), 403-447.

[22] Z.C.Han, Yan Yan Li, *The Yamabe problem on manifolds with boundaries: existence and compactness results*, Duke Math. J. **99** (1999), 489-542.

[23] Z.C. Han, Yan Yan Li, *The existence of conformal metrics with constant scalar curvature and constant boundary mean curvature*, Comm. Anal. Geom., **8** (2000), 809-869 .

[24] E. Hebey , M. Vaugon , *Le problème de Yamabe équivariant*, Bull. Sci. Math. **117** (1993), 241-286.

[25] J. Lee ,T.H. Parker , *The Yamabe problem*, Bull. Amer. Math. Soc. **17** (1987), 37-91.

[26] P. L. Li, J. Q. Liu, *Nirenberg's problem on the 2-dimensional hemisphere*, Int.J.Math. **4** (1993), 927-939

[27] Y.Y. Li , *The Nirenberg problem in a domain with boundary*, Top. Meth. Nonlin. Anal. **6** (1995), 309-329.

[28] Y.Y. Li , M. Zhu, , *Uniqueness theorems through the method of moving spheres*, Duke Math. J. **80** (1995), 383-417.

[29] L. Ma, , *Conformal metrics with prescribed men curvature on the boundary of the unit ball*, to appear

[30] M. Morse , G. Van Schaak , *The critical point theory under general boundary conditions*, Annals of Math. **35** (1934), 545-571.

[31] R. Schoen , *Conformal deformation of a Riemannian metric to constant scalar curvature*, J. Diff. Geom. **20** (1984), 479-495.

[32] R. Schoen , S.T. Yau , *Conformally flat manifolds, Kleinian groups, and scalar curvature*, Invent. Math. **92** (1988), 47-71.

[33] R. Schoen , S.T. Yau , *On the proof of Positive Mass Conjecture in General Relativity*, Comm.Math.Phys. **65** (1979),45-76

[34] Spivak M., A comprehensive introduction to differential geometry, 2nd edition, Houston, Publish or Perish (1970)

RHEINISCHE FRIERICH-WILHELMS- UNIVERSITÄT BONN, MATHEMATISCHES INSTITUT, BERINGSTRASSE 4, D- 53115 BONN, GERMANY

E-mail address: `ahmedou@math.uni-bonn.de`

Contemporary Mathematics
Volume **350**, 2004

Cubic Quasilinear Wave Equation
and Bilinear Estimates

Hajer BAHOURI and Jean-Yves CHEMIN

This paper is dedicated to Haim Brezis and Felix Browder

ABSTRACT. In this text, we shall give an outline of some recent results of local wellposdness for cubic quasilinear wave equation for initial data less regular than what is required by the energy method. To go below the regularity prescribed by the classical theory of strictly hyperbolic equations, we prove bilinear estimate for the product of two solutions for wave operators whose coefficients are not very regular.

Introduction

In this paper, our interest is to prove local solvability for equations of the type

$$(EC) \begin{cases} \partial_t^2 u - \Delta u - \displaystyle\sum_{1 \leq j,k \leq d} g^{j,k} \partial_j \partial_k u &= 0 \\ \Delta g^{j,k} &= Q_{j,k}(\partial u, \partial u) \\ (u, \partial_t u)_{|t=0} &= (u_0, u_1). \end{cases}$$

where $Q_{j,k}$ are quadratic forms on $\mathbb{R}^{d+1}$. In all this work, we shall state, for a function u on $[0,T] \times \mathbb{R}^d$,

$$\nabla u \stackrel{\text{def}}{=} (\partial_1 u, \cdots, \partial_d u), \ \partial u \stackrel{\text{def}}{=} (\partial_t u, \partial_1 u, \cdots, \partial_d u) \text{ and } g \cdot \nabla^2 u \stackrel{\text{def}}{=} \sum_{1 \leq j,k \leq d} g^{j,k} \partial_j \partial_k u.$$

When no confusion is possible, we shall also state $\gamma \stackrel{\text{def}}{=} (\nabla u_0, u_1)$. This problem of course is a model one. The general problem consists in considering equations of the type

$$\begin{cases} \partial_t^2 u - \Delta u - \displaystyle\sum_{1 \leq j,k \leq d} g^{j,k} \partial_j \partial_k u &= \displaystyle\sum_{1 \leq j,k \leq d} \widetilde{Q}_{j,k}(\partial g^{j,k}, \partial u) \\ \Delta g^{j,k} &= Q_{j,k}(\partial u, \partial u) \\ (u, \partial_t u)_{|t=0} &= (u_0, u_1). \end{cases}$$

where $\widetilde{Q}_{j,k}$ are quadratic form on $\mathbb{R}^{d+1}$ and where all the quadratic forms are supposed to be smooth functions of u. This simply complicates a little the estimates

---

1991 *Mathematics Subject Classification.* Primary 35J10.

without any relevant new phenomenon. In the frame work of (EC), it makes sense to work with small data and this simplifies the proofs.

Energy methods allow to prove local wellposedness for initial data (u_0, u_1) in $H^{\frac{d}{2}+\frac{1}{2}} \times H^{\frac{d}{2}-\frac{1}{2}}$. More precisely, we have the following theorem.

THEOREM 0.1. *If $d \geq 3$, let (u_0, u_1) be in $H^{\frac{d}{2}+\frac{1}{2}} \times H^{\frac{d}{2}-\frac{1}{2}}$ such that $\|\gamma\|_{\dot{H}^{\frac{d}{2}-1}}$ is small enough. Then, a positive times T exists such that a unique solution u of (EC) exists in $C([0,T]; H^{\frac{d}{2}+\frac{1}{2}}) \cap C^1([0,T]; H^{\frac{d}{2}-\frac{1}{2}})$. Moreover, a constant C exists (which of course does not depend on the initial data) such that $T \geq C\|\gamma\|_{\dot{H}^{\frac{d}{2}-\frac{1}{2}}}^{-2}$.*

Let us recall that H^s is the usual Sobolev space on $\mathbb{R}^d$ and that $\dot{H}^s$ is the homogeneous one and we shall state

$$\|f\|_s^2 \overset{\text{def}}{=} \int_{\mathbb{R}^d} |\xi|^{2s} |\widehat{f}(\xi)|^2 d\xi.$$

This is an Hilbert space when $s < d/2$.

The goal of this paper is to go below the regularity $H^{d/2+1/2}$ for the initial data. Let us have a look to the scaling properties of equation (EC). If u is a solution of (EC), then $u_\lambda(t,x) \overset{\text{def}}{=} u(\lambda t, \lambda x)$ is also a solution of (EC). The space which is invariant under this scaling is $\dot{H}^{\frac{d}{2}}$. So the above theorem appears to require $1/2$ derivative more than the scaling. The goal of this work is to try to go as closed as possible to the scaling invariant regularity.

Some results in that direction have been proved by the authors (see [1] and [2]) and also by D. Tataru (see [22]) for quasilinear wave equations of the following type

$$(E) \begin{cases} \partial_t^2 u - \Delta u - G(u) \cdot \nabla^2 u &= F(u)Q(\partial u, \partial u) \\ (u, \partial_t u)_{|t=0} &= (u_0, u_1) \end{cases}$$

where G is a smooth function vanishing at 0 with value in K such that $\mathrm{Id} + K$ is a convex compact subset of the set of positive symmetric matrices. Let us recall this results. Let us notice that the scaling of the two equations (E) and (EC) is the same.

THEOREM 0.2. *If $d \geq 3$, let (u_0, u_1) be in $H^s \times H^{s-1}$ for $s > s_d$ with $s_d = \dfrac{d}{2} + \dfrac{1}{2} + \dfrac{1}{6}$. Then, a positive time T exists such that a unique solution u exists such that*

$$\partial u \in C([0,T]; H^{s-1}) \cap L^2([0,T]; L^\infty).$$

Moreover, a constant C exists such that $T^{\frac{2}{3}+(s-s_d)} \geq C\|\gamma\|_{\dot{H}^{s-1}}^{-1}$.

This theorem has been proved with $1/4$ instead than $1/6$ in [1] and then improved a little bit in [2] and proved with $1/6$ by D. Tataru in [22]. Strichartz estimates for quasilinear equations are the key point of the proofs. Recently, S. Klainerman and S. Rodnianski have announced a better index. Their proof is based on very different methods. In this case, the energy methods give the classical index $s > d/2 + 1$ and $T \geq C\|\gamma\|_{H^{s-1}}^{-1+s-s_d}$.

The goal of this work is to do the analogous in the case of Equation (EC). The result will be the following.

THEOREM 0.3. *If $d \geq 5$, let (u_0, u_1) be in $H^s \times H^{s-1}$ with $s > \dfrac{d}{2} + \dfrac{1}{6}$ such that $\|\gamma\|_{\dot{H}^{\frac{d}{2}-1}}$ is small enough. Then, a positive times T exists such that a unique* *solution u of (EC) exists such that $\partial u \in C([0,T]; H^{s-1}) \cap L_T^2(\dot{B}_{4,2}^{\frac{d}{4}-\frac{1}{2}})$ where $\dot{B}_{4,2}^{\frac{d}{4}-\frac{1}{2}}$ is the Besov space defined in Definition 1.1. Moreover, for any positive α, we have that $T^{\frac{1}{6}+\alpha} \geq C_\alpha \|\gamma\|_{\dot{H}^{\frac{d}{2}-\frac{5}{6}+\alpha}}^{-1}$.*

The case of dimension 4 is a little bit different. The theorem is the following.

THEOREM 0.4. *If $d = 4$, let (u_0, u_1) be in $H^s \times H^{s-1}$ with $s > 2 + \dfrac{1}{6}$ such that $\|\gamma\|_{\dot{H}^1}$ is small enough. Then, a positive times T exists such that a unique solution u of (EC) exists such that*

$$\partial u \in C([0,T]; H^{s-1}) \cap L_T^2(\dot{B}_{6,2}^{\frac{1}{6}}) \quad \text{and} \quad \partial g \in L_T^1(L^\infty).$$

where $\dot{B}_{6,2}^{\frac{d}{6}-\frac{1}{2}}$ denotes the Besov space defined in Definition 1.1. Moreover, for any positive α, a constant C_α exists such that $T^{\frac{1}{6}+\alpha} \geq C_\alpha \|\gamma\|_{\dot{H}^{\frac{d}{2}-\frac{5}{6}+\alpha}}^{-1}$.

Remarks

- If we think in term of small data (i.e. of initial data of the type $\epsilon(u_0, u_1)$), then energy methods give a life span in ϵ^{-2}. The above theorem gives a life span of order $\epsilon^{-6+\alpha}$ for any positive α.
- As we shall see, the case when $d \geq 5$ can be treated only with Strichartz estimates simply because if ∂u belongs to $L_T^2(\dot{B}_{4,2}^{\frac{d}{4}-\frac{1}{2}})$ then ∂g is in $L_T^1(L^\infty)$.
- The case when $d = 4$ requires bilinear estimates. This fact appears in the statement of Theorem 0.4 through the following phenomenon: the fact that ∂u is in $L_T^2(\dot{B}_{6,2}^{\frac{1}{6}})$ does not imply that the time derivative of g belongs to $L_T^1(L^\infty)$. Of course this condition is crucial in particular to get the basic energy estimate. But we have been unable to exibit a Banach space $\mathcal{B}$ which contains the solution u and such that if a function a is in $\mathcal{B}$, then $\partial \Delta^{-1}(a^2)$ belongs to $L_T^1(L^\infty)$.

Acknowledgments We want to thank S. Klainerman for introducing us to this problem and also for fruitful discussions. We thank J.-M. Bony for very important discussions about the concept of microlocalized functions.

1. Method of the proof

1.1. Some basic facts in Littlewood-Paley theory. Let us begin by recalling the basis of Littlewood-Paley theory. Let us denote by $\mathcal{C}$ the ring of center 0, of small radius $3/4$ and of big radius $8/3$. Let us choose two non negative radial functions χ and φ belonging respectively to $\mathcal{D}(B(0, 4/3))$ and $\mathcal{D}(\mathcal{C})$ such that

$$(1) \qquad \chi(\xi) + \sum_{q\in\mathbb{N}} \varphi(2^{-q}\xi) = \sum_{q\in\mathbb{Z}} \varphi(2^{-q}\xi) = 1,$$

$$(2) \qquad |p-q| \geq 2 \Rightarrow \text{Supp } \varphi(2^{-q}\cdot) \cap \text{Supp } \varphi(2^{-p}\cdot) = \emptyset,$$

$$(3) \qquad q \geq 1 \Rightarrow \text{Supp } \chi \cap \text{Supp } \varphi(2^{-q}\cdot) = \emptyset,$$

and if $\widetilde{\mathcal{C}} = B(0, 2/3) + \mathcal{C}$, then $\widetilde{\mathcal{C}}$ is a ring and we have

$$(4) \qquad |p-q| \geq 5 \Rightarrow 2^p\widetilde{\mathcal{C}} \cap 2^q\mathcal{C} = \emptyset.$$

Notations

$$h = \mathcal{F}^{-1}\varphi \quad \text{and} \quad \widetilde{h} = \mathcal{F}^{-1}\chi,$$

$$\Delta_q u = \varphi(2^{-q}D)u = 2^{qd}\int h(2^q y)u(x-y)dy,$$

$$S_q u = \sum_{p \le q-1} \Delta_p u = \chi(2^{-q}D)u = 2^{qd}\int \widetilde{h}(2^q y)u(x-y)dy.$$

We shall often denote $\Delta_q u$ by u_q. Let us recall the definition of Besov spaces.

DEFINITION 1.1. *Let s be a real number, and (p,r) in $[1,\infty]^2$. Let us state*

$$\|u\|_{\dot{B}^s_{p,r}} \overset{\mathrm{def}}{=} \left\|\left(2^{qs}\|\Delta_q u\|_{L^p}\right)_{q\in\mathbb{Z}}\right\|_{\ell^r(\mathbb{Z})}$$

If $s < d/p$ then the closure of the compactly smooth functions with respect to this norm is a Banach space and we have that $\dot{H}^s = \dot{B}^s_{2,2}$ and the norm $\|\cdot\|_{\dot{B}^s_{2,2}}$ is equivalent to $\|\cdot\|_{\dot{H}^s}$.

Notations We shall also state

$$\|a\|_s \overset{\mathrm{def}}{=} \|a\|_{\dot{B}^s_{2,2}}, \quad \|b\|_{L^p_I(E)} \overset{\mathrm{def}}{=} \|b\|_{L^p(I;E)}, \quad \|b\|_{L^p_T(E)} \overset{\mathrm{def}}{=} \|b\|_{L^p([0,T];E)}$$

$$\text{and} \quad \|b\|_{T,s} \overset{\mathrm{def}}{=} \|b\|_{L^\infty_T(\dot{B}^s_{2,2})}.$$

Here we want to explain the problems we have to solve to prove Theorem 0.4. As in the case of Equation (E), the basic fact is energy estimates. This implies the control of

$$\int_0^T \|\partial g(t,\cdot)\|_{L^\infty}dt.$$

In the case of Equation (E), it is obtained by Strichartz estimates. This will be the case here when $d \ge 5$ not when $d = 4$. Let us follow now ideas of S. Klainerman and D. Tataru (see [**17**]). If u is the solution of the constant coefficient wave equation, let us estimate

$$\int_0^T \|\partial\Delta^{-1}\big(\partial_j u(t,\cdot)\partial_k u(t,\cdot)\big)\|_{L^\infty}dt.$$

As $\partial_t\Delta^{-1}\big(\partial_j u(t,\cdot)\partial_k u(t,\cdot)\big) = \Delta^{-1}\big(\partial_t\partial_j u\partial_k u(t,\cdot)\big) + \Delta^{-1}\big(\partial_j u\partial_t\partial_k u(t,\cdot)\big)$, we have to control expression of the type

$$\int_0^T \|\Delta^{-1}\big(\partial_t\partial_j u\partial_k u(t,\cdot)\big)\|_{L^\infty}dt.$$

When $d \ge 4$, we have that $\|\Delta^{-1}\big(\partial_t\partial_j u\partial_k u(t,\cdot)\big)\|_{B^{\frac{d}{2}}_{2,1}} \le C\|\partial u(t,\cdot)\|^2_{\frac{d}{2}-\frac{1}{2}}$. So we get that

$$\int_0^T \|\Delta^{-1}\big(\partial_t\partial_j u\partial_k u(t,\cdot)\big)\|_{L^\infty}dt \le T\|\partial u\|^2_{T,\frac{d}{2}-\frac{1}{2}}.$$

Then the proof of Theorem 0.1 is routine. If we want to go below this $H^{\frac{d}{2}+\frac{1}{2}}$ regularity of the initial data, we shall use Strichartz estimates. Let us introduce Bony's decomposition which consists in writing

$$ab = \sum_q S_{q-1}a\Delta_q b + \sum_q S_{q-1}b\Delta_q a + \sum_{\substack{-1\le j\le 1 \\ q}} \Delta_q a\Delta_{q-j}b.$$

When $d \ge 4$, we have $\|\partial^k u_q\|_{L^2_T(L^\infty)} \le C2^{q(\frac{d}{2}-\frac{1}{2}+k-1)}\|\gamma_q\|_{L^2}$. Then it is easy to prove that

$$\left\|\Delta^{-1}\left(\sum_q S_{q-1}\partial^2 u\partial u_q\right)\right\|_{L^1_T(L^\infty)} \le C\|\gamma\|^2_{\frac{d}{2}-1}.$$

The symmetric term can be treated exactly along the same lines. The remainder term

$$\Delta^{-1}\Big(\sum_{\substack{-1\leq j\leq 1 \\ q}} \partial^2 u_q \partial u_{q-j}\Big)$$

is much more difficult to treat particulary in dimension 4. The reason why is the following. When d is greater or equal to 5, the Strichartz estimates tells us that

$$\|\partial^k u_q\|_{L^2_T(L^4)} \leq 2^{q\left(\frac{d}{4}-\frac{1}{2}+k-1\right)}\|\gamma_q\|_{L^2}.$$

So thanks to Bernstein inequality, we infer that

$$\Big\|\Delta_p \Delta^{-1}\Big(\sum_{\substack{-1\leq j\leq 1 \\ q\geq p-N_0}} \Delta_q \partial^2 u \Delta_{q-j}\partial u\Big)\Big\|_{L^1_T(L^\infty)}$$

$$\leq C 2^{p\left(\frac{d}{2}-2\right)}\sum_{\substack{-1\leq j\leq 1 \\ q\geq p-N_0}} 2^{q\frac{d}{2}}\|\gamma_q\|_{L^2}\|\gamma_{q-j}\|_{L^2}$$

$$\leq C \sum_{\substack{-1\leq j\leq 1 \\ q\geq p-N_0}} 2^{-(q-p)\left(\frac{d}{2}-2\right)} 2^{q(d-2)}\|\gamma_q\|_{L^2}\|\gamma_{q-j}\|_{L^2}.$$

Convolution and Cauchy-Schwarz inequalities implies that $\|\Delta^{-1}\big(\partial^2 u \partial u\big)\|_{L^1_T(L^\infty)} \leq C\|\gamma\|^2_{\frac{d}{2}-1}$. The case of dimension 4 is much more delicate. In dimension 4, the Strichartz estimate is

$$\|\partial^k u_q\|_{L^2_T(L^6)} \leq 2^{q\left(\frac{4}{3}-\frac{1}{2}+k-1\right)}\|\gamma_q\|_{L^2}.$$

So the series $\partial^2 u_q \partial u_{q-j}$ does not converge in $L^1_T(L^3)$ because the only estimate we have is

$$\|\partial^2 u_q \partial u_{q-j}\|_{L^1_T(L^3)} \leq C 2^{q\frac{8}{3}}\|\gamma_q\|^2_{L^2} \leq C 2^{q\frac{2}{3}} d_q\|\gamma\|^2_1 \quad\text{with}\quad \sum_q d_q = 1.$$

To overcome this difficulty, we follow an idea of D. Tataru and S. Klainerman (see [**17**]).

1.2. Bilinear estimates and precised Strichartz estimates. To explain the basic ideas of bilinear estimates, let us consider the case of constant coefficient case.

PROPOSITION 1.1. *Let u_1 and u_2 two solutions of $\partial_t^2 u_j - \Delta u_j = 0$ and $(\partial u_j)_{|t=0} = \gamma_j$. Then, if $d \geq 4$,*

$$\|\partial\Delta^{-1}Q(\partial u, \partial u)\|_{L^1_T(L^\infty)} \leq C_{\epsilon,T}\|\gamma_1\|_{\frac{d}{2}-1+\epsilon}\|\gamma_2\|_{\frac{d}{2}-1+\epsilon}.$$

The precised Strichartz estimates are described by the following proposition.

PROPOSITION 1.2. *If $d \geq 3$, a constant C exists such that for any T and any $h \leq 1$, if Supp $\widehat{u}_j$ and Supp $\mathcal{F}(\Box u(t,\cdot))$ are included in a ball of radius h and in the ring $\mathcal{C}$, we have*

$$\|u\|_{L^2_T(L^\infty)} \leq C\big(h^{d-2}\log(e+T)\big)^{\frac{1}{2}}\big(\|u(0)\|_{L^2} + \|\partial_t u(0)\|_{L^2} + \|\Box u\|_{L^1_T(L^2)}\big).$$

As usual it is deduced with the $TT^\star$ argument from the following dispersive inequality.

LEMMA 1.1. *Let $\mathcal{C}$ be a ring of $\mathbb{R}^d$. A constant C exists such that if u_0 and u_1 are functions in $L^1(\mathbb{R}^d)$ such that*

$$\text{Supp } (\widehat{u}_j) \subset \mathcal{C} \quad \text{and} \quad \max\{\delta(\text{Supp } (\widehat{u}_0)), \delta(\text{Supp } (\widehat{u}_1))\} \leq h,$$

then, for any $\widetilde{d}$ between 0 and $d-1$, we have

$$\|u(t,\cdot)\|_{L^\infty} \leq \frac{Ch^{d-\widetilde{d}}}{t^{\frac{\widetilde{d}}{2}}}\left(\|u_0\|_{L^1} + \|u_1\|_{L^1}\right),$$

where u denotes the solution of $\partial_t^2 u - \Delta u = 0$ and $\partial_t^j u_{|t=0} = u_j$.

This inequality is proved in [**17**] in the case when $\widetilde{d} = d - 1$. The general case is obtained by interpolation with the classical Sobolev embedding.

Let us recall that we want to estimate the

$$\left\|\Delta_p \Delta^{-1}\left(\sum_{\substack{-1 \leq j \leq 1 \\ q \geq p - N_0}} \Delta_q \partial^2 u \Delta_{q-j} \partial u\right)\right\|_{L_T^1(L^\infty)}$$

With scaling, we can assume that $q = 1$ and let us state $h = 2^{p-q}$. Let us define $(\phi_\nu)_{1 \leq \nu \leq N_h}$ a partition of unity of the ring $\mathcal{C}$ such that Supp $\phi_\nu \subset B(\xi_\nu, h)$. Then, using the fact that the support of the Fourier transform of the product of two functions is included in the sum of the support of their Fourier transform, a family of functions $(\widetilde{\phi}_\nu)_{1 \leq \nu \leq N_h}$ exists such that Supp $\widetilde{\phi}_\nu \subset B(-\xi_\nu, 2h)$ and

$$(5) \qquad \chi(h^{-1}D)(\partial^2 v \partial v)) = \sum_{\nu=1}^{N_h} \chi(h^{-1}D)\left(\partial^2 \widetilde{\phi}_\nu(D) v \partial \phi_\nu(D) v\right).$$

Applying Proposition 1.2 gives

$$\|\chi(h^{-1}D)(\partial^2 v \partial v))\|_{L_T^1(L^\infty)} \leq Ch^{d-2} \log(e + T) \sum_{\nu=1}^{N_h} \|\widetilde{\phi}_\nu(D)\gamma\|_{L^2} \|\phi_\nu(D)\gamma\|_{L^2}.$$

The Cauchy Schwarz inequality implies that

$$\|\chi(h^{-1}D)(\partial^2 v \partial v))\|_{L_T^1(L^\infty)}$$

$$\leq Ch^{d-2} \log(e + T)\left(\sum_{\nu=1}^{N_h} \|\widetilde{\phi}_\nu(D)\gamma\|_{L^2}^2\right)^{\frac{1}{2}}\left(\sum_{\nu=1}^{N_h} \|\phi_\nu(D)\gamma\|_{L^2}^2\right)^{\frac{1}{2}}.$$

The almost orthogonality of $(\widetilde{\phi}_\nu(D)\gamma_1)_{1 \leq \nu \leq N_h}$ and $(\phi_\nu(D)\gamma_2)_{1 \leq \nu \leq N_h}$ implies that

$$(6) \qquad \|\chi(h^{-1}D)(\partial^2 v \partial v))\|_{L_T^1(L^\infty)} \leq Ch^{d-2} \log(e + T)\|\gamma\|_{L^2}\|\gamma\|_{L^2}.$$

So after rescaling, we get that

$$\left\|\Delta_p \Delta^{-1}\left(\sum_{\substack{-1 \leq j \leq 1 \\ q \geq p - N_0}} \Delta_q \partial^2 u \Delta_{q-j} \partial u\right)\right\|_{L_T^1(L^\infty)}$$

$$\leq 2^{p(d-4)} \sum_{\substack{-1 \leq j \leq 1 \\ q \geq p - N_0}} \log(e + 2^q T) 2^{2q} \|\gamma_q\|_{L^2} \|\gamma_{q-j}\|_{L^2}.$$

If $\gamma \in H^{\frac{d}{2}-1+\frac{\epsilon}{2}}$ then we have

$$\left\| \Delta_p \Delta^{-1}\left(\sum_{\substack{-1 \leq j \leq 1 \\ q \geq p - N_0}} \Delta_q \partial^2 u \Delta_{q-j} \partial u \right) \right\|_{L_T^1(L^\infty)} \leq (2^p T)^{-\epsilon} \sum_{\substack{-1 \leq j \leq 1 \\ q \geq p - N_0}} 2^{-(q-p)(d-4+\epsilon)}$$

$$\times 2^{q\left(\frac{d}{2}-1\right)} (2^q T)^{\frac{\epsilon}{2}} \|\gamma_q\|_{L^2} 2^{(q-j)\left(\frac{d}{2}-1\right)} (2^q T)^{\frac{\epsilon}{2}} \|\gamma_{q-j}\|_{L^2}.$$

So the series convergences in $L_T^1(L^\infty)$ for large p. The case when p is small (low frequencies) is nothing but Sobolev embeddings.

The problem we have to solve in this work is to prove this bilinear estimate in the context of quasilinear wave equation. To do this, we follow the lines of [1] and [2]. As we shall use geometrical optics technics, we need to deal with smooth functions in time also. This leads to the following iterative scheme introduced in [2]. Let us define the sequence $(u^{(n)})_{n\in\mathbb{N}}$ by the first term $u^{(0)}$ satisfying

$$\begin{cases} \partial_t^2 u^{(0)} - \Delta u^{(0)} &= 0 \\ (u^{(0)}, \partial_t u^{(0)})_{|t=0} &= (S_0 u_0, S_0 u_1), \end{cases}$$

and by the following induction

$$(\mathcal{R}_n) \begin{cases} \partial_t^2 u^{(n+1)} - \Delta u^{(n+1)} - G_{n,T} \cdot \nabla^2 u^{(n+1)} &= 0 \\ (u^{(n+1)}, \partial_t u^{(n+1)})_{|t=0} &= (S_{n+1} u_0, S_{n+1} u_1) \end{cases}$$

with $G_{n,T} \overset{\text{def}}{=} \theta(T^{-1}) G_n$ with $G_n^{j,k} \overset{\text{def}}{=} \Delta^{-1} Q_{j,k}(\partial u^{(n)}, \partial u^{(n)})$ where θ is a function of $\mathcal{D}(]-1,1[)$ whose value is 1 near 0. Let us point out that the sequence $(u^{(n)})_{n\in\mathbb{N}}$ does depend on T. We introduce some notations which will be used all along this work. If α is a (small) positive number, let us define

$$s_\alpha \overset{\text{def}}{=} \frac{d}{2} + \frac{1}{6} + \alpha \quad \text{and} \quad N_T^\alpha(\gamma) \overset{\text{def}}{=} T^{\frac{1}{6}+\alpha} \|\gamma\|_{s_\alpha - 1}.$$

Let us introduce the assertions we are going to prove by induction.

- If $d \geq 5$,

$$(\mathcal{P}_n) \begin{cases} \|\partial u^{(n)}\|_{L^2([0,T];\dot{B}_{4,2}^{\frac{d}{4}-\frac{1}{2}})} &\leq C_0 N_T^\alpha(\gamma) \\ \|\partial u^{(n)}\|_{T,s-1} &\leq e^3 \|\gamma\|_{s-1} \quad \text{for any} \quad s \in \left[s_\alpha - 1, \frac{d}{2} + \frac{1}{2}\right]; \end{cases}$$

- if $d = 4$,

$$(\mathcal{P}_n) \begin{cases} \|\partial u^{(n)}\|_{L^2([0,T];\dot{B}_{6,2}^{\frac{d}{6}-\frac{1}{2}})} &\leq C_0 N_T^\alpha(\gamma) \\ \|\partial G_{n,T}\|_{L^1([0,T];L^\infty)} &\leq 2 \\ \|\partial u^{(n)}\|_{T,s-1} &\leq e^3 \|\gamma\|_{s-1} \quad \text{for any} \quad s \in \left[\frac{3}{2} + \alpha, \frac{d}{2} + \frac{1}{2}\right]. \end{cases}$$

This paper consists in proving that if $\|\gamma\|_{\dot{H}^{\frac{d}{2}-1}} + N_T^\alpha(\gamma)$ is small enough, $(\mathcal{P}_0)$ is true and $(\mathcal{P}_n)$ implies $(\mathcal{P}_{n+1})$. The remainder is routine of non linear partial differential equations.

2. Reduction to microlocalized estimates

By microlocalization of the estimates, we mean that we shall prove estimates that are valid on time intervals whose length depend on the frequency parameter. These techniques have been introduced in [1] and used in [2] and improved by D. Tataru in [22].

For $\Lambda_0 > 0$, we shall consider a family of smooth functions $\mathcal{G} = (G_\Lambda)_{\Lambda \geq \Lambda_0}$ defined on $I_\Lambda \times \mathbb{R}^d$ such that G_Λ is small enough in L^∞ norm and for any $k \geq 0$, the following quantities

$$(7) \qquad \|\mathcal{G}\|_0 \overset{\text{def}}{=} \sup_{\Lambda \geq \Lambda_0} \|\nabla G_\Lambda\|_{L^1_{I_\Lambda}(L^\infty)} + |I_\Lambda| \|\nabla^2 G_\Lambda\|_{L^1_{I_\Lambda}(L^\infty)} \quad \text{and}$$

$$(8) \qquad \|\mathcal{G}\|_k \overset{\text{def}}{=} \sup_{\Lambda \geq \Lambda_0} |I_\Lambda| \Lambda^k \|\nabla^{k+2} G_\Lambda\|_{L^1_{I_\Lambda}(L^\infty)} \quad \text{for} \quad k \geq 1.$$

are finite. Let us denote by P_Λ the operator $P_\Lambda v \overset{\text{def}}{=} \partial_\tau^2 v - \Delta v - \sum_{j,k} G_\Lambda^{j,k} \partial_j \partial_k v.$

THEOREM 2.1. *Let $\mathcal{C}$ be a ring of $\mathbb{R}^d$ and ϵ_0 a positive real number. Let us assume that $\|\mathcal{G}\|_0$ is small enough. For any positive real number $\epsilon \leq \epsilon_0$, a constant C_ϵ exists which satisfies the following properties. Let f_1 and f_2 two functions in $L^1_{I_\Lambda}(L^2)$ and γ_1 and γ_2 two functions of L^2; let us assume that all the spectrum of those functions is included in $\mathcal{C}$. Let us assume that $|I_\Lambda| \leq \Lambda^{2-\epsilon}$. Then if $v_{1,\Lambda}$ and $v_{2,\Lambda}$ are solutions of*

$$(E_\Lambda) \begin{cases} P_\Lambda v_{j,\Lambda} &= f_j \\ \partial v_{j,\Lambda|\tau=0} &= \gamma_j \end{cases}$$

we shall have the following properties:

- *if $d \geq 5$, we have $\|\partial v_{j,\Lambda}\|_{L^2_{I_\Lambda}(L^4)} \leq C(\|\gamma_j\|_{L^2} + \|f_j\|_{L^1_{I_\Lambda}(L^2)}).$*
- *if $d = 4$, we have $\|\partial v_{j,\Lambda}\|_{L^2_{I_\Lambda}(L^6)} \leq C(\|\gamma_j\|_{L^2} + \|f_j\|_{L^1_{I_\Lambda}(L^2)}).$*
- *if $d \geq 3$, then we have, for any $h \leq 1$ and any $\epsilon > 0$,*

$$\|\chi(h^{-1}D)(\partial v_{1,\Lambda} \partial v_{2,\Lambda})\|_{L^1_{I_\Lambda}(L^\infty)} \leq C_\epsilon h^{d-2-\epsilon} \log(e + |I_\Lambda|)$$
$$\times \left(\|\gamma_1\|_{L^2} + \|f_1\|_{L^1_{I_\Lambda}(L^2)}\right)\left(\|\gamma_2\|_{L^2} + \|f_2\|_{L^1_{I_\Lambda}(L^2)}\right).$$

If h is small enough, this is Sobolev embedding. Using Bernstein inequality, we can write

$$\begin{aligned} \|\chi(h^{-1}D)(\partial v_{1,\Lambda}\partial v_{2,\Lambda})\|_{L^1_{I_\Lambda}(L^\infty)} &\leq h^d \|\partial v_{1,\Lambda}\partial v_{2,\Lambda}\|_{L^1_{I_\Lambda}(L^1)} \\ &\leq h^d |I_\Lambda| \|\partial v_{1,\Lambda}\|_{L^\infty_{I_\Lambda}(L^2)} \|\partial v_{2,\Lambda}\|_{L^\infty_{I_\Lambda}(L^2)} \\ &\leq h^d |I_\Lambda| \left(\|\gamma_1\|_{L^2} + \|f_1\|_{L^1_{I_\Lambda}(L^2)}\right) \\ &\qquad \times \left(\|\gamma_2\|_{L^2} + \|f_2\|_{L^1_{I_\Lambda}(L^2)}\right). \end{aligned}$$

So when $h^d |I_\Lambda| \leq h^{d-2-\epsilon}$, the above bilinear estimate is proved. From now on, we assume that $|I_\Lambda| \geq h^{-2-\epsilon}.$

3. Approximation of the solution and geometrical optics

3.1. The approximation of the solution. The key point is the use of Hamilton-Jacobi equation. The following proposition (and its proof) is a small modification of Proposition 6.1 of [1].

PROPOSITION 3.1. *Let F be a real valued smooth function on $\mathbb{R}^d \times \mathbb{R}^N$ bounded as all its derivatives such that*

$$F(\zeta, G) = \pm\big(|\zeta|^2 + G(\zeta, \zeta)\big)^{\frac{1}{2}} \quad \text{for all} \quad \zeta \in \widetilde{\mathcal{C}}.$$

For any positive real number ϵ, a positive real number α exists such that, if $\|\mathcal{G}\|_0 \leq \alpha$, then, for any $\Lambda \geq \Lambda_0$, for any η, a solution Φ_Λ of the equation

$$(\widetilde{HJ_\Lambda}) \begin{cases} \partial_\tau \Phi_\Lambda(\tau, y, \eta) = F_\Lambda(\tau, y, \partial_y \Phi_\Lambda(\tau, y, \eta)) \\ \Phi_\Lambda(0, y, \eta) = (y|\eta) \end{cases} \text{with } F_\Lambda(\tau, z, \zeta) \overset{\text{def}}{=} F(\zeta, G_\Lambda(\tau, z)).$$

exists and is smooth on $I_\Lambda \times \mathbb{R}^d \times \mathbb{R}^d$. Moreover, the family defined by $\Phi \overset{\text{def}}{=} (\Phi_\Lambda)_{\Lambda \geq \Lambda_0}$ satisfies the following properties: For any couple of integer (k, ℓ), a constant $C_{k,\ell}$ (indepedant of ϵ) exists such that

$$\tag{9} \sup_{\Lambda \geq \Lambda_0} \|\partial_y \partial_\eta \Phi_\Lambda - \mathrm{Id}\|_{L^\infty(I_\Lambda \times \mathbb{R}^{2d})} \leq \epsilon C,$$

$$\tag{10} \sup_{\Lambda \geq \Lambda_0} |I_\Lambda| \Lambda^k \|\partial_\eta^\ell \nabla^{2+k} \Phi_\Lambda\|_{L^\infty(I_\Lambda \times \mathbb{R}^{2d})} \leq \epsilon C_{k,\ell} \quad \text{et}$$

$$\tag{11} \sup_{\Lambda \geq \Lambda_0} \|\partial_\eta^{\ell+2} \Phi_\Lambda\|_{L^\infty(I_\Lambda \times \mathbb{R}^{2d})} \leq \epsilon C_\ell |I_\Lambda|.$$

The link between the solution of the Hamilton-Jacobi equation and the hamiltonian flow is decribed by the two following lemmas.

LEMMA 3.1. *Let Φ_Λ be the solution of the above Hamilton-Jacobi equation $(\widetilde{HJ_\Lambda})$ and Ψ_Λ the Hamiltonian flow of $-F_\Lambda(\tau, Y)$ i.e. the solution of*

$$\begin{cases} \dfrac{d\Psi_\Lambda}{d\tau}(\tau, y, \eta) = -H_{F_\Lambda}(\tau, \Psi_\Lambda(Y)) \\ \Psi_\Lambda(0, y, \eta) = (y, \eta). \end{cases}$$

Then we have

$$\begin{aligned} (\partial_\eta \Phi_\Lambda)(\tau, \Psi_\Lambda^y(\tau, y, \eta), \eta) &= y \quad \text{and} \\ (\partial_y \Phi_\Lambda)(\tau, \Psi_\Lambda^y(\tau, y, \eta), \eta) &= \Psi_\Lambda^\eta(\tau, y, \eta). \end{aligned}$$

LEMMA 3.2. *A constant C_0 exists such that for any couple of positive number (r, h) such that $|I_\Lambda| \geq h^{-2}$ then if*

$$g_a(dy^2, d\eta^2) \overset{\text{def}}{=} \frac{dy^2}{K^2} + \frac{d\eta^2}{h^2} \quad \text{with} \quad K = C|I_\Lambda|h$$

then, we have the following two properties.

- *For any couple (Y, Z) and for any $\tau \in I_\Lambda$, we have*

$$\tag{12} \frac{1}{C_0} g_a(Y - Z) \leq g_a(\Psi_\Lambda(\tau, Y) - \Psi_\Lambda(\tau, Z)) \leq C_0 \, g_a(Y - Z).$$

- *For any couple of points (Y_0, Z_τ) of $T^\star\mathbb{R}^d$ such that $g_a(Z_\tau, \Psi_\Lambda(\tau, Y_0))^{\frac{1}{2}} \geq C_0 r$, if we have $(z, \eta) \in B_{g_a}(Y_0, r)$ and $(y, \zeta) \in B_{g_a}(Z_\tau, r)$ then*

$$g_a(\nabla_\eta \Phi_\Lambda(\tau, y, \eta) - z, \nabla_y \Phi_\Lambda(\tau, y, \eta) - \zeta) \geq frac1C_0 g_a(Z_\tau, \Psi_\Lambda(\tau, Y_0)).$$

Let us state the approximation theorem which tells us that the solution can be represented by a Fourier integral operator up to arbitrary small error term (i.e. terms which are smaller than any given power of Λ^{-1}).

THEOREM 3.1. *Let us assume that $\|\mathcal{G}\|_0$ is small enough and that $|I_\Lambda| \leq \Lambda^{2-\epsilon}$. Then, for any integer N, two families of functions $(\sigma_\Lambda^\pm)$ (with value in $\mathbb{R}^2$) on $I_\Lambda \times \mathbb{R}^{2d}$ and a constant C exist such that the following properties are satisfied.*

- Let $(v_\Lambda)_{\Lambda \geq \Lambda_0}$ be the family of solutions of (E_Λ) with $f = 0$ and with initial data $\gamma = (\gamma^0, \gamma^1)$; if we state

$$(13) \qquad \mathcal{I}_\Lambda^\pm(\gamma) \stackrel{def}{=} \int_{\mathbb{R}^d} e^{i\Phi_\Lambda^\pm(\tau,y,\eta)} \sigma_\Lambda^\pm(\tau,y,\eta) \cdot \widehat{\gamma}_\pm(\eta) d\eta \quad then$$

$$(14) \qquad \|\nabla(v_\Lambda - \mathcal{I}_\Lambda^+(\gamma) - \mathcal{I}_\Lambda^-(\gamma))\|_{L^\infty_{I_\Lambda}(L^2)} \leq C\Lambda^{-N}\|\gamma\|_{L^2}.$$

- We have

$$\|\partial_{\tau,y}^\alpha \partial_\eta^\beta \sigma^\pm\|_{L^\infty(I_\Lambda \times \mathbb{R}^{2d})} \leq C_{\alpha,\beta}\Lambda^{-|\alpha|}.$$

The proof of this is done in [**1**] and [**2**].

3.2. The precised Strichartz estimate. The theorem is the following.

THEOREM 3.2. *Let $\mathcal{C}$ be a ring of $\mathbb{R}^d$ and let us assume that $\mathcal{G}_0$ is small enough. For any positive real number ϵ, a constant C_ϵ exists which satisfies the following properties. Let f be a function in $L^1_{I_\Lambda}(L^2)$ and γ a function of L^2; let us assume that those two functions have their support included in $\mathcal{C}$ and of diameter less than h. Let us assume $|I_\Lambda| \leq \Lambda^{2-\epsilon}$. Then if v_Λ is the solution of*

$$(E_\Lambda) \left\{ \begin{array}{rcl} P_\Lambda v_\Lambda & = & f \\ \partial v_{\Lambda|\tau=0} & = & \gamma. \end{array} \right.$$

we have

$$\|v_\Lambda\|_{L^2_{I_\Lambda}(L^\infty)} \leq Ch^{\frac{d-2}{2}} \left(\log(e + |I_\Lambda|)\right)^{\frac{1}{2}} \left(\|\gamma\|_{L^2} + \|f\|_{L^1_{I_\Lambda}(L^2)}\right).$$

The proof of this theorem consits in an adaptation of [**17**] associated to the classical $TT^\star$ argument as explained in [**2**].

4. The concept of microlocalized functions

In this section, we present the concept of microlocalized functions introduced by J.-M. Bony in [**5**]. This is related to the Weyl-Hörmander calculus (see [**10**], [**7**]).

4.1. A simplified version of pseudo-differential calculus. In this paragraph, we shall consider a positive quadratic form g on $T^\star\mathbb{R}^d$ such that the symplectic conjugate defined by

$$g^\sigma(T) \stackrel{def}{=} \sup_{W \neq 0} \frac{[T, W]}{g(W)}$$

satisfies the uncertainty principle $g^\sigma \geq g$. Here $[\cdot, \cdot]$ denotes the basic symplectic form on $T^\star\mathbb{R}^d$

$$[(x,\xi),(y,\eta)] = \sum_{j=1}^{d}\left(\xi^j y_j - \eta^j x_j\right).$$

In all this paper, we are going to be in the case when

$$g(dx, d\xi) = \frac{dx^2}{K^2} + \frac{d\xi^2}{h^2}.$$

In this case, we have $g^\sigma = \lambda^2 g$ with $\lambda = Kh$. The uncertainty principle means that $\lambda \geq 1$.

We shall mesure the length of derivatives of smooth functions on $T^\star\mathbb{R}^d$ with respect to this metric g. More precisely, let us define, for any smooth function φ on $T^\star\mathbb{R}^d$,

$$\|\varphi\|_{j,g} \stackrel{\text{def}}{=} \sup_{\substack{k \leq j \\ X \in T^\star\mathbb{R}^d}} \sup_{\substack{(T_\ell)_{1 \leq \ell \leq k} \\ g(T_\ell) \leq 1}} |D^k\varphi(X)(T_1,\cdot,T_k)|.$$

Now, to a function φ in $\mathcal{D}(T^\star\mathbb{R}^d)$, we associate the operator φ^D defined by

$$(\varphi^D u)(x) = (2\pi)^{-d} \int_{T^\star\mathbb{R}^d} e^{i(x-y|\xi)}\varphi(y,\xi)u(y)dyd\xi.$$

This choice of the quantization process makes the computation of section 5 simpler. If the function $\varphi(x,\xi)$ is equal to $\varphi_1(x)\varphi_2(\xi)$, then $\varphi^D u = \mathcal{F}^{-1}\big(\varphi_2(\mathcal{F}(\varphi_1 u))\big)$. Moreover we have

$$\mathcal{F}(\varphi^D u)(\xi) = \int_{\mathbb{R}^d} e^{-i(y|\xi)}\varphi(y,\xi)u(y)dy.$$

Later on in this paper we shall need to decompose L^2 functions whose Fourier transform is supported in the ring $\mathcal{C}$ using these operators φ^D. These two lemmas are proved in [6].

LEMMA 4.1. *A sequence* $(X_\nu)_{\nu \in \mathcal{Z}}$ *exists such that two sequencies* $(\varphi_\nu)_{\nu \in \mathcal{Z}}$ *and* $(\psi_\nu)_{\nu \in \mathcal{Z}}$ *which satisfy the following properties.*

- *the support of φ_ν is included in a ball* $B_\nu \stackrel{\text{def}}{=} B_g(X_\nu, r)$,
- *A sequence $(C_j)_{j \in \mathbb{N}}$ exists (which depends only on r and not in K and h) such that*

$$\forall \nu \in \mathcal{Z}, \ \|\varphi_\nu\|_{j,g} \leq C_j,$$

- *the functions ψ_ν are not supported in B_ν but confined, i.e. a sequence $(C_N)_{N \in \mathbb{N}}$ exists such that for all $\nu \in \mathcal{Z}$*

$$\|\psi_\nu\|_{N,g,X} \stackrel{\text{def}}{=} \sup_{\substack{k \leq N \\ X \in T^\star\mathbb{R}^d}} (1 + \lambda^2 g(X - B_\nu))^N \sup_{\substack{(T_\ell)_{1 \leq \ell \leq k} \\ g(T_\ell) \leq 1}} |D^k\psi_\nu(X)(T_1,\cdot,T_k)|$$
$$\leq \ C_N,$$

- *For any function u of L^2 whose Fourier transform has a support included in $\mathcal{C}$, we have*

$$\sum_{\nu \in \mathcal{Z}} \varphi_\nu^D \psi_\nu^D u = \sum_{\nu \in \mathcal{Z}} \varphi_\nu^D u = u.$$

Such partitions of unity are "compatible" with L^2 in the following sense.

LEMMA 4.2. *A constant C exists such that*

$$C^{-1}\|u\|_{L^2}^2 \leq \sum_\nu \|\varphi_\nu^D u\|_{L^2}^2 \leq C\|u\|_{L^2}^2 \quad \text{and} \quad \sum_\nu \|\psi_\nu^D u\|_{L^2}^2 \leq C\|u\|_{L^2}^2.$$

LEMMA 4.3. *The operator φ^D maps L^p into L^p for any $p \in [1,\infty]$. More precisely, a constant C and an integer N exists such that*

$$\forall X_0 \in T^\star\mathbb{R}^d, \ \|\varphi^D a\|_{L^p} \leq C\|\varphi\|_{N,g,X_0}\|a\|_{L^p}.$$

Now we can define the concept of microlocalized function.

DEFINITION 4.1. *Let X_0 be a point of $T^\star \mathbb{R}^d$ and (C_0, r) a couple of positive real numbers. A function u in $L^2(\mathbb{R}^d)$ is said to be (C_0, r)-microlocalized in X_0 if a sequence of integer $(k_N)_{N \in \mathbb{N}}$ exists such that, for any N, the quantities*

$$\mathcal{M}_{X_0, N}^{C_0, r}(u) \overset{\text{def}}{=} \sup_{g(X - X_0)^{\frac{1}{2}} \geq C_0 r} \lambda^{2N} g(X - X_0)^N \sup_{\substack{\varphi \in \mathcal{D}(B_g(X, r)) \\ \|\varphi\|_{k_N, g} \leq 1}} \|\varphi^D u\|_{L^2}$$

are finite ($B_g(X, r)$ denotes the set of points of $T^\star \mathbb{R}^d$ such that $g(Y - X)^{\frac{1}{2}} \leq r$).

A basic example of microlocalized functions is given by the following proposition which is a corollary of Theorem 2.2.1. of [**7**].

PROPOSITION 4.1. *A sequence of integers $(k_N)_{N \in \mathbb{N}}$ and a sequence of positive real numbers $(C_N)_{N \in \mathbb{N}}$ exist such that the following properties are satisfied. Let X_0 be a point of $T^\star \mathbb{R}^d$, φ_0 a function in $\mathcal{D}(B_g(X_0, r))$ and u a function of $L^2(\mathbb{R}^d)$. Then the function $\varphi_0^D u$ is $(3, r)$-microlocalized in X_0 and, for any N, we have*

$$\mathcal{M}_{X_0, N, g}^{3, r}(u) \leq C_N \|\varphi_0\|_{k_N, g} \|u\|_{L^2}.$$

The concept of uniformly microlocalized families of functions will be a basic tool.

DEFINITION 4.2. *Let $g \overset{\text{def}}{=} (g_a)_{a \in A}$ be a family of metrics, $\mathcal{X} \overset{\text{def}}{=} (X_a)_{a \in A}$ a family of points of $T^\star \mathbb{R}^d$ and (C_0, r) a pair of positive real numbers. A family of functions $U \overset{\text{def}}{=} (u_a)_{a \in A}$ of $L^2(\mathbb{R}^d)$ is said to be uniformly (C_0, r)-microlocalized in $\mathcal{X}$ with respect to g if, for any integer N,*

$$\mathcal{M}_{N, \mathcal{X}, g}^{C_0, r}(U) \overset{\text{def}}{=} \sup_{a \in A} \mathcal{M}_{X_a, N, g_a}^{C_0, r}(u_a) < \infty.$$

4.2. A lemma about the product. Using suitable integrations by part, we prove in [**3**] the following lemma.

LEMMA 4.4. *A constant C_0 exists such that, for any integer N, a constant C_N and an integer k_N exist which satisfy the following properties.*

If u_1 and u_2 are two L^2 functions on $\mathbb{R}^d$, if χ is a function of $\mathcal{D}(\mathbb{R}^d)$ supported in an euclidian ball of radius r, if φ_1 and φ_2 are two functions of $\mathcal{D}(T^\star \mathbb{R}^d)$ respectively supported in $B_g(Y_1, r)$ and in $B_g(Y_2, r)$, then if $g(\check{Y}_1 - Y_2)^{\frac{1}{2}} \geq C_0 r$, for any N, we have

$$\left\| \chi(h^{-1} D)(\varphi_1^D u_1 \varphi_2^D u_2) \right\|_{L^1}$$
$$\leq C_N \|\varphi_1\|_{k_N, g} \|\varphi_2\|_{k_N, g} \left(1 + \lambda^2 g(\check{Y}_1 - Y_2)\right)^{-N} \|u_1\|_{L^2} \|u_2\|_{L^2}$$

where we are defined $\check{Y} \overset{\text{def}}{=} (y, -\eta)$ if $Y = (y, \eta)$.

5. The propagation theorem

Let us prove that microlocalization properties propagates along the flow of P_Λ.

THEOREM 5.1. *A constant C_0 exists which satisfies the following property.*

Let us consider a point $Y_0 = (y_0, \eta_0)$ of $T^\star \mathbb{R}^d$ such that η_0 belongs to $\mathcal{C}$, a smooth function ϕ supported in $B_{g_a}(Y_0, r)$ and a function γ of L^2. Then $\mathcal{I}_\Lambda^{\pm}(\phi^D \gamma)(\tau, \cdot)$ is (C_0, r)-microlocalized near $\Psi_\Lambda^{\pm}(\tau, Y_0)$. Moreover, for any integer N, a constant C and an integer k exist (which depend only on N) such that

$$\mathcal{M}_{\Psi_\Lambda^{\pm}(\tau, Y_0), N, g_a}^{C_0, r}\left(\mathcal{I}_\Lambda^{\pm}(\phi^D \gamma)(\tau, \cdot)\right) \leq C \|\phi\|_{k, g_a} \|\gamma\|_{L^2}.$$

In the following proof of this theorem, we shall drop the exponent $\pm$ for sake a simplicity of the notations. By definition of the microlocalized functions, we have to estimate

$$\mathcal{J} \overset{\text{def}}{=} \mathcal{F}\big(\varphi^D_{Z_\tau} \mathcal{I}_\Lambda(\phi^D \gamma)(\tau, \cdot)\big)$$

where Z_τ is a point of $T^\star \mathbb{R}^d$ such that $g_a(Z_\tau - \Psi_\Lambda(\tau, Y_0))^{\frac{1}{2}} \geq C_0 r$. By definition, we have

$$\mathcal{J}(\zeta) \;=\; \int_{\mathbb{R}^d} \mathcal{K}(\zeta, z)\gamma(z)dz \quad \text{with}$$

$$\mathcal{K}(\zeta, z) \overset{\text{def}}{=} \int_{\mathbb{R}^{2d}} e^{-i(y|\zeta)+i\Phi_\Lambda(\tau,y,\eta)-i(z|\eta)} \varphi_{Z_\tau}(y, \zeta)\sigma_\Lambda(\tau, y, \eta)\phi(z, \eta)dyd\eta.$$

The proof consists in integrations by parts in the above integral with respect to $\mathcal{L}$ defined by

$$\mathcal{L}f \overset{\text{def}}{=} \frac{1}{1+|\Theta|^2}\left(f - i|I_\Lambda|^{-\frac{1}{2}}\Theta_y \partial_\eta f - i|I_\Lambda|^{\frac{1}{2}}\Theta_\eta \partial_y f\right) \quad \text{with}$$

$$\Theta \overset{\text{def}}{=} (\Theta^y, \Theta^\eta) \overset{\text{def}}{=} \left(|I_\Lambda|^{-\frac{1}{2}}\big((\nabla_\eta \Phi_\Lambda(\tau, y, \eta) - z\big), |I_\Lambda|^{\frac{1}{2}}\big(\nabla_y \Phi_\Lambda(\tau, y, \eta) - \zeta\big)\right).$$

It is obvious that $\mathcal{L}\big(e^{-i(y|\zeta)+i\Phi_\Lambda(\tau,y,\eta)-i(z|\eta)}\big) \;=\; e^{-i(y|\zeta)+i\Phi_\Lambda(\tau,y,\eta)-i(z|\eta)}$. So as usual, we have

$$\mathcal{K}(\zeta, z) = \int_{\mathbb{R}^{2d}} e^{-i(y|\zeta)+i\Phi_\Lambda(\tau,y,\eta)-i(z|\eta)} (^t\mathcal{L})^N \big(\varphi_{Z_\tau}(y, \zeta)\sigma_\Lambda(\tau, y, \eta)\phi(z, \eta)\big)dyd\eta$$

Let us state the following technical lemma.

LEMMA 5.1. *For any integer N, a family of functions $(L_{\alpha,N})_{|\alpha|\leq N}$ exists such that $L_{\alpha,N}(Y, \mathcal{Y})$ is a smooth function from $T^\star\mathbb{R}^d \times (T^\star\mathbb{R}^d)^{M_N}$ and such that*

$$(15) \qquad \|\partial^\beta_Y L_{\alpha,N}(Y, \cdot)\|_{L^\infty((T^\star\mathbb{R}^d)^{M_N})} \leq C_{N,|\beta|}(1 + |Y|^2)^{-\frac{N+|\beta|}{2}}.$$

Moreover, they satisfy

$$(^t\mathcal{L})^N f = \sum_{|\alpha|\leq N} L_{\alpha,N}(\Theta, (\partial^\beta \Theta)_{|\beta|\leq N})\widetilde{\partial}^\alpha f$$

where $\widetilde{\partial}$ denotes differentiation of length 1 for the metric $\widetilde{g}_a$ defined by

$$\widetilde{g}_a(dy^2, d\eta^2) \overset{\text{def}}{=} |I_\Lambda|^{-1}dy^2 + |I_\Lambda|d\eta^2 = \lambda g_a(dy^2, d\eta^2).$$

Between g_a and $g_a^\sigma = \lambda^2 g_a$ and as the metric $\widetilde{g}_a$ is greater than $g_\Lambda \overset{\text{def}}{=} \Lambda^{-2}dx^2 + d\eta^2$, we have thanks to Leibniz formula that derivatives of $\widetilde{g}_a$-length 1 of $\varphi_{Z_\tau}(y, \zeta)\sigma_\Lambda(\tau, y, \eta)\phi(z, \eta)$ are bounded uniformly with respect to the involved parameters. So using Lemma 5.1, we have

$$\forall N, \; \exists C_N \;/\; \left|(^t\mathcal{L})^N\big(\varphi_{Z_\tau}(y, \zeta)\sigma_\Lambda(\tau, y, \eta)\phi(z, \eta)\big)\right| \leq C_N(1 + |\Theta|^2)^{-\frac{N}{2}}.$$

So by definition of Θ and $\widetilde{g}_a$, we infer that, for any integer N, we have

$$|\mathcal{K}(\zeta, z)| \leq C_N \int_{\substack{|y-z_\tau|\leq rK \\ |\eta-\eta_0|\leq rh}} \left(1 + \lambda g_a\big(\nabla_\eta \Phi_\Lambda(\tau, y, \eta) - z, \nabla_y \Phi_\Lambda(\tau, y, \eta) - \zeta\big)\right)^{-N} dyd\eta.$$

Now let us apply Lemma 3.2. Using the fact that $g_a(Z_\tau, \Psi_\Lambda(\tau, Y_0))^{\frac{1}{2}}$ is greater than $C_0 r$, that (z, η) belongs to $B_{g_a}(Y_0, r)$ and (y, ζ) to $B_{g_a}(Z_\tau, r)$ we infer that

$$|\mathcal{K}(\zeta, z)| \leq C_N \left(1 + \lambda g_a\big(Z_\tau - \Psi_\Lambda(\tau, Y_0)\big)\right)^{-N} \left(1 + \lambda g_a\big((z, \zeta) - (y_0, \Psi_\Lambda^\eta(\tau, Y_0))\big)\right)^{-N}$$

$$\times \int_{\substack{|y - z_\tau| \leq rK \\ |\eta - \eta_0| \leq rh}} \frac{dy\,d\eta}{\left(1 + \lambda g_a\big(\nabla_\eta \Phi_\Lambda(\tau, y, \eta) - z, \nabla_y \Phi_\Lambda(\tau, y, \eta) - \zeta\big)\right)^N} \cdot$$

Let us state the change of variables

$$y' = |I|^{-\frac{1}{2}} \big(\nabla_\eta \Phi_\Lambda(\tau, y, \eta) - z\big) \quad \text{and} \quad \eta' = |I|^{\frac{1}{2}} \big(\nabla_y \Phi_\Lambda(\tau, y, \eta) - \zeta\big).$$

As the jacobian of this change of variables is closed to 1, it turns out that

$$|\mathcal{K}(\zeta, z)| \leq C_N \left(1 + \lambda g_a\big(Z_\tau - \Psi_\Lambda(\tau, Y_0)\big)\right)^{-N} \left(1 + \lambda g_a\big((z, \zeta) - (y_0, \Psi_\Lambda^\eta(\tau, Y_0))\big)\right)^{-N}.$$

Immediate integrations imply that

$$|I_\Lambda|^{-\frac{d}{2}} \int |\mathcal{K}(\zeta, z)|dz + |I_\Lambda|^{\frac{d}{2}} \int |\mathcal{K}(\zeta, z)|d\zeta \leq C_N \left(1 + \lambda g_a(Z_\tau - \Psi_\Lambda(\tau, Y_0))\right)^{-N}.$$

By Schur's lemma we get that, for any N

$$\|\mathcal{J}\|_{L^2} \leq C_N \left(1 + \lambda g_a(Z_\tau - \Psi_\Lambda(\tau, Y_0))\right)^{-N} \|\gamma\|_{L^2}.$$

As $g_a(Z_\tau - \Psi_\Lambda(\tau, Y_0))^{\frac{1}{2}} \geq C_0 r$, then $\lambda g_a(Z_\tau - \Psi_\Lambda(\tau, Y_0)) \geq C_0 r \lambda g_a(Z_\tau - \Psi_\Lambda(\tau, Y_0))^{\frac{1}{2}}$. So Theorem 5.1 is proved.

6. The conclusion of the proof

To conclude the proof of theorem 2.1 let us first apply Lemma 4.4 and Theorem 5.1 to concentrate on real interaction. Because variable coefficents do not respect the localization in frequency space, we shall need to decompose the interval I_Λ. In this section, we shall state

$$\mathcal{J}(\tau, y) \overset{\text{def}}{=} \chi(h^{-1}D)\big(\mathcal{I}_\Lambda(\gamma_1)(\tau, y)\mathcal{I}_\Lambda(\gamma_2)(\tau, y)\big).$$

The equivalent of Identity (5) of the constant coefficient case is the following lemma.

LEMMA 6.1. *Let $J = (\tau_J, \tau_J^+)$ be a subinterval of I_Λ such that*

$$|J| \leq h|I_\Lambda| \quad \text{and} \quad \|\nabla G_\Lambda\|_{L^1_J(L^\infty)} \leq h\|\nabla G_\Lambda\|_{L^1_{I_\Lambda}(L^\infty)}.$$

Then two families (ϕ_μ) and (θ_μ) of confined symbols exists such that, for any integer N,

$$\forall \mu, \ \|\phi_\mu\|_{N, g_a, \Psi_\Lambda(\tau_J, Y_\mu)} + \|\theta_\mu\|_{N, g_a, \check{\Psi}_\Lambda(\tau_J, Y_\mu)} \leq C_N$$

and, for any N, a constant C_N exists such that

$$\|\mathcal{J} - \underline{\mathcal{J}}\|_{L^1_J(L^\infty)} \leq C_N h\lambda^{-N}(|I_\Lambda|h^2)h^{d-2}\|\gamma_1\|_{L^2}\|\gamma_2\|_{L^2} \quad \text{with}$$

$$\underline{\mathcal{J}}(\tau) \overset{\text{def}}{=} \sum_{\substack{\mu \\ \mu' \in A_\mu}} \chi(h^{-1}D)\Big(\phi_\mu^D \mathcal{I}_\Lambda\big(\varphi_\mu^D \psi_\mu^D \gamma_1\big)(\tau, \cdot) \times \theta_\mu^D \mathcal{I}_\Lambda\big(\varphi_{\mu'}^D \psi_{\mu'}^D \gamma_2\big)(\tau, \cdot)\Big)$$

and

$$A_\mu \subset \big\{\mu' \,/\, g_a(Y_{\mu'} - \Psi_\Lambda(\tau_J, \check{\Psi}_\Lambda(\tau_J, Y_\mu))) \leq Cr\big\}.$$

Now we shall decompose the interval I_Λ on subintervals J such that the above lemma can be applied on J. Let us introduce the following function on the interval I_Λ

$$H(\tau) \overset{\text{def}}{=} \left(\sum_\mu \|\mathcal{I}_\Lambda(\varphi_\mu^D \psi_\mu^D \gamma_1)(\tau, \cdot)\|_{L^\infty}^2 \right)^{\frac{1}{2}} \left(\sum_\mu \|\mathcal{I}_\Lambda(\varphi_\mu^D \psi_\mu^D \gamma_2)(\tau, \cdot)\|_{L^\infty}^2 \right)^{\frac{1}{2}}.$$

Using precised Strichartz estimates, we get that

$$\|\mathcal{I}_\Lambda(\varphi_\mu^D \psi_\mu^D \gamma_j)\|_{L^2_{I_\Lambda}(L^\infty)} \le C\big(\log(e + |I_\Lambda|)\big)^{\frac{1}{2}} h^{\frac{d-2}{2}} \|\psi_\mu^D \gamma_j\|_{L^2}.$$

So, using Cauchy-Schwartz inequality and Lemma 4.2, we get that

$$\begin{aligned}
\int_{I_\Lambda} H(\tau)d\tau &\le& C\big(\log(e + |I_\Lambda|)\big)h^{d-2}\left(\sum_\mu \|\psi_\mu^D \gamma_1\|_{L^2}^2 \right)^{\frac{1}{2}} \left(\sum_\mu \|\psi_\mu^D \gamma_2\|_{L^2}^2 \right)^{\frac{1}{2}} \\
&\le& C\big(\log(e + |I_\Lambda|)\big)h^{d-2}\|\gamma_1\|_{L^2}^2 \|\gamma_2\|_{L^2}^2.
\end{aligned}$$

As in section 2, we decompose I_Λ in intervals J such that

$$|J| \le h|I_\Lambda|, \quad \|\nabla G_\Lambda\|_{L^1_J(L^\infty)} \le h\|\nabla G_\Lambda\|_{L^1_{I_\Lambda}(L^\infty)} \quad \text{and} \quad \int_J H(\tau)d\tau \le h\int_{I_\Lambda} H(\tau)d\tau.$$

Let us estimate $\|\underline{\mathcal{J}}\|_{L^1_J(L^\infty)}$. Lemma 4.3 implies that

$$\|\underline{\mathcal{J}}(\tau)\|_{L^\infty} \le \sum_{\substack{\mu \\ \mu' \in A_\mu}} \left\|\mathcal{I}_\Lambda\big(\varphi_\mu^D \psi_\mu^D \gamma_1\big)(\tau, \cdot)\right\|_{L^\infty} \left\|\mathcal{I}_\Lambda\big(\varphi_{\mu'}^D \psi_{\mu'}^D \gamma_2\big)(\tau, \cdot)\right\|_{L^\infty}.$$

Because the number of elements of A_μ is bounded uniformly with respect to μ, by Cauchy-Schwarz inequality, we infer that $\|\underline{\mathcal{J}}(\tau)\|_{L^\infty} \le H(\tau)$. So by construction of J,

$$\|\underline{\mathcal{J}}\|_{L^1_J(L^\infty)} \le Ch\big(\log(e + |I_\Lambda|)\big)h^{d-2}\|\gamma_1\|_{L^2}^2 \|\gamma_2\|_{L^2}^2.$$

As in section 2, the number of intervals J is less than Ch^{-1}. As $\lambda = |I_\Lambda|h^2$, the theorem is proved if we apply Lemma 6.1 with N large enough.

References

[1] H. Bahouri and J.-Y. Chemin, Équations d'ondes quasilinéaires et inégalités de Strichartz, *American Journal of Mathematics*, **121**, 1999, pages 1337–1377.

[2] H. Bahouri and J.-Y. Chemin, Équations d'ondes quasilinéaires et effet dispersif, *International Mathematical Research News*, **21**, 1999, pages 1141–1178.

[3] H. Bahouri and J.-Y. Chemin, Microlocal analysis, bilinear estimates and cubic quasilinear wave equation, *in preparation*.

[4] J.-M. Bony, Calcul symbolique et propagation des singularités pour les équations aux dérivées partielles non linéaires, *Annales de l'École Normale Supérieure*, **14**, 1981, pages 209–246.

[5] J.-M. Bony, personnal communication.

[6] J.-M. Bony and J.-Y. Chemin, Espaces fonctionnels associés au calcul de Weyl-Hörmander, *Bulletin de la Société Mathématique de France*, **122**, 1994, pages 77–118.

[7] J.-M.Bony and N.Lerner, Quantification asymptotique et microlocalisation d'ordre supérieur, *Annales de l'École Normale Supérieure*, **22**, 1989, pages 377–433.

[8] J.-Y. Chemin and C.-J. Xu, Inclusions de Sobolev en calcul de Weyl-Hörmander et systèmes sous-elliptiques, *Annales de l'École Normale Supérieure*, **30**, 1997, pages 719–751.

[9] J. Ginibre and G. Velo, Generalized Strichartz inequalities for the wave equation, *Journal of Functional Analysis*, **133**, 1995, page 50–68.

[10] L. Hörmander, *The analysis of linear partial differential equations*, Springer Verlag, 1983.

[11] L. Kapitanski, Some generalization of the Strichartz-Brenner inequality, *Leningrad Mathematical Journal*, **1**, 1990, pages 693–721.

[12] S. Klainerman, The null condition and global existence to non linear wave equations, *Communications in Pure and Applied Mathematics*, **38**, 1985, pages 631–641.

[13] S. Klainerman and M. Machedon, Space-time estimates for null forms and the local existence theorem, *Communications in Pure and Applied Mathematics*, **46**, 1993, pages 1221–1268.

[14] S. Klainerman and M. Machedon, Smoothing estimates for null forms and applications, *Duke Mathematical Journal*, **81**, 1995, pages 99–133.

[15] S. Klainerman and M. Machedon, On the regularity properties of a model problem relates to wave maps, *Duke Mathematical Journal*, **87**, 1997, pages 553–589.

[16] S. Klainerman and M. Machedon, Estimates for null forms and the spaces $H_{s,\delta}$, *International Mathematical Research News*, **15**, 1998, pages 756–774.

[17] S. Klainerman and D. Tataru, On the optimal local regularity for the Yang-Mills equations in $\mathbf{R}^{4+1}$, Journal of the American Mathematical Society, **12**, 1999, pages 93–116.

[18] H. Lindblad, A sharp couternexample to local existence of low regularity solutions to non linear wave equations, *Duke Mathematical Journal*, **72**, 1993, pages 503–539.

[19] G. Ponce and T. Sideris, Local regularity of non linear wave equations in three space dimensions, *Communications in Partial Differential Equations*, **18**, 1993, pages 169–177.

[20] H. Smith, A parametrix construction for wave equation with $C^{1,1}$ coefficients, *Annales de l'Institut Fourier*, **48**, 1998, pages 797–835.

[21] D. Tataru, Local and global results for wave maps I, *Communications in Partial Differential Equations*, **23**, 1998, pages 1781–1793.

[22] D. Tataru, Strichartz estimates for second order hyperbolic operators with nonsmooth coefficients III, *preprint*

UNIVERSITÉ DE TUNIS, DÉPARTEMENT DE MATHMATIQUES, 1060 TUNIS, TUNISIA
E-mail address: `Hajer.Bahouri@fst.rnu.tn`

UNIVERSITÉ PIERRE-ET-MARIE-CURIE, ANALYSE NUMÉRIQUE, 4, PLACE JUSSIEU, 75230 PARIS CEDEX 05, FRANCE
E-mail address: `chemin@ann.jussieu.fr`

Contemporary Mathematics
Volume **350**, 2004

Ginzburg-Landau Functionals, Phase Transitions and Vorticity

F. Bethuel, G. Orlandi

This paper is dedicated to Haim Brezis and Felix Browder

ABSTRACT. We review some old and new results concerning asymptotics for Ginzburg-Landau functionals, both in the scalar and in the complex case. In particular we try to emphasize some analogies and differences between the two theories.

I. GINZBURG-LANDAU FUNCTIONALS

Let Ω be a smooth open domain in $\mathbb{R}^N$, $N \geq 1$. Consider functions

$$u : \Omega \to \mathbb{R}^d, \qquad d \geq 1.$$

Ginzburg-Landau energies are families of functionals involving a non-convex potential. A typical example is given by

$$E_\varepsilon(u) = \int_\Omega e_\varepsilon(u) = \frac{1}{2} \int_\Omega |\nabla u|^2 + \frac{1}{4\varepsilon^2} \int_\Omega W(u),$$

where the potential W is a non-convex smooth function from Ω into $\mathbb{R}^+$, and $\varepsilon > 0$ denotes a parameter which will be assumed small throughout this paper. In particular we are interested in the asymptotic limit $\varepsilon \to 0$.

The set

$$\Sigma = \{y \in \mathbb{R}^d, \; W(y) = 0\},$$

which is sometimes called the vacuum manifold in the physical literature plays an important role in this analysis. As we will see, its topology will enter in an essential way.

1991 *Mathematics Subject Classification.* 35J60, 58E50, 49J10.

Key words and phrases. Ginzburg-Landau, phase transitions, vortices.

Supoort first author Institut Universitaire de France, European grant RTN HPRN-CT-2002-00274 "Front-Singularities".

Support second author MURST Research Project "Calcolo delle Variazioni" European grant RTN HPRN-CT-2002-00274 "Front-Singularities".

These functionals have been introduced in the early fifties by Ginzburg and Landau in order to describe phase transitions in condensed matter Physics (more precisely, at low temperature). The nature of the predicted defects (e.g. points, lines, walls) depends crucially on d and Σ (see [23]). Among the many variants of Ginzburg-Landau functionals, there are in particular those including electro-magnetic effects, as for instance in superconductivity. Related models have been developed in particle physics (as for examples, Yang-Mills-Higgs theory).

In this paper we will focus on the cases $d = 1$ and $d = 2$ (i.e. u real or complex-valued). Moreover, we assume the potential is

$$W(u) = (1 - |u|^2)^2\,,$$

often called two-well potential for $d = 1$. This potential is used in large number of physical models. Note that

$$\Sigma = \{-1, 1\} \quad \text{if } d = 1\,, \qquad\qquad \Sigma = S^1 \quad \text{if } d = 2\,,$$

where S^1 is the unit circle in $\mathbb{R}^2$. In the first case, i.e. $d = 1$, the non-connectedness of Σ yields typically codimension one defects, whereas in the second case, i.e. $d = 2$, Σ is not simply connected and allows for defects of codimension two.

Our aim is to survey recent results concerning the asymptotic behavior, as $\varepsilon \to 0$, of minimizers as well as critical points of E_ε. Various boundary conditions on $\partial\Omega$ are of interest (in particular for Physics). Here for simplicity we will restrict ourselves to Dirichlet boundary conditions such as $u = g_\varepsilon$ on $\partial\Omega$, with $|g_\varepsilon| = 1$. The Euler-Lagrange equation satisfied by critical points of E_ε is then

$$(\text{GL})_\varepsilon \qquad \begin{cases} -\Delta u_\varepsilon = \frac{1}{\varepsilon^2} u_\varepsilon (1 - |u_\varepsilon|^2) & \text{in } \Omega \\ u_\varepsilon = g_\varepsilon & \text{on } \partial\Omega. \end{cases}$$

The associated function space is

$$H^1_{g_\varepsilon}(\Omega; \mathbb{R}^d) = \left\{ u \in L^2(\Omega; \mathbb{R}^d)\,, \ \frac{\partial u}{\partial x_i} \in L^2(\Omega; \mathbb{R}^d)\,, \ u = g_\varepsilon \text{ on } \partial\Omega \right\}.$$

For this type of boundary data it is easy to prove existence of minimizers, i.e. there exists $u_\varepsilon \in H^1_{g_\varepsilon}(\Omega; \mathbb{R}^d)$ such that

$$E_\varepsilon(u_\varepsilon) = \inf_{u \in H^1_{g_\varepsilon}} E_\varepsilon(u)\,.$$

In the study of the asymptotics of minimizers (or critical points) it is of course useful to derive some compactness. However in general, sequences of minimizers u_ε will diverge in the natural energy norm H^1, i.e.

$$\int_\Omega |\nabla u_\varepsilon|^2 \to +\infty \qquad\qquad \text{as } \varepsilon \to 0\,.$$

On the other hand we expect that, as $\varepsilon \to 0$, possible limiting maps u_* of the sequence u_ε will be forced by the potential term W to take values in Σ, i.e. $|u_*| = 1$ a.e. in Ω. We will see that one can obtain compactness in larger spaces than H^1, whose definition depends on the nature of Σ.

A second class of problems of physical interest are the associated evolution problems, such as the nonlinear heat flow equation

$$(\text{PGL})_\varepsilon \qquad \begin{cases} \dfrac{\partial u_\varepsilon}{\partial t} - \Delta u_\varepsilon = \dfrac{1}{\varepsilon^2} u_\varepsilon (1 - |u_\varepsilon|^2) & \text{in } \Omega \times (0, +\infty), \\[2mm] u_\varepsilon(x,0) = u_\varepsilon^0(x) & \text{for a.e. } x \in \Omega, \\[2mm] u_\varepsilon(x,t) = g_\varepsilon(x) & \text{for a.e. } (x,t) \in \partial\Omega \times (0, +\infty). \end{cases}$$

and the nonlinear Schrödinger equation

$$(\text{NLS})_\varepsilon \qquad \begin{cases} i\dfrac{\partial u_\varepsilon}{\partial t} - \Delta u_\varepsilon = \dfrac{1}{\varepsilon^2} u_\varepsilon (1 - |u_\varepsilon|^2) & \text{in } \Omega \times (0, +\infty), \\[2mm] u_\varepsilon(x,0) = u_\varepsilon^0(x) & \text{for a.e. } x \in \Omega, \\[2mm] u_\varepsilon(x,t) = g_\varepsilon(x) & \text{for a.e. } (x,t) \in \partial\Omega \times (0, +\infty). \end{cases}$$

We will not discuss here the results obtained for these problems (except the existence problem for travelling waves in $(\text{NLS})_\varepsilon$). Let us mention however, that their study relies heavily, for some aspects, on the analysis developed for the elliptic case.

II. THE SCALAR CASE

The case $d = 1$, i.e. u is real-valued, was extensively studied in the 80's, in particular by the De Giorgi school (starting with the seminal work by Modica-Mortola [26], and [25]) and also, motivated by physical questions, since the works by Gurtin and Sternberg [32,17].

As mentioned, in many interesting situations, for a sequence of minimizers u_ε, the H^1 norm will diverge as $\varepsilon \to 0$. However (for suitable boundary conditions), the bound

$$E_\varepsilon(u_\varepsilon) \le \frac{K}{\varepsilon},$$

where K is independent on ε, can be easily established. We will see that this estimate is sufficient to derive some compactness. More precisely, given a sequence v_ε verifying the bound $E_\varepsilon(v_\varepsilon) \le K\varepsilon^{-1}$, there exists a subsequence v_{ε_n} such that

$$v_{\varepsilon_n} \to v_* \quad \text{in } L^1(\Omega), \qquad v_*(x) \in \{-1,1\} \qquad \text{for a.e. } x \in \Omega.$$

Moreover, $v_* \in BV(\Omega)$.

In order to get some insight for this type of result, let us first consider the one dimensional case, which captures already some of the essential features of the problem.

II.1 The case $N = 1$

Take $\Omega = (-1,1)$ and, for example, as boundary condition

$$g : \partial\Omega \to \mathbb{R} \qquad g(-1) = -1, \qquad g(1) = 1.$$

There exists a unique minimizer u_ε in $H^1_g(\Omega)$ of

$$E_\varepsilon(u) = \int_{-1}^{1} \frac{\dot u^2}{2} + \frac{1}{4\varepsilon^2}(1 - |u|^2)^2,$$

moreover, u_ε is odd and verifies the equation

$$\begin{cases} -\ddot{u}_\varepsilon = \dfrac{1}{\varepsilon^2} u_\varepsilon(1 - |u_\varepsilon|^2) & \text{in } (-1,1), \\ u_\varepsilon(-1) = -1 \,, u_\varepsilon(1) = 1 \,. \end{cases}$$

Elementary calculations show that $E_\varepsilon(u_\varepsilon) \simeq c_0/\varepsilon$ for a constant c_0, the energy balance

$$\int_{-1}^{1} \frac{\dot{u}^2}{2} \simeq \int_{-1}^{1} \frac{1}{4\varepsilon^2}(1 - |u|^2)^2 \,,$$

and the convergence in rescaled units to the optimal profile

$$u_\varepsilon\left(\frac{x}{\varepsilon}\right) \to \tanh\left(\frac{x}{2}\right) \qquad \text{as } \varepsilon \to 0 \,.$$

In particular, $u_\varepsilon(x) \to u_*(x)$, where

$$\begin{cases} u_*(x) = -1 & \text{if } x < 0 \,, \\ u_*(x) = 1 & \text{if } x > 0 \,. \end{cases}$$

It turns therefore out that the cost of an interface is of the order of $1/\varepsilon$. Hence, interesting configurations with a finite number of interfaces will correspond to maps v_ε such that

$$E_\varepsilon(v_\varepsilon) \le \frac{K}{\varepsilon} \,.$$

In this context, the following result, which holds for an arbitrary N-dimensional domain Ω, is available.

PROPOSITION 1. *Let $(v_\varepsilon)_{\varepsilon>0}$ a sequence such that*

(II.1) $$E_\varepsilon(v_\varepsilon) \le \frac{K}{\varepsilon} \,.$$

Then, for a subsequence $\varepsilon_n \to 0$,

$$v_{\varepsilon_n} \to v_* \qquad \text{in } L^1(\Omega) \,,$$

where $v_*(x) \in \{-1, 1\}$ for a.e. $x \in \Omega$, and $v_* \in BV(\Omega)$.

Sketch of proof: we have

$$\frac{\varepsilon}{2} \int_\Omega |\nabla v_\varepsilon|^2 + \frac{1}{4\varepsilon} \int_\Omega (1 - |v_\varepsilon|^2)^2 = \varepsilon E_\varepsilon(v_\varepsilon) \le K \,.$$

Hence, from the inequality $ab \le \frac{1}{2}(a^2 + b^2)$, it holds

$$\int_\Omega |\nabla v||1 - |v_\varepsilon|^2| \le \frac{\varepsilon}{2} \int_\Omega |\nabla v_\varepsilon|^2 + \frac{1}{2\varepsilon} \int_\Omega (1 - |v_\varepsilon|^2)^2 \le 2K \,,$$

that is

$$\int_\Omega |\nabla \chi(v_\varepsilon)| \le 2K \,,$$

where $\chi(t) = t - t^3/3$. This yields a uniform bound in $W^{1,1}$ for $\chi(v_\varepsilon)$, and a subsequence $\chi(v_{\varepsilon_n})$ converges therefore weakly in $BV(\Omega)$, hence strongly in L^1. So does $v_{\varepsilon_n} = \chi^{-1}(\chi(v_{\varepsilon_n}))$.

Note that if $N = 1$ and $\Omega = (-1, 1)$, then it is an elementary exercise to prove that the condition $v_* \in BV(\Omega)$ and $|v_*| = 1$ a.e. in Ω implies that v_* has only a finite numbers of jumps, depending on K. In view of this remark and Proposition 1, in the case $N = 1$ the limiting map v_* will have a finite number of jumps, of cardinality less than K/μ, where the absolute constant μ represents the minimal cost in energy to perform one jump from -1 to 1 (or vice versa).

One therefore obtain the following corollary

COROLLARY 1. *Let $K > 0$ and $\Omega = (-1, 1)$, and let v_ε be a sequence such that $E_\varepsilon(v_\varepsilon) \leq \frac{K}{\varepsilon}$. Then, there exists a subsequence $\varepsilon_n \to 0$, $\ell \in \mathbb{N}$, ℓ depending only on K, ℓ points $a_1, \ldots, a_\ell \in (-1, 1)$, and $\chi \in \{-1, 1\}$, such that*

$$v_{\varepsilon_n}(x) \to \chi \prod_{i=1}^{\ell} \left(\frac{x - a_i}{|x - a_i|} \right) \qquad in \; L^1(\Omega).$$

Remark II.1. i) Let us emphasize that condition (II.1) does not imply that the sequence v_ε is bounded in BV. A simple example is given by

$$v_\varepsilon(x) = 1 + \varepsilon^{1/2} \sin(\frac{x}{\varepsilon}), \qquad \text{for } -1 \leq x \leq 1.$$

Clearly the maps v_ε satisfy (II.1) but they are not equibounded in BV.

ii) On the other hand, one may prove that given any sequence v_ε satisfying (II.1) there exists another sequence $\tilde{v}_\varepsilon$ verifying also (II.1) which is equibounded in BV and which is close to the original sequence v_ε in the following sense:

$$\|v_\varepsilon - \tilde{v}_\varepsilon\|_{L^1(\Omega)} \to 0 \qquad \text{as } \varepsilon \to 0.$$

The main point is to get rid of the possible small oscillations of v_ε on the set where it takes values close to $+1$ and -1. This is achieved by a composition with a suitable projection on $\Sigma = \{-1, 1\}$.

II.2. The case $N \geq 2$

As in dimension one, the fact that $v_* \in BV(\Omega)$ and $|v_*| = 1$ a.e. in Ω allows to deduce regularity properties for the jump set of v_*, which are best expressed in the language of Geometric Measure Theory. This step is however far from being elementary, and relies on De Giorgi's theory of finite perimeter sets. More precisely, let $w_* \in BV(\Omega)$ (so that Dw_* is a measure), and $|w_*| = 1$ a.e.. Let $\Omega_*^{\pm} = \{x \in \Omega, w_*(x) = \pm 1\}$. Then Dw_* is supported on the $(N-1)$-rectifiable set $\partial^*\Omega_*^{\pm}$, the reduced boundary of $\Omega_*^{\pm}$. [For the definition of reduced boundary, see e.g. [31]; the reduced boundary is included of the usual topological boundary. In the smooth case they actually coincide, but in general they may be different].

Using minimality (or stationarity) of the sequence v_ε one may deduce additional properties on the interface, i.e. the jump set of v_*. For instance, if $u_\varepsilon \to u_*$ is a sequence of minimizers, then the interface, will be an area minimizing $(N-1)$-rectifiable set [26], while if $v_\varepsilon \to v_*$ is a sequence of critical points in E_ε^K, then the interface will be a stationary varifold [18]. Here $E_\varepsilon^K = \{v, \; E_\varepsilon(v) \leq \frac{K}{\varepsilon}\}$ denotes the level set of E_ε in the natural energy space $H^1(\Omega)$.

Remark II.2. In many physical problems, the parameter ε is so small that very little information is lost by considering v_* instead of v_ε. This means that one may replace the variational problem for the functions by the variational problem for the interface. This statement can be made rigourous in the sense of Γ-convergence.

If one is interested at fine properties at the level ε, one may wonder how the interface behaves for v_ε. This is a difficult question for stationary points. However, for minimizers u_ε the interface behaves basically as the one dimensional optimal profile $\tanh(x/2)$ matched at the scaling ε on the interface of u_*.

III. THE COMPLEX CASE: VORTICITY

Here we will consider $u : \Omega \to \mathbb{C} \simeq \mathbb{R}^2$, and

$$W(u) = \frac{1}{4}(1 - |u|^2)^2 \,,$$

so that $\Sigma = \{y \in \mathbb{C}, \ W(y) = 0\} = \{y \in \mathbb{C}, \ |y| = 1\} = S^1$. A new type of singularity can appear here, due to the fact that $\pi_1(S^1) = \mathbb{Z} \neq 0$. Interesting new cases of topological defects appear therefore for planar Ω, i.e. for $N = 2$ (this is somewhat similar to the one dimensional case for scalar problems).

III.1 The case $N = 2$

III.1.1 A minimization problem

Let $\Omega = D^2 = \{z \in \mathbb{C} \simeq \mathbb{R}^2, \ |z| = 1\}$, and consider

$$g : \partial\Omega = S^1 \to S^1$$

defined by $g(z) = z$, i.e. $g = Id_{S^1}$. Let us consider the minimization problem

$$\mu_\varepsilon = \inf\{E_\varepsilon(v), \ v \in H_g^1(D^2; \mathbb{C})\} \,.$$

It is natural, due to the symmetries in the problem, to seek solutions of the form

$$w_\varepsilon(z) = f_\varepsilon(r) \exp(i\theta) = f_\varepsilon(r)\frac{z}{|z|}$$

where $z = r\exp(i\theta)$ (in polar coordinates), and $f_\varepsilon : \mathbb{R}^+ \to \mathbb{R}$ is smooth, with $f_\varepsilon(0) = 0$. Choosing f_ε such that $f_\varepsilon(r) = 1$ for $r \geq \varepsilon$, and $|f'| \leq 2$, a simple computation shows that

$$\mu_\varepsilon \leq E_\varepsilon(w_\varepsilon) \leq K|\log \varepsilon| \,.$$

Actually, it has been proved that the minimizers u_ε, for small ε, do have radial symmetry [24,27]. Moreover, as in the scalar case, we may define an optimal profile (although it is not given by an explicit formula). More precisely, there exists a unique function $f : \mathbb{R}^+ \to \mathbb{R}$ satisfying

$$\begin{cases} -f'' - \frac{1}{r}f' + \frac{1}{r^2}f = f(1 - f^2) & \text{on } [0, +\infty) \\ f(0) = 0 \end{cases} \,.$$

Then we have

$$u_\varepsilon(z) \simeq f(\frac{|z|}{\varepsilon}) \exp(i\theta)$$

and

$$u_\varepsilon(z) \to u_*(z) = \frac{z}{|z|} \qquad \text{as } \varepsilon \to 0, \qquad \text{in } W^{1,p}, \ p < 2 \,,$$

and in $C_{loc}^k(D^2 \setminus \{0\})$. The map $u_*(z) = z/|z|$ realizes thus the prototypical singularity that can appear in the asymptotics for minimization problems.

III.1.2 Asymptotics for solutions to $(\mathbf{GL})_\varepsilon$

In contrast to the scalar case, there is of course a large choice of boundary conditions verifying $|g| = 1$. The general situation where Ω is a smooth bounded starshaped domain in $\mathbb{R}^2$ was analyzed in [8].

THEOREM 1, ([8]). *Let $\Omega \subset \mathbb{R}^2$ be a smooth bounded starshaped domain, $g : \partial\Omega \to S^1$ and $w_\varepsilon \in H^1_g(\Omega; \mathbb{R}^2)$ a critical point of the Ginzburg-Landau functional E_ε. Then there exists $K > 0$ such that*

$$E_\varepsilon(w_\varepsilon) \le K|\log \varepsilon|, \qquad \frac{1}{4\varepsilon^2} \int_\Omega (1 - |w_\varepsilon|^2)^2 \le K.$$

Moreover, there exists points $a_1, \ldots, a_\ell$ in Ω, integers $d_1, \ldots, d_\ell \ne 0$, with $\ell \le K$, and a harmonic function $\varphi : \Omega \to \mathbb{R}$ such that $w_\varepsilon \to w_*$ in $W^{1,p}(\Omega)$ for any $p < 2$ and in $C^k_{loc}(\Omega \setminus \{a_1, \ldots, a_\ell\})$, where

$$w_*(z) = \exp(i\varphi(z)) \prod_{i=1}^{\ell} \left(\frac{z - a_i}{|z - a_i|} \right)^{d_i}.$$

The points a_i are usually called "vortices" (in analogy with the terminology of fluid dynamics), the integers d_i their multiplicity. Since φ is harmonic, it is completely determined by the boundary condition, the location of the points a_i and their multiplicities. As a matter of fact it can be proved that the configuration $(a_1, d_1), \ldots, (a_\ell, d_\ell)$ is not arbitrary, but has to be critical for a suitable renormalized energy (i.e. independent of ε). Again, the boundary condition enters in an essential way in the definition of this energy.

As the reader might already have noticed, there are strong analogies between the 1-dimensional scalar case and the planar complex case: clearly vortices and jumps play a somewhat similar role. Let us stress however a few differences:

i) the typical energy necessary to the formation of a vortex is of order $|\log \varepsilon|$, whereas for jumps it is ε^{-1};
ii) the multiplicity of a vortex is an integer, whereas the jumps can be labeled by coefficients in $\mathbb{Z}_2$.
iii) in a (vague) sense, jumps do not "interact", whereas vortices do. Their interaction is governed by the renormalized energy.

Another striking difference concerns the way the theory has been developed in both cases. Indeed, PDE techniques have played an important role in the starting development for the complex case, while the emphasis was put first, for the scalar case, on variational methods (e.g. compactness, Γ-convergence...).

III.1.3 Compactness for level sets of E_ε

In order to handle the two theories from a more homogeneous point of view, the following question appears naturally:

(**Q**) what kind of compactness can be expected for sequences of maps $w_\varepsilon \in W^K_\varepsilon$, where

$$W^K_\varepsilon = \{v \in H^1(\Omega), \, E_\varepsilon(v) \le K|\log \varepsilon|\} ?$$

A simple example shows that no general compactness result for reasonable norms can be derived, due to possible divergences in the phase. Take, for instance

$$w_\varepsilon(z) = \exp(i\varphi(z)\sqrt{|\log\varepsilon|})\,,$$

with $\varphi : \Omega \to \mathbb{R}$ a non-constant smooth function. We have $|w_\varepsilon| = 1$, hence

$$E_\varepsilon(w_\varepsilon) = \frac{1}{2}\int_\Omega |\nabla w_\varepsilon|^2 = \frac{|\log\varepsilon|}{2}\int_\Omega |\nabla\varphi|^2 \le K|\log\varepsilon|\,.$$

On the other hand, $|\nabla w_\varepsilon| = O(|\log\varepsilon|^{1/2})$, so that any norm of the gradient will diverge as $\varepsilon \to 0$. Actually, even for solutions of the Ginzburg-Landau equation, no compactness has to be expected even in L^1 (see [15]).

However, one may split the contribution of the "topological" part from the rest of the phase to assert, in analogy with Corollary 1,

THEOREM 2 ([5,4]). *Let $K > 0$ and $(v_\varepsilon)_{\varepsilon>0}$, $v_\varepsilon : \Omega \to \mathbb{C}$ such that*

$$E_\varepsilon(v_\varepsilon) \le K|\log\varepsilon|\,.$$

Let $G \subset\subset \Omega$ be a smooth open simply connected set. Then, there exists a subsequence $\varepsilon_n \to 0$, ℓ points $a_1,\ldots,a_\ell \in G$, integers $d_1,\ldots,d_\ell \ne 0$, with $\sum_1^\ell |d_i| \le K'$, for some constant K' depending only on K, and functions $\varphi_{\varepsilon_n} : G \to \mathbb{R}$ such that

$$\int_G |\nabla\varphi_{\varepsilon_n}|^2 \le K|\log\varepsilon|$$

and

$$v_{\varepsilon_n} \cdot \exp(-i\varphi_{\varepsilon_n}) \to \prod_{i=1}^\ell \left(\frac{z - a_i}{|z - a_i|}\right)^{d_i} \qquad in\ H^s(G)\,,\ s < 1.$$

Notice that in the previous example, $\ell = 0$ (i.e. there are no vortices) and taking $\varphi_\varepsilon = \varphi \cdot \sqrt{|\log\varepsilon|}$, one may write, as above,

$$w_{\varepsilon_n} \cdot \exp(-i\varphi_{\varepsilon_n}) \equiv 1\,.$$

Sketch of proof: the idea is to introduce a regularization of v_ε in order to get rid of possible "small dipoles" (i.e. pairs of vortices having opposite multiplicities and whose distance is say $o(\varepsilon^{1/2})$), and to keep only the "relevant" part of the vorticity of v_ε.

Assume for simplicity that $|v_\varepsilon| \le 2$, and consider a minimizer w_ε of

$$F_\varepsilon(u) = \frac{1}{2}\int_\Omega \frac{|u - v_\varepsilon|^2}{\varepsilon} + E_\varepsilon(u)\,, \qquad u \in H^1(\Omega;\mathbb{R}^2)\,.$$

Then w_ε verifies the perturbed Ginzburg-Landau equation

$$\frac{w_\varepsilon - v_\varepsilon}{\varepsilon} = \Delta w_\varepsilon + \frac{1}{\varepsilon^2} w_\varepsilon(1 - |w_\varepsilon|^2)\,.$$

One can easily show that $E_\varepsilon(w_\varepsilon) \le E_\varepsilon(v_\varepsilon) \le K|\log\varepsilon|$, and

$$\int_\Omega |w_\varepsilon - v_\varepsilon|^2 \le 2K\varepsilon|\log\varepsilon|\,.$$

Performing a change of scale, and denoting

$$\tilde{\varepsilon} = \varepsilon^{1/2}$$
$$\tilde{w}(x) = w(\tilde{\varepsilon}x),$$

we are then led to the equation

(III.1)
$$\tilde{w}_\varepsilon - \tilde{v}_\varepsilon = \Delta\tilde{w}_\varepsilon + \frac{1}{\tilde{\varepsilon}^2}\tilde{w}_\varepsilon(1 - |\tilde{w}_\varepsilon|^2),$$

and the left hand side in (III.1) is bounded in L^∞. Many techniques developed in the context of the standard Ginzburg-Landau equation (see [8,12,33]) apply to (III.1). In particular, on G, the maps w_ε will have a finite number of vortices, bounded independently of ε. More precisely, there exists points $a_1^\varepsilon, \ldots, a_\ell^\varepsilon$, integers $d_1^\varepsilon, \ldots, d_\ell^\varepsilon$, and a constant $\lambda > 0$ such that $|w_\varepsilon| \geq 1/2$ on $G \setminus \cup_1^\ell B(a_i^\varepsilon, \lambda\varepsilon)$, and

$$\frac{w_\varepsilon(z)}{|w_\varepsilon(z)|} = \exp(i\varphi_\varepsilon(z))\prod_{i=1}^{\ell}\left(\frac{z - a_i^\varepsilon}{|z - a_i^\varepsilon|}\right)^{d_i^\varepsilon} \qquad \text{on } G \setminus \cup_{i=1}^{\ell}B(a_i^\varepsilon, \lambda\varepsilon),$$

where $\varphi_\varepsilon : G \to \mathbb{R}$ are suitable functions. Moreover, we have

$$\|w_\varepsilon - v_\varepsilon\|_{L^2} \leq C\varepsilon^{1/2}|\log\varepsilon|^{1/2}$$
$$\|\nabla(w_\varepsilon - v_\varepsilon)\|_{L^2} \leq C|\log\varepsilon|^{1/2},$$

so that, for $s < 1$, $\|w_\varepsilon - v_\varepsilon\|_{H^s} \leq C\varepsilon^\alpha$, for some $0 < \alpha < 1$, and after a few simple computations the conclusion follows.

Comment. i) Theorem 2 shows that the possible lack of compactness is merely due to the phase (which is a real-valued function). On the other hand, the "topological" contribution due to the vortices is essentially compact.

ii) In view of the previous remark, some topological properties of the level sets of E_ε can be reduced the properties of the level sets of the renormalized energy on the space of configurations of vortices (which is finite dimensional). This fact has been used in [5,30,34,13] in order to find solutions to $(GL)_\varepsilon$ by variational methods (mountain pass, Ljusternik-Schnirelman theory, etc...).

III.1.4 Compactness for Jacobians

A related but conceptually different approach for locating the vorticity for maps $v_\varepsilon \in W_\varepsilon^K$ has been proposed first in [20] and, independently, in [2]. This approach was motivated in particular by the study via Γ-convergence of complex (and more generally, vector-valued) Ginzburg-Landau functionals (also in arbitrary dimensions).

The main idea here is to look at the Jacobians of v_ε, which allows to characterize its topological part. More precisely, for $v = (v^1, v^2) : \Omega \to \mathbb{R}^2$ a smooth map, its Jacobian Jv is the 2-form defined by

$$Jv = dv^1 \wedge dv^2 = \frac{1}{2}d(v^1 dv^2 - v^2 dv^1).$$

In two dimensions, it may be identified with a scalar function, namely

$$Jv = \det(\nabla v) = v_x \times v_y,$$

where, for $a, b \in \mathbb{R}^2$, $a \times b = a^1 b^2 - a^2 b^1$. Note that $v_x \times v_y = 0$ whenever v_x and v_y are colinear. Hence, when $|v| = 1$, we have $Jv \equiv 0$. In particular, oscillations in the phase of v are not "seen" by its Jacobian Jv.

It is then proved that

PROPOSITION 2. *Let* $v_\varepsilon : \Omega \to \mathbb{R}^2$ *such that* $E_\varepsilon(v_\varepsilon) \leq K|\log \varepsilon|$. *Then there exists a subsequence* $\varepsilon_n \to 0$, ℓ *points* $a_1, \ldots, a_\ell \in \Omega$, *and integers* $d_1, \ldots, d_\ell \neq 0$, *with* $\sum_1^\ell |d_i| \leq K'$, *for some constant* K' *depending only on* K, *such that*

$$Jv_{\varepsilon_n} \to \pi \sum_{i=1}^{\ell} d_i \delta_{a_i} \qquad in \ [C_c^{0,\alpha}(\Omega)]^*, \ for \ any \ \alpha > 0 \,.$$

Remark III.1. i) the corresponding result in the one dimensional scalar case would be $\dot{v}_{\varepsilon_n} \to 2\chi \sum_{i=1}^{\ell} (-1)^i \delta_{a_i}$ (see Corollary 1).

ii) Proposition 2 could also be derived using Theorem 2. However, the approaches in [20,2] are more complete and give also interesting results for higher energy levels than the ones considered here.

III.2 The case $N \geq 3$

Since in dimension two vortices are points, and therefore codimension two defects, one expects, likewise, that in higher dimensions defects for the complex Ginzburg-Landau functional will concentrate on sets of codimension two.

In this direction, the theory for **solutions** to the Ginzburg-Landau equation was developed during the last decade in [28,21,22,14,9,10,7,11]. In particular, the main results in [7,10,11] apply to the following situation. Let Ω be a smooth bounded, simply connected domain in $\mathbb{R}^N$, $N \geq 3$. For $\varepsilon > 0$ a small parameter, consider solutions $w_\varepsilon : \Omega \to \mathbb{C}$ of the Ginzburg-Landau equation with Dirichlet data g_ε in $H^{1/2}(\partial\Omega; \mathbb{C})$. We assume moreover that there exist positive constants M_0, M_1, M_2 such that w_ε and g_ε verify conditions (H1), (H2), (H3) or (H3bis) below.

(H1) $$E_\varepsilon(w_\varepsilon) \leq M_0 |\log \varepsilon| \,,$$

(H2) $$\|g_\varepsilon\|^2_{H^{1/2}(\partial\Omega)} \leq M_1 \,,$$

(H3) $$|g_\varepsilon| = 1 \qquad \text{a.e. in } \partial\Omega \,,$$

(H3bis) $$\frac{1}{2} \int_{\partial\Omega} |\nabla g_\varepsilon|^2 + \frac{1}{4\varepsilon^2} \int_{\partial\Omega} (1 - |g_\varepsilon|^2)^2 \leq M_2 |\log \varepsilon| \,.$$

THEOREM 3, ([7,10,11]). *Let* w_ε *be a solution of* (GL_ε) *satisfying* (H1), (H2), (H3) *or* (H3bis). *Then, for a subsequence* $\varepsilon_n \to 0$, *there exist a map* $w_* \in W^{1,p}(\Omega)$, $\forall \, 1 \leq p < \frac{N}{N-1}$, *and a map* $g_* \in H^{1/2}(\partial\Omega)$ *such that*

i) $|w_*| = 1$ *on* Ω, $|g_*| = 1$, $w_* = g_*$ *on* $\partial\Omega$;

ii) $w_{\varepsilon_n} \to w_*$ *in* $W^{1,p}(\Omega)$, $g_{\varepsilon_n} \rightharpoonup g_*$ *in* $H^{1/2}(\partial\Omega)$;

iii) $\mathrm{div}(w_* \times \nabla w_*) = 0$ *in* Ω;

iv) $\dfrac{e_{\varepsilon_n}(w_{\varepsilon_n})}{|\log \varepsilon_n|} \rightharpoonup \mu_*$ *as measures, where* μ_* *is a bounded measure on* $\bar{\Omega}$.

Set $\mathcal{S} = supp(\mu_*)$.

v) $\mathcal{S}$ *is a closed subset of* $\bar{\Omega}$ *with* $\mathcal{H}^{N-2}(\mathcal{S}) < +\infty$;

vi) $w_* \in C^\infty(\Omega \backslash S)$ *and for any ball* $B(x_0, r)$ *included in* $\Omega \backslash S$ *there exists a function* $\varphi_* \in C^\infty(B(x_0, r))$, *such that* $\Delta \varphi_* = 0$, $w_* = \exp(i\varphi_*)$;

vii) $w_{\varepsilon_n} \to w_*$ *in* $C^k(K)$, *for any compact subset* K *of* $\Omega \setminus S$;

viii) S *is* $\mathcal{H}^{N-2}$*-rectifiable;* μ_* *is a stationary varifold.*

Remark III.2. i) Statement viii) uses crucially the analysis of [6].

ii) In case w_ε is minimizing, then S is an area-minimizing integer multiplicity rectifiable current (see [21,29,20,2]).

iii) More generally, the following equation has been considered:

$$(III.2) \qquad i\,|\log \varepsilon| \vec{c}(x) \cdot \nabla w = \Delta w + \frac{1}{\varepsilon^2} w(1 - |w|^2) + |\log \varepsilon|^2 d(x) w,$$

where $\vec{c}$ and d are smooth functions on $\bar{\Omega}$. The Ginzburg-Landau equation for superconductivity, as well as the travelling wave equation for $(NLS)_\varepsilon$ are included as particular cases of this equation. In [3], an analysis similar to the analysis in Theorem 3 was carried out for equation (III.2). The main difference is that the varifold associated to μ_* is no longer stationary, but satisfies the curvature equation

$$\vec{H}(x) = \star \left(\vec{c}(x) \wedge \star \frac{dJ_*}{d\mu_*} \right), \qquad \text{for } \mu_*\text{-a.e. } x \in S,$$

where $\vec{H}(x)$ denotes the generalized mean curvature of the varifold associated to μ_* at x, the measure $J_* = d(w_* \times dw_*)$ is the Jacobian of w_*, and $\frac{dJ_*}{d\mu_*}$ is the Radon-Nykodim derivative of J_* with respect to μ_*.

Concerning the **compactness** for level sets of E_ε, the results in Section III.1.3 can be extended as follows (see [20,2,4]). Let v_ε be a sequence such that

$$E_\varepsilon(v_\varepsilon) \leq K |\log \varepsilon|.$$

Then, there exists a subsequence ε_n, functions $\varphi_{\varepsilon_n} : \Omega \to \mathbb{R}$ such that

$$\int_\Omega |\nabla \varphi_{\varepsilon_n}|^2 \leq K |\log \varepsilon|,$$

and $v_* \in W^{1,p}(\Omega; S^1)$, $p < \frac{N}{N-1}$, such that

$$v_{\varepsilon_n} \cdot \exp(-i\varphi_{\varepsilon_n}) \to v_* \qquad \text{in } H^s_{loc}(\Omega), \ s < s_N.$$

Moreover, Jv_* is a measure, and therefore, by results in [16,19], it is a codimension two integer multiplicity rectifiable current. A "canonical" construction of v_* as in Theorem 2, relying on the Biot-Savart law, can be found in [1].

References

[1] G. Alberti, S. Baldo, G. Orlandi, Functions with prescribed singularities, J. Eur. Math. Soc., to appear.

[2] G. Alberti, S. Baldo, G. Orlandi, Variational convergence for functionals of Ginzburg-Landau type, preprint (2002).

[3] F. Bethuel, G. Orlandi, D. Smets, Vortex rings for the Gross-Pitaevskii equation, J. Eur. Math. Soc., to appear.

[4] F. Bethuel, G. Orlandi, D. Smets, Approximations with vorticity bounds for the Ginzburg-Landau functional, preprint (2003).

[5] L. Almeida, F. Bethuel, Topological methods for the Ginzburg-Landau equation, J. Math Pures Appl. **11** (1998), 1–49.

[6] L. Ambrosio, H.M. Soner, A measure theoretic approach to higher codimension mean curvature flow, Ann. Sc. Norm. Sup. Pisa, Cl. Sci. (4), **25** (1997), 27–49.

[7] F. Bethuel, J. Bourgain, H. Brezis, G. Orlandi, $W^{1,p}$ estimates for solutions to the Ginzburg-Landau equation with boundary data in $H^{1/2}$, C.R. Acad. Sc. Paris, Série 1, **333** (2001), 1069–1076.

[8] F. Bethuel, H. Brezis, F. Hélein, *Ginzburg-Landau Vortices*, Birkhäuser, Boston, (1994).

[9] F. Bethuel, H. Brezis, G. Orlandi, Small energy solutions to the Ginzburg-Landau equation, C.R. Acad. Sc. Paris, Série 1 **331** (2000), 763–770.

[10] F. Bethuel, H. Brezis, G. Orlandi, Asymptotics for the Ginzburg-Landau equation in arbitrary dimensions, J. Funct. Anal. **186** (2001), 432–520. Erratum, *Ibid.* **188** (2002), 548–549. [11] F. Bethuel, G. Orlandi, Uniform estimates for the parabolic Ginzburg-Landau equation, ESAIM, C.O.C.V. **8** (2002), 219–238.

[12] F. Bethuel, T. Rivière, Vortices for a variational problem related to superconductivity, Annales I.H.P. Anal. Nonlin. **12** (1995), 243–303.

[13] F. Bethuel, J.C. Saut, Travelling waves for the Gross-Pitaevskii equation, Annales IHP, Phys. Théor. **70**, (1999).

[14] J. Bourgain, H. Brezis, P. Mironescu, On the structure of the Sobolev space $H^{1/2}$ with values into the circle. C.R. Acad. Sc. Paris, Série I **331** (2000), 119–124.

[15] H. Brezis, P. Mironescu, Sur une conjecture de E. De Giorgi relative a l'energie de Ginzburg-Landau, C.R. Acad. Sci Paris (I) **319** (1994), 167–170.

[16] M. Giaquinta, L. Modica, J. Souček, Cartesian Currents and the Calculus of Variations, Springer, (1998).

[17] M. Gurtin, On a theory of phase transitions with interfacial energy, Arch. Rat. Mech. Anal. **87** (1985), 187–212.

[18] J. Hutchinson, Y. Tonegawa, Convergence of phase interfaces in the van der Walls-Cahn-Hilliard theory, Calc. Var. and P.D.E. **10** (2000), 49–84.

[19] R. Jerrard, H.M. Soner, Functions of bounded higher variation, preprint (1999).

[20] R. Jerrard, H.M. Soner, The Jacobian and the Ginzburg-Landau energy, Calc. Var. PDE **14** (2002), 151–191.

[21] F.H. Lin, T. Rivière, Complex Ginzburg-Landau equations in high dimensions and codimension two area minimizing currents, J. Eur. Math. Soc. **1** (1999), 237–311. Erratum, *Ibid.*

[22] F.H. Lin, T. Rivière, A quantization property for static Ginzburg-Landau vortices, Comm. Pure Appl. Math. **54** (2001), 206–228.

[23] N.D. Mermin, The topological theory of defects in ordered media, Rev. Mod. Phys. **51** (1979), 592–649

[24] P. Mironescu, Les minimiseurs locaux pour l'équation de Ginzburg-Landau sont à symétrie radiale, C.R. Acad. Sci. Paris (I) **323** (1996), 593-598.

[25] L. Modica, The gradient theory of phase transitions and the minimal interface criterion, Arch. Rat. Mech. Anal. **98** (1987), 123–142.

[26] L. Modica, S. Mortola, Un esempio di Γ-convergenza, Boll. Un. Mat. Ital. B **14** (1977), 285–299.

[27] F. Pacard, T. Rivière, *Linear and non linear aspects of vortices*, Birkhäuser, (2001).

[28] T. Rivière, Line vortices in the $U(1)$-Higgs model, ESAIM, C.O.C.V. **1** (1996), 77–167.

[29] E. Sandier, Ginzburg-Landau minimizers from $\mathbb{R}^{N+1}$ to $\mathbb{R}^N$ and minimal connections, Indiana Univ. Math. J. **50** (2001), 1807–1844. (2002).

[30] S. Serfaty, Local minimizers for the Ginzburg-Landau energy near critical magnetic fields, I and II, Comm. Contemp. Math. **1** (1999), 213-254 and 295-333.

[31] L.Simon, *Lectures on Geometric Measure Theory*, proceedings of the Centre for Math. Analysis, Australian Nat. Univ., Canberra (1983).

[32] P. Sternberg, The effect of a singular perturbation on nonconvex variational problems, Arch. Rat. Mech. Anal. **101** (1988), 209–260.

[33] M. Struwe, On the asymptotic behavior of the Ginzburg-Landau model in 2 dimensions, J. Diff. Int. Equ. **7** (1994), 1613–1624. Erratum *Ibid.* **8** (1995), 224.

[34] F. Zhou, Q. Zhou, A remark on the multiplicity of solutions for the Ginzburg-Landau equation, Ann. IHP, Analyse Nonlinéaire **16** (1999), 255-267.

ANALYSE NUMERIQUE, UNIVERSITÉ P. ET M. CURIE, BC 187, 4, PL. JUSSIEU 75252 PARIS CEDEX 05, FRANCE

E-mail address: bethuel@ann.jussieu.fr

Dipartimento di Informatica, Università di Verona, strada le Grazie, 37134 Verona, Italy

E-mail address: orlandi@sci.univr.it

Contemporary Mathematics
Volume **350**, 2004

On the Topology of Conformally Compact Einstein 4-manifolds

Sun-Yung A. Chang, Jie Qing and Paul Yang

In honor of Haim Brezis and Felix Browder

ABSTRACT. In this paper we study the topology of conformally compact Einstein 4-manifolds. When the conformal infinity has positive Yamabe invariant and the renormalized volume is also positive we show that the conformally compact Einstein 4-manifold will have at most finite fundamental group. Under the further assumption that the renormalized volume is relatively large, we conclude that the conformally compact Einstein 4-manifold is diffeomorphic to B^4 and its conformal infinity is diffeomorphic to S^3.

0. Introduction

Conformally compact Einstein manifolds play an important part in the AdS/CFT correspondence, a promising new area under extensive development in string theory [Mc] [GKP] [W]. Mathematically the study of conformally compact Einstein structures began with the work of Fefferman and Graham ([FG1]) in which they gave a procedure to find local conformal invariants and the associated conformally covariant operators. It has been a challenging problem to give general existence theory. Recently M. Anderson ([A2]) has shown the existence of conformally compact Einstein metrics whose boundary is the 3-sphere with arbitrary conformal structure in the connected component of the round one, thus extending the perturbation existence result of Graham and Lee ([GL]). On the other hand, using several conformally covariant differential equations and their ties to Chern-Gauss-Bonnet formula, two of the coauthors with Gursky studied the conformal geometry of closed 4-manifolds and made progress in understanding of the topology of 4-manifolds. In our view, conformally compact Einstein 4-manifolds provide a platform to study conformal geometry of closed 3-manifolds. In this note we will apply the results in four dimension to 3-manifolds.

1991 *Mathematics Subject Classification.* 53C25, 53C80, 58J60.
Key words and phrases. Conformally compact Einstein metrics, positive Yamabe constant, renormalized volume, vanishing theorem.
First author supported by NSF Grant DMS–0070542.
Second author supported in part by NSF Grant DMS–0103160.
Third author supported by NSF Grant DMS–0070526.

An oriented manifold (X, g) with boundary M is a conformally compact Einstein manifold if there is a complete Einstein metric g in the interior of X, and a smooth defining function s for the boundary $M = \partial X = \{s = 0\}$ so that $(X, s^2 g)$ is a compact Riemannian manifold with boundary. Since defining functions are not unique the data (X, g) determines a conformal structure $(M, [\hat{g}])$ which is called the conformal infinity of (X, g). It will be advantageous to find a defining function s whose associated conformal metric enjoys optimal positivity property. To this end, one of us first observe ([Q]),

THEOREM. *Suppose that (X^{n+1}, g) is a conformally compact Einstein manifold, and that $\hat{g}$ is a Yamabe metric for $(M, [\hat{g}])$. Then there exists a conformal compactification $(X, u^{-2}g)$, which has a totally geodesic boundary M and whose scalar curvature $R[u^{-2}g] \geq \frac{n+1}{n-1} R[\hat{g}]$.*

An important invariant of the conformally compact Einstein structure is the renormalized volume (see section one for the precise definition). The volume of a conformally compact Einstein manifold is infinite. But an appropriate normalization gives rise to the new invariants as suggested in the works of Maldacena [Mc], Witten [W], and Gubser, Klebanov and Polyakov [GKP]. The renormalization was carried out by Henningson and Skenderis in [HS] and later also by Graham in [Gr]. Anderson in [A1] observed that the renormalized volume for a conformally compact Einstein 4-manifold appears in the Chern-Gauss-Bonnet formula

$$(0.1) \qquad 8\pi^2 \chi(X) = \frac{1}{4} \int_X |\mathcal{W}|^2 dv_g + 6V,$$

where $\mathcal{W}$ is the Weyl curvature and the norm of the Weyl tensor is given by $|\mathcal{W}|^2 = \mathcal{W}_{ijkl}\mathcal{W}^{ijkl}$; i.e., the usual definition when W is viewed as a section of $\otimes^4 T^* M^4$ and V is the renormalized volume of a conformally compact Einstein 4-manifold (X^4, g). It follows that the renormalized volume is an invariant of the underlying conformal structure of the conformally compact Einstein manifold. On the other hand, we recall that for a closed 4-manifold (Y, g), we have the Chern-Gauss-Bonnet formula

$$(0.2) \qquad 8\pi^2 \chi(Y) = \int_Y (\frac{1}{4}|\mathcal{W}|^2 + \sigma_2(A_g)) dv_g,$$

where $A_g = Rc - \frac{R}{6}g$ is the Weyl Schouten tensor and $\sigma_2(A_g)$ is the second symmetric function of the eigenvalues of A_g. Thus the renormalized volume is the Ricci part of the Chern-Gauss-Bonnet integral. According to the above cited Theorem, the given compactification X has a totally geodesic boundary. It is then natural to consider the doubling manifold Y. In this setting, we will show that the renormalized volume plays an important role in restricting the topology and geometry of the conformally compact Einstein manifolds. In particular, we have

THEOREM A. *Suppose that (X, g) is a conformally compact Einstein 4-manifold with its conformal infinity of positive Yamabe constant and suppose the renormalized volume V satisfies*

$$(0.3) \qquad V > \frac{1}{3}\frac{4\pi^2}{3}\chi(X),$$

Then X is homeomorphic to the 4-ball B^4 up to a possible finite cover.

This result uses the vanishing theorems of Gursky [Gu2] to show that the doubling manifold Y is a homology sphere, and the main result of [CGY1] to show that Y admits a conformal metric of positive Ricci tensor, and hence has at most finite fundamental group. The homeomorphism classification theory of Donaldson and Freedman then implies that Y is homeomorphic to the 4-sphere up to a finite cover. On the other hand, under a stronger hypothesis, we can conclude that Y is diffeomorphic to the 4-sphere.

THEOREM B. *Suppose that (X, g) is a conformally compact Einstein 4-manifold with conformal infinity of positive Yamabe constant. Then*

$$(0.4) \qquad V > \frac{1}{2} \frac{4\pi^2}{3} \chi(X)$$

implies that X is diffeomorphic to B^4 and M is diffeomorphic to S^3.

These considerations demonstrate that the renormalized volume is an important conformal invariant of the conformally compact Einstein structure. The recent work of Fefferman and Graham [FG2] showed that it may be interpreted as a Q curvature integral associated to a natural boundary operator. It would be of interest to understand this invariant. One may ask whether the positivity of the renormalized volume is an intrinsic property of the conformal structure of the boundary M alone. This is related to the uniqueness question of the conformally compact Einstein structure. There is some partial answer to this question in the article of Anderson [A3] (cf. [Wa]).

The paper is organized as follows: In section 1 we introduce conformally compact Einstein manifolds and relevant properties. In section 2 we construct positive eigenfunctions and present a conformal compactification of a conformally compact Einstein manifolds. In section 3 we recall the results in [CGY1] [CGY2] [Gu1] [Gu2] and prove our main theorems.

1. Conformally compact Einstein manifolds

Let us start with the definition of a conformally compact Einstein manifold. Suppose X^{n+1} is a compact oriented (n+1)-manifold with boundary $\partial X = M^n$. A Riemannian metric g in the interior of X is said to be a $C^{m,\alpha}$ conformally compact if $\bar{g} = s^2 g$ extends as a $C^{m,\alpha}$ metric on $\bar{X}$, where s is a defining function of the boundary in the sense that: $s > 0$ in X, $s = 0$ and $ds \neq 0$ on M. Clearly defining functions are not unique. For a given defining function, the metric $\bar{g}$ restricted to TM induces a metric $\hat{g}$ on M. $\hat{g}$ rescales upon changing the defining function r, therefore defines a conformal structure on M. We call $(M, [\hat{g}])$ the conformal infinity of the conformally compact manifold (X, g). Therefore (X^{n+1}, g) is a conformally compact Einstein manifold if it is conformally compact and $\mathrm{Ric}(g) = -ng$. A good reference for basic properties of conformally compact Einstein manifolds is the paper of Graham [Gr].

The boundary regularity of the conformally compact Einstein metric is an important issue. Thanks to M. Anderson [A2], in 4 dimension, we know that $C^{2,\alpha}$ would imply the full smoothness of the conformally compact Einstein metric. In other words, the conformally compact Einstein 4-manifolds we will be focus on in the last section are, always smooth when the conformal infinity is smooth, therefore, the local expansions of Einstein metrics near the infinity are always available. Sufficient boundary regularity is assumed in higher dimension to ensure the expansion though.

Solving a nonlinear first order PDE by the method of characteristics near the boundary, one has (cf. Lemma 2.1 and Lemma 2.2 in [Gr]):

LEMMA 1.1. *Given any conformally compact manifold* (X, g), *a metric* $\hat{g}$ *in the conformal class* $[\hat{g}]$ *determines a unique defining function* s *in a neighborhood of* M *such that*

$$(1.1) \qquad g = s^{-2}((ds)^2 + g_s),$$

where, when n *is odd,*

$$(1.2) \qquad g_s = \hat{g} + g^{(2)}s^2 + \text{even powers of } s + g^{(n-1)}s^{n-1} + g^{(n)}s^n + \cdots ;$$

$g^{(i)}$ *for* $i < n$ *are determined by local geometry of* $(M, \hat{g})$ *and* $g^{(n)}$ *is trace-free, when* n *is even,*

$$(1.3) \qquad g_s = \hat{g} + g^{(2)}s^2 + \text{even powers of } s + hs^n \log s + g^{(n)}s^n + \cdots ,$$

$g^{(i)}$ *for* $i < n$, h *and the trace of* $g^{(n)}$ *are determined by local geometry of* $(M, \hat{g})$, *and* h *is trace-free.*

For example, one can calculate that

$$(1.4) \qquad g^{(2)} = -\frac{1}{n-2}(\hat{R}_{ij} - \frac{\hat{R}}{2(n-1)}\hat{g}_{ij}).$$

As a realization of the so-called AdS/CFT correspondence, Henningson and Skenderis in [HS] and Graham in [Gr] defined and calculated the renormalized volume for a conformally compact Einstein manifold. Namely, they considered the expansions, when n is odd,

$$(1.5) \quad \mathrm{Vol}(\{s > \epsilon\}) = c_0\epsilon^{-n} + c_2\epsilon^{-n+2} + \text{ odd powers of } \epsilon + c_{n-1}\epsilon^{-1} + V + o(1),$$

when n is even,

(1.6)
$$\mathrm{Vol}(\{s > \epsilon\}) = c_0\epsilon^{-n} + c_2\epsilon^{-n+2} + \text{ even powers of } \epsilon + c_{n-2}\epsilon^{-2} + L\log\frac{1}{\epsilon} + V + o(1),$$

where

$$(1.7) \qquad c_i = \int_M v^{(i)} dv_{\hat{g}},$$

and $v^{(i)}$ are local scalar invariants of $(M, \hat{g})$ of order i. More importantly he showed the renormalized volume V for n odd and L for n even are independent of the choice of defining functions. Incidentally when n is even the quantity $L = \int_M v^{(n)} dv_{\hat{g}}$ is a conformal invariant of M, and when n is odd, V is only an invariant of the ambient structure (X, g).

Particularly, when $n = 3$, Anderson in [A] relates the renormalized volume to the Euler number of a conformally compact Einstein 4-manifold,

$$(1.8) \qquad 8\pi^2\chi(X) = \frac{1}{4}\int_X |\mathcal{W}|^2 dv_g + 6V.$$

2. Conformal compactifications

In this section we present a conformal compactification of a general conformally compact Einstein manifold. As observed in [Q], when one considers the hemisphere compactification of the hyperbolic space, the conformal factor is a positive eigenfunction. Namely, recall

$$(H^{n+1}, g_H) = (B^{n+1}, (\frac{2}{1 - |y|^2})^2 |dy|^2),$$

and

$$(S^{n+1}_+, g_S) = (B^{n+1}, (\frac{2}{1 + |y|^2})^2 |dy|^2),$$

that is $g_S = t^{-2} g_H$, where we denote by

$$t = \frac{1 + |y|^2}{1 - |y|^2}.$$

One finds after some calculation (cf. [Q]),

$$(2.1) \qquad \Delta_{g_H} t = (n+1)t.$$

Given a general conformally compact Einstein manifold, using the theory of uniformly degenerate elliptic linear PDE developed in [M] [GL] [L], one can construct positive eigenfunctions. A good reference for our discussions here is a paper of Lee [L]. The following lemma is a slight improvement of Lemma 5.2 in [L].

LEMMA 2.1. *Suppose that (X^{n+1}, g) is a conformally compact Einstein manifold. Then there exists a positive function u such that*

$$(2.2) \qquad \Delta u = (n+1)u \quad in \ X.$$

In addition, when n is odd,

$$(2.3) \quad u = \frac{1}{s} + \frac{\hat{R}}{4n(n-1)} s + w^{(4)} s^3 + \ odd \ powers \ of \ s + w^{(n-1)} s^{n-2} + O(s^n)$$

and when n is even,

$$(2.4) \quad u = \frac{1}{s} + \frac{\hat{R}}{4n(n-1)} s + w^{(4)} s^3 + \ odd \ powers \ of \ s + w^{(n)} s^{n-1} + O(s^n),$$

where $w^{(i)}$ are all local invariants of Riemannian geometry of $(M, \hat{g})$ of order i.

PROOF. To apply the theory of uniformly degenerate elliptic PDE (cf. [L], for example) we need to calculate

$$\Delta \frac{1}{s} = \frac{s^{n+1}}{\sqrt{\det g_s}} \partial_s (s^{1-n} \sqrt{\det g_s} \partial_s \frac{1}{s}) = \frac{n+1}{s} - \frac{1}{2} \mathrm{Tr}_{\hat{g}} \partial_s g_s.$$

Then from the expansion of g_s obtained by Graham as stated in Lemma 1.1 in the previous section, when n is odd,

$$\Delta \frac{1}{s} = \frac{n+1}{s} + p^{(2)} s + \ odd \ powers \ of \ s \ + p^{(n-1)} s^{n-2} + O(s^n),$$

and when n is even,

$$\Delta \frac{1}{s} = \frac{n+1}{s} + p^{(2)} s + \ odd \ powers \ of \ s \ + p^{(n)} s^{n-1} + O(s^n),$$

where $p^{(k)} = -\frac{k}{2}\mathrm{Tr}_{\hat{g}}g^{(k)}$ are local invariants of Riemannian geometry of $(M, \hat{g})$ of order k, and

$$p^{(2)} = \frac{\hat{R}}{2(n-1)},$$

for example. Now let us calculate the key ingredient in the theory of uniformly degenerate elliptic PDE: the indicial equation. For that purpose we first calculate with separating variables for any smooth function ϕ defined on $(M, \hat{g})$,

$$\Delta(\phi s^k) = \frac{\phi s^{n+1}}{\sqrt{\det g_s}}\partial_s(s^{1-n}\sqrt{\det g_s}\partial_s s^k) + \frac{s^{k+2}}{\sqrt{\det g_s}}\partial_\alpha(\sqrt{\det g_s}g_s^{\alpha\beta}\partial_\beta\phi),$$

where

$$\frac{\phi s^{n+1}}{\sqrt{\det g_s}}\partial_s(s^{1-n}\sqrt{\det g_s}\partial_s s^k) = k(k-n)\phi s^k + \frac{k}{2}\phi s^{k+1}\mathrm{Tr}_{\hat{g}}\partial_s g_s$$

and

$$\frac{s^{k+2}}{\sqrt{\det g_s}}\partial_\alpha(\sqrt{\det g_s}g_s^{\alpha\beta}\partial_\beta\phi) = s^{k+2}\tilde{\Delta}\phi + s^{k+2}(\cdots \text{even powers of } s \cdots),$$

ϕ is independent of s, and $\tilde{\Delta}$ is the Laplacian of $(M, \hat{g})$. Therefore the indicial equation for us is

$$(\Delta - (n+1))(\phi s^k) = (k(k-n) - (n+1))\phi s^k + s^{k+2}(\cdots \text{ even powers of } s \cdots)$$

Then we may write, when n is odd

$$(\Delta - (n+1))(\frac{1}{s} + w^{(2)}s + \text{ odd powers of } s + w^{(n-1)}s^{n-2}) = O(s^n)$$

and when n is even

$$(\Delta - (n+1))(\frac{1}{s} + w^{(2)}s + \text{ odd powers of } s + w^{(n)}s^{n-1}) = O(s^n).$$

Notice that $w^{(i)}$ are all local invariants of Riemannian geometry of $(M, \hat{g})$, for example for $i = 2$ we have

$$w^{(2)} = \frac{\hat{R}}{4n(n-1)}.$$

Then, the lemma follows from Proposition 3.3 in the paper of Lee [L]. Note that positivity of u is a simple consequence of the maximum principle.

We remark that for the standard hyperbolic space we have

$$u = t = \frac{1}{s} + \frac{1}{4}s.$$

We can now calculate the scalar curvature for the metric $u^{-2}g$ on X as in [Q]. We find, for u obtained in Lemma 2.1,

$$(2.5) \qquad -\Delta u^{-\frac{n-1}{2}} - \frac{n^2-1}{4}u^{-\frac{n-1}{2}} = \frac{n^2-1}{4}(u^2 - |du|^2)u^{-\frac{n+3}{2}}.$$

Thus the scalar curvature of $u^{-2}g$ is

$$(2.6) \qquad R[u^{-2}g] = n(n+1)(u^2 - |du|^2).$$

Notice that u is determined by the choice of a defining function, therefore the choice of a metric of the conformal infinity by Lemma 1.1. So we may choose a Yamabe metric $\hat{g}$ for $(M, [\hat{g}])$ in the following theorem.

THEOREM 2.2. *Suppose that (X^{n+1}, g) is a conformally compact Einstein manifold, and that u is the eigenfunction obtained in Lemma 2.1 for a Yamabe metric $\hat{g}$ of the conformal infinity $(M, [\hat{g}])$. Then $(X^{n+1}, u^{-2}g)$ is a compact manifold with totally geodesic boundary M and the scalar curvature is greater than or equal to $\frac{n+1}{n-1}\hat{R}$.*

PROOF. First by Lemma 2.1, one can estimate

$$u^2 - |du|^2 = \frac{1}{s^2} + \frac{\hat{R}}{2n(n-1)} + O(s^2) - s^2(\partial_s u)^2 - s^2 g_s^{\alpha\beta} u_\alpha u_\beta$$

$$(2.7) \qquad = \frac{1}{s^2} + \frac{\hat{R}}{2n(n-1)} - \frac{1}{s^2} + \frac{\hat{R}}{2n(n-1)} + O(s^2)$$

$$= \frac{\hat{R}}{n(n-1)} + O(s^2).$$

Thus in light of (2.6), the theorem follows from a maximum principle and the following Bochner formula,

$$(2.8) \qquad -\Delta(u^2 - |du|^2) = 2|Ddu - ug|^2,$$

which has also been used in [L]. The smoothness assertion follows from the expansion (2.3).

3. Topology of conformally compact Einstein 4-manifolds

In this section we apply the vanishing theorems of Gursky and the deformation theory of [CGY] of conformal metrics on the doubling 4-manifold Y in order to draw conclusions about the topology of the conformally compact manifold X. Under increasingly stringent conditions on the renormalized volume V, we demonstrate the vanishing of $H^1(Y)$, $H^2(Y)$. This is a reflection of the increasing constraint placed on the Weyl-Schouten tensor of the compactified structure.

According to the compactification obtained in Theorem 2.2 we have a compact 4-manifold $(X, \tilde{g})$ with totally geodesic boundary M. In addition the scalar curvatures satisfies $R[\tilde{g}] \geq 2R[\hat{g}]$, where $\hat{g}$ is the Yamabe metric for the conformal infinity $(M, [\hat{g}])$. Recall the Chern-Gauss-Bonnet formula for 4-manifolds with totally geodesic boundary

$$(3.1) \qquad 8\pi^2\chi(X) = \int_X (\frac{1}{4}|\tilde{W}|^2 + \sigma_2(A_{\tilde{g}}))dv_{\tilde{g}}$$

where

$$\sigma_2(A) = \frac{1}{24}R^2 - \frac{1}{2}|E|^2$$

and E is the traceless Ricci curvature. It follows from the conformal invariance

$$(3.2) \qquad |\tilde{W}|^2 dv_{\tilde{g}} = |W|^2 dv_g$$

and Anderson's formula (1.8) that

$$(3.3) \qquad \int_X \sigma_2(A_{\tilde{g}}) dv_{\tilde{g}} = 6V.$$

Consider the doubling manifold $(Y, \bar{g})$ of $(X, \tilde{g})$. The metric $\bar{g}$ on Y belongs to $C^{2,1}$, i.e. its second derivative is of Lipschitz class, therefore the curvature tensor is of Lipschitz class. Consequently we have

PROPOSITION 3.1. *If the conformal infinity has positive Yamabe invariant, then the renormalized volume satisfies*

$$(3.4) \qquad V \leq \frac{4}{3}\pi^2.$$

Equality in (3.4) holds if and only if (X, g) is the hyperbolic space, therefore $(M, [\hat{g}])$ is the round sphere.

PROOF. According to Gursky ([Gu1]) if a 4-manifold $(Y, \bar{g})$ is of positive Yamabe constant, then

$$(3.5) \qquad \int_Y \sigma_2(A_{\bar{g}}) dv_{\bar{g}} \leq 16\pi^2,$$

and equality holds if and only if $(Y, \bar{g})$ is the round 4-sphere. In the situation at hand, the doubling space Y has a metric of positive scalar curvature according to Theorem 2.2, therefore,

$$(3.6) \qquad V \leq \frac{1}{6} \int_X \sigma_2(\tilde{g}) dv_{\tilde{g}} \leq \frac{1}{6} 8\pi^2 = \frac{4}{3}\pi^2$$

and equality holds if and only if

$$\int_X |\mathcal{W}|^2 dv_g = 0.$$

In this case the simply connected hyperbolic manifold is the hyperbolic space, and the conformal infinity is the round sphere. The proof is complete.

Let us assume that (X, g) is a conformally compact Einstein 4-manifold whose conformal infinity $(M, [\hat{g}])$ has positive Yamabe invariant. According to the spectral result of Lee in [L] and a vanishing result of Wang [Wa], there are no harmonic one forms in L^2. Therefore it follows from the isomorphism established by Mazzeo [M], the space of L^2 harmonic one forms is isomorphic to the relative cohomology $H^1(X, M)$, hence the latter has to vanish.

We have the long exact sequence in cohomology for the pair (X, M):

$$(3.7) \qquad \ldots H^k(X, M) \xrightarrow{j^*} H^k(X) \xrightarrow{i^*} H^k(M) \xrightarrow{\delta^*} H^{k+1}(X, M) \ldots$$

and the Mayer-Vietoris sequence for the double $Y = X \cup X'$ where X' is the second copy of X:

$$(3.8) \qquad \ldots H^{k-1}(M) \xrightarrow{\gamma} H^k(Y) \xrightarrow{\psi^*} H^k(X) \oplus H^k(X') \xrightarrow{\phi^*} H^k(M) \ldots$$

Since $H^1(X, M) = 0$, it follows from the long exact sequence that $i^* H^1(X) = H^1(X)$. The Mayer-Vietoris sequence then implies that $H^1(Y) = H^1(X)$.

We wish to impose a second positivity assumption on the conformally compact Einstein structure in addition to the positivity of Yamabe invariant of the boundary: the renormalized volume V should be positive. We remark that this implies in view of the formula (0.1) that $\chi(X) > 0$. In fact we have a stronger consequence according to the first vanishing theorem of Gursky [Gu2]:

THEOREM 3.2. *[Gu2] If (N, g) is a closed 4-manifold with positive Yamabe invariant, and*

$$(3.9) \qquad \int_N \sigma_2(A_{\bar{g}})dv_{\bar{g}} > 0,$$

then $H^1(N) = 0$.

In order to apply the vanishing theorem, we need to assure ourselves that the doubling metric is smooth. Since the metric on X is of the class $C^{2,1}$ and the boundary is totally geodesic, we may (if necessary) smooth out the metric by a small perturbation without changing the two positivity assumptions: positive Yamabe invariant and $\int_Y \sigma_2(A_{\bar{g}})dv_{\bar{g}} > 0$. We conclude thus $H^1(X) = H^1(Y) = 0$.

Next we formulate a stronger positivity condition in order to control the second homology. Recall for a closed 4-manifold $(Y, \bar{g})$, we have the Chern-Gauss-Bonnet formula (0, 2). On the other hand we also have the Hirzebruch signature formula

$$(3.10) \qquad 12\pi^2 \tau(Y) = \int_Y \frac{1}{4}(|\bar{W}^+|^2 - |\bar{W}^-|^2)dv_{\bar{g}}$$

Combining the two formulae we have

$$(3.11) \qquad 4\pi^2(2\chi(Y) + 3\tau(Y)) = \frac{1}{4}\int_Y |\bar{W}^+|^2 dv_{\bar{g}} + \int_Y \sigma_2(A_{\bar{g}})dv_{\bar{g}}$$

and

$$(3.12) \qquad 4\pi^2(2\chi(Y) - 3\tau(Y)) = \frac{1}{4}\int_Y |\bar{W}^-|^2 dv_{\bar{g}} + \int_Y \sigma_2(A_{\bar{g}})dv_{\bar{g}}.$$

THEOREM 3.3. *[Gu2] Suppose that $(Y, \bar{g})$ is a closed oriented 4-manifold with positive Yamabe constant. Then*

$$(3.13) \qquad \frac{1}{4}\int_Y |\bar{W}^+|^2 dv_{\bar{g}} < \int_Y \sigma_2(A_{\bar{g}})dv_{\bar{g}}$$

implies that the self-dual part of the second homology vanishes; similarly,

$$(3.14) \qquad \frac{1}{4}\int_Y |\bar{W}^-|^2 dv_{\bar{g}} < \int_Y \sigma_2(A_{\bar{g}})dv_{\bar{g}}$$

implies that the anti-self-dual part of the second homology vanishes.

PROPOSITION 3.4. *Assume that (X, g) is a conformally compact Einstein 4 manifold whose conformal infinity has positive Yamabe invariant and that*

$$(0.3) \qquad V > \frac{1}{3}\frac{4\pi^2}{3}\chi(X)$$

then $V > 0$ and $H^2(X, \mathbb{R}) = 0$.

PROOF. According to (1.8),

$$\frac{1}{4}\int_X |\tilde{W}|^2 dv_{\tilde{g}} = 8\pi^2\chi(X) - 6V < 12V.$$

Therefore $V > 0$. If we translate the assumption (0.3) into assumption on the doubling manifold $(Y, \bar{g})$, we have

$$(3.15) \qquad \int_Y \sigma_2(A_{\bar{g}})dv_{\bar{g}} > \frac{8}{3}\pi^2\chi(Y)$$

here we note that $\chi(Y) = 2\chi(X)$. It follows from (0.2) that (3.15) is equivalent to

$$(3.16) \qquad \frac{1}{4}\int_Y |\bar{W}|^2 dv_{\bar{g}} < 2\int_Y \sigma_2(A_{\bar{g}})dv_{\bar{g}}.$$

At this point, we remark once more that we may smooth the doubling metric if necessary while preserve the condition (3.16).

We can now assume without loss of generality that

$$(3.17) \qquad \frac{1}{4}\int_Y |\bar{W}^+|^2 dv_{\bar{g}} < \int_Y \sigma_2(A_{\bar{g}})dv_{\bar{g}}.$$

and apply the vanishing theorem Theorem 3.2 of Gursky, and conclude that $H_+^2(Y) = 0$. From (3.11) we also have

$$(3.18) \qquad 4\pi^2(2\chi(Y) + 3\tau(Y)) \geq \int_Y \sigma_2(A_{\bar{g}})dv_{\bar{g}} > \frac{8}{3}\pi^2\chi(Y).$$

A short computation with the two long exact sequences (3.7) and (3.8) and using the fact $H^1(X) = H^1(Y) = 0$ will yield that $H^2(Y) = H^2(X) \oplus H^2(X')$ and hence $dim H^2(Y)$ is even. Hence we we may assume $dim H_-^2(Y) = 2k$. Then (3.18) yields

$$(3.19) \qquad 2(2 + 2k) + 3(-2k) > \frac{2}{3}(2 + 2k)$$

which is impossible for any positive integer k. Therefore $k = 0$ and $H^2(X, \mathbb{R}) = H^2(Y, \mathbb{R}) = 0$ and we have finished the proof of the proposition.

So far we have applied the vanishing theorems of Gursky to conformally compact Einstein manifolds. Next we develop the implication the recent work of Chang-Gursky-Yang [CGY1] [CGY2] on the doubling manifold Y.

THEOREM 3.5. *[CGY1] Suppose that $(Y, \bar{g})$ is an oriented closed 4-manifold with positive Yamabe constant and that*

$$\int_Y \sigma_2(A_{\bar{g}})dv_{\bar{g}} > 0.$$

Then there exists a conformal metric $g' = e^{2w}\bar{g}$ with positive Ricci tensor.

Proof of Theorem A

As a consequence of Theorem 3.5, we can conclude that the doubling manifold Y has a finite fundamental group. If we consider its simply connected covering $\tilde{Y}$, the homeomorphic classification theory of Donaldson and Freedman for simply connected 4-manifolds implies that the manifold $\tilde{Y}$ is determined by its intersection form. It follows from Proposition 3.4 that $H^2(Y, \mathbb{R}) = 0$ under the assumption (0.3) of Theorem A. The vanishing of $H^2(Y, \mathbb{R})$ in turn implies that $\tilde{Y}$ is homeomorphic to S^4. We have thus established Theorem A.

We will now discuss the condition (0.4) in Theorem B which is stronger than the condition (0.3) in the statement of Theorem A. To see the implication of (0.4), we first recall a main result in [CGY2].

THEOREM 3.6. *[CGY2] Suppose that $(Y, \bar{g})$ is an oriented closed 4-manifold with positive Yamabe constant and that*

$$\frac{1}{4} \int_Y |\bar{\mathcal{W}}|^2 dv_{\bar{g}} < \int_Y \sigma_2(A_{\bar{g}}) dv_{\bar{g}}.$$

Then Y is diffeomorphic to S^4.

Proof of Theorem B

Again we will use the doubling manifold $(Y, \bar{g})$. We know by our assumption (0.4) that $(Y, \bar{g})$ is a $C^{2,1}$ closed doubling 4-manifold with positive scalar curvature and

$$(3.20) \qquad \frac{1}{4} \int_Y |\bar{\mathcal{W}}|^2 dv_{\bar{g}} < \int_Y \sigma_2(A_{\bar{g}}) dv_{\bar{g}}.$$

We observe that the smoothness condition $\bar{g} \in C^{2,1}$ means that we can find a nearby smooth metric g_0 of $\bar{g}$ such that the above mentioned three properties all preserved. Now we are ready to apply the argument in [CGY1] [CGY2] to deform g_0 into the metric of the standard 4-sphere with the doubling property preserved all the way. For this purpose we want to sketch the ideas given in [CGY1] [CGY2] in the following.

First, one considers the following functional $F : W^{2,2}(Y) \longrightarrow \mathbb{R}$:

$$(3.21) \qquad F[\omega] = \tilde{\gamma}_1 \tilde{I}[\omega] + \gamma_1 I[\omega] + \gamma_2 II[\omega] + \gamma_3 III[\omega].$$

For our situation we need to consider an even nowhere-vanishing symmetric $(0, 2)$ tensor η which may be taken as the metric g_0 for example. For particular choice of the coefficients

$$(3.22) \qquad \begin{aligned} \tilde{\gamma}_1 &= -\frac{\int_Y \sigma_2(A_0)dv_0 - \int_Y |\mathcal{W}_0|^2 dv_0}{2 \int_Y |\eta_0|^2 dv_0} < 0, \\ \gamma_1 &= -\frac{1}{8}, \\ \gamma_2 &= 1, \\ \gamma_3 &= \frac{3\delta - 2}{24} > 0, \end{aligned}$$

following the work of [CY], it is proved in [CGY1] [CGY2] that there exists smooth function ω which achieves the infimum of $F[\omega]$ over $W^{2,2}(Y)$ and satisfies

$$(3.23) \qquad \sigma_2(A) - |\mathcal{W}|^2 = \frac{\delta}{4}\Delta R - \tilde{\gamma}_1 |\eta|^2$$

for all $\delta > \frac{2}{3}$, here we need to emphasize that terms in (3.23) are taken with respect to the metric $e^{2\omega}g_0$. Second, by a priori estimate and non-degeneracy of the linearization of (3.23) (cf. [CGY1] [CGY2]) one may conclude that there exists metric $e^{2\omega_\delta}g_0$ which satisfies (3.23) for all $\delta > 0$. Third, take a metric $e^{2\omega_\delta}g_0$ obtained by the previous steps for sufficient small δ and deform it by Yamabe flow in the conformal class. It was proved in [CGY1] [CGY2] that, after some time, the metrics along the Yamabe flow satisfy

$$(3.24) \qquad\qquad \sigma_2(A) - \frac{1}{4}|\mathcal{W}|^2 > 0.$$

Finally, one applies a weak pinching result of Margerin [Ma]. Namely, with (3.24), one can deform again the metric by Ricci flow into the standard metric for the round 4-sphere (cf. Section 2 in [CGY2]).

Finally we need to make sure that in each of the above steps the doubling symmetry is preserved. For this purpose we simply make sure that the solution in each step is unique, at least locally. For the first step, one may use the uniqueness of the extremal of F, because of $\kappa = 0$ (cf. Theorem 2.1 in [CY]), to conclude that the extremal of F has to be an even function on the doubling manifold (Y, g_0). In second step, we have a local uniqueness as the linearized equation is non-degenerate, which also assures the solution ω_δ obtained in the argument in [CGY1] [CGY2] is even for all $\delta > 0$. The last two steps are clear since the uniqueness of the two geometric flows guarantees that all metrics along the flows respect the doubling symmetry if the initial metrics are. Thus the proof is completed.

References

[A1] M. Anderson, L^2 curvature and volume renormalization of the AHE metrics on 4-manifolds, Math. Res. Lett., 8 (2001) 171-188.

[A2] M. Anderson, Boundary regularity, uniqueness and non-uniqueness for AH Einstein metrics on 4-manifolds, preprint.

[A3] M. Anderson, Einstein metrics with prescribed conformal infinity on 4-manifolds, preprint.

[CY] S.-Y. A. Chang and P. Yang, Extremal metrics of zeta functional determinants on 4-manifolds, Ann. of Math. 142(1995) 171-212.

[CGY1] S.-Y. A. Chang, M. Gursky and P. Yang, An equation of Monge-Ampere type in conformal geometry and 4-manifolds of positive Ricci curvature, Annals of Math. 155 (2002), 709-787.

[CGY2] S.-Y. A. Chang, M. Gursky and P. Yang, A conformally invariant Sphere theorem in four dimension, Preprint 2002.

[FG1] C. Fefferman, and C.R. Graham, Conformal invariants, in *The mathematical heritage of Elie Cartan*, Asterisque, 1985, 95-116.

[FG1] C. Fefferman, and C.R. Graham, Q-curvature and Poincare metrics, Math. Res. Lett., 9 (2002), no. 2-3, 139–151.

[Gr] C. R. Graham, Volume and Area renormalizations for conformally compact Einstein metrics. The Proceedings of the 19th Winter School "Geometry and Physics" (Srni, 1999). Rend. Circ. Mat. Palermo (2) Suppl. No. 63 (2000), 31–42.

[GL] C.R. Graham and J. Lee, Einstein metrics with prescribed conformal infinity on the ball. Adv. Math. 87 (1991), no. 2, 186–225.

[GKP] S.S.Gubser, I.R.Klebanov and A.M.Polyakov, Gauge theory correlators from non-critical string theory, Phys Lett. B 428 (1998) 105-114, hep-th/9802109

[Gu1] M. Gursky, The principal eigenvalue of a conformally invariant differential operator, with application to semilinear elliptic PDE, Comm. Math. Phys. 207 (1999), 131-143.

[Gu2] M. Gursky, The Weyl functional, de Rham cohomology, and Kähler-Einstein metrics. Ann. of Math. (2) 148 (1998), no. 1, 315–337.

[HS] M. Henningson and K. Skenderis, The holographic Weyl anomaly, J. High. Energy Phys. 07 (1998) 023 hep-th/9806087; Holography and the Weyl anomaly hep-th/9812032

[L] J. Lee, The spectrum of an asymptotically hyperbolic Einstein manifold. Comm. Anal. Geom. 3 (1995), no. 1-2, 253–271.

[Mc] J. Maldacena, The large N limit of superconformal field theories and supergravity, Adv. Theo. Math. Phy. 2 (1998) 231-252, hep-th/9711200

[Ma] C. Margerin, A sharp characterization of the smooth 4-sphere in curvature forms; CAG, 6 (1998), no.1, 21-65.

[M] R. Mazzeo. The Hodge cohomology of a conformally compact metric, J. Diff. Geom. 28(1988) 309-339.

[Q] J. Qing, On the rigidity for conformally compact Einstein manifolds, Preprint 2002.

[Wa] X. Wang, On conformally compact Einstein manifolds, Math. Res. Lett. 8(2001), no. 5-6, 671–688.

[W] E. Witten, Anti-de Sitter space and holography, Adv. Theo. Math. Phy. 2 (1998) 253-290, hep-th/9802150

PRINCETON UNIVERSITY, DEPARTMENT OF MATHEMATICS, PRINCETON, NJ 08544-1000
E-mail address: chang@math.princeton.edu

UNIVERSITY OF CALIFORNIA, DEPARTMENT OF MATHEMATICS, SANTA CRUZ, CA 95064
E-mail address: qing@math.ucsc.edu

PRINCETON UNIVERSITY, DEPARTMENT OF MATHEMATICS, PRINCETON, NJ 08544-1000
E-mail address: yang@math.princeton.edu

Contemporary Mathematics
Volume **350**, 2004

On Loop Spaces of Configuration Spaces, and Related Spaces

F. R. Cohen

This paper is dedicated to H. Brezis and F. Browder

ABSTRACT. Loop spaces of configuration spaces may be regarded as spaces of continuous loops of particles moving through time t for which the particles are not allowed to collide at any fixed time t, and where the particles at time $t = 1$ have returned to their original positions at time $t = 0$. The purpose of this article is to give an exposition of properties of these loop spaces, related spaces, and applications together with a smattering of new results.

1. Introduction

Consider the configuration space

$$\mathrm{Conf}(M, k) = \{(m_1, \ldots, m_k) \in M^k \mid m_i \neq m_j \ \text{if} \ i \neq j\}$$

where M is a manifold. The (pointed) loop space of $\mathrm{Conf}(M, k)$,

$$\Omega\mathrm{Conf}(M, k),$$

is the space of ordered k-tuples of functions $(f_1, f_2, \cdots, f_k)$ such that

(1) each f_i is a continuous function

$$f_i : [0, 1] \to M,$$

with

$$f_i(0) = f_i(1) = p_i$$

for specified points $p_1, p_2, \cdots, p_k$, and

(2) $f_i(t) \neq f_j(t)$ if $i \neq j$.

The purpose of this article is to give an exposition of features of loop spaces of configuration spaces. In addition, there are other natural versions of these spaces given by total spaces of fibre bundles over $\Omega\mathrm{Conf}(M, k)$. These versions arise as paths of particles for which each particle has a spin or for which the manifold M has additional natural symmetries. The main example here of this last case is given by a discrete subgroup of $PSL(2, \mathbb{R})$ acting freely, and properly discontinuously on

1991 *Mathematics Subject Classification.* Primary: 55R80, 20F14; Secondary: 55N25, 55Q52.

Key words and phrases. Descending central series, loop space homology, surface groups.

*Partially supported by the NSF grant number 9704410.

the upper $1/2$-plane $\mathbb{H}^2$ as well as where the configuration space is replaced by a more general complement of a complex hyperplane arrangement. Some of these cases are addressed below.

In addition, the homology groups of loop spaces of configuration spaces are sometimes given by universal enveloping algebras of certain Lie algebras which connect to geometric invariants, as well as homotopy groups, and combinatorial group theory. For example, these Lie algebras have admitted applications with the Kohno-Drinfel'd monodromy theorem, as well as the Knishnik-Zamolodchikov equations [**4, 20, 21, 22, 23, 24**], descriptions of Vassiliev invariants of pure braids via iterated integals [**21, 22, 23, 24**], as well as in homotopy theory concerning the interplay between Artin's braid groups, the homotopy groups of the 2-sphere [**3, 14**], and certain double suspensions [**8**].

More specifically, let M be a manifold of dimension m with Q_i a finite subset of M having cardinality i a fixed positive integer. The loop space of the configuration space of points in M is frequently homotopy equivalent to a product of more familiar spaces $\Omega(M - Q_i)$ as given next.

EXAMPLE 1.1. If M is a manifold given by the complement of a point in a manifold N without boundary, then there is a homotopy equivalence

$$\Omega\mathrm{Conf}(M, k) \to \prod_{0 \leq i \leq k-1} \Omega(M - Q_i)$$

as pointed out in [**9**]. The proof of this remark is described in section 1 here.

However, this product decomposition fails to capture the natural loop multiplication given by concatenation of loops. The main work is to describe how these component pieces $\Omega(M - Q_i)$ of the homotopy type of $\Omega(M \vee (\vee_i S^{m-1}))$ assemble into $\Omega\mathrm{Conf}(M, k)$.

In case M is closed, there are related fibrations, as well as product decompositions. The case for which M is a sphere is special, and close to the Euclidean case in the case of odd dimensional spheres. The cases for which M is a real, complex, or quaternionic projective space are again special, and are addressed in work of [**27, 28**]. The case for which the rational cohomology of M has at least two algebraically independent cohomology algebra generators, multiplicative properties of $\Omega\mathrm{Conf}(M, k)$ are given in [**9**] by applying striking results of Félix, and Thomas [**18**].

A related example arises via paths of particles in Euclidean 3-space $\mathbb{R}^3$ modulo the action of rotation group $SO(3)$ acting naturally on the configuration space with the spinor group

$$Spin(3) = SU(2)$$

acting with the center of $Spin(3)$ fixing every point. Thus there is a projection map with singularities (not the projection in a fibre bundle)

$$\mathrm{Conf}(\mathbb{R}^3, k) \to \mathrm{Conf}(\mathbb{R}^3, k)/Spin(3).$$

The Borel construction, the homotopy orbit space,

$$ESpin(3) \times_{Spin(3)} \mathrm{Conf}(\mathbb{R}^3, k)$$

yields a fibre bundle

$$ESpin(3) \times_{Spin(3)} \mathrm{Conf}(\mathbb{R}^3, k) \to BSpin(3)$$

with fibre $\mathrm{Conf}(\mathbb{R}^3, k)$. This bundle does not admit either a (non-trivial) product decomposition or a cross-section. However, after passage to loop spaces, there are associated product decompositions:

EXAMPLE 1.2. If $k \geq 2$, then there is a homotopy equivalence

$$\Omega \mathrm{Conf}(\mathbb{R}^3, k) \to \Omega(ESpin(3) \times_{Spin(3)} \mathrm{Conf}(\mathbb{R}^3, k)) \times \Omega S^3.$$

Thus the homology of $\Omega(ESpin(3) \times_{Spin(3)} \mathrm{Conf}(\mathbb{R}^3, k))$ is a quotient Hopf algebra of the homology of $\Omega \mathrm{Conf}(\mathbb{R}^3, k)$. Furthermore, $H_* \Omega(ESpin(3) \times_{Spin(3)} \mathrm{Conf}(\mathbb{R}^3, k))$

(1) is given by the tensor algebra generated by elements $B_{i,j}$ of degree 1 for $1 \leq j < i \leq k$ modulo the 2-sided ideal generated by the "infinitesimal braid relations" (also called the "Yang-Baxter Lie relations") as listed below together with the relation $\sum_{1 \leq j < i \leq k} B_{i,j}{}^2 + \sum_{1 \leq s < t < i \leq k} [B_{i,s}, B_{i,t}] = 0$, and

(2) has Euler-Poincaré series given by $(1 + t) \cdot [(1 - 2t) \cdots (1 - (k-1)t)]^{-1}$.

Further information is given in Theorem 8.2.

On the other-hand, the homology of $\Omega \mathrm{Conf}(\mathbb{R}^m, k)$ is isomorphic to the universal enveloping algebra of a certain Lie algebra which satisfies the "infinitesimal braid relations", or "Yang-Baxter Lie" relations arising from both classical constructions within combinatorial group theory, quantum invariants, as well as homotopy theory [**9**, **15**, **14**, **20**, **21**, **22**, **23**, **24**]. Work of T. Kohno [**20**, **21**, **22**, **23**] gives iterated integrals representing the cohomology of these loop spaces in the context of work of K. T. Chen [**5**]. These integrals represent the rational cohomology of $\Omega \mathrm{Conf}(\mathbb{R}^m, n)$. A short discussion of how these homology groups admit natural structures which occur within quantum groups, low dimensional topology, as well as homotopy theory is given in Section 9.

Loop spaces of configuration spaces may be regarded as "intermediate spaces" between spaces of all continuous functions from a circle, $\Lambda(M)$, and subspaces given by spaces of continuous embeddings of a circle in M given by $Emb(S^1, M)$. There is an evaluation map

$$\theta : Emb(S^1, M) \to \Lambda \mathrm{Conf}(M, k)$$

which factors the natural evaluation map at a fixed set of k points in S^1,

$$Emb(S^1, M) \to \mathrm{Conf}(M, k)$$

described in [**9**] defined by the equation $\theta(f)(z) = (f(z), f(\lambda \cdot z), f(\lambda^2 \cdot z), \cdots, f(\lambda^{k-1} \cdot z))$ where λ is a primitive k-th root of unity. In addition, there are evaluation maps

$$\Lambda \mathrm{Conf}(M, k) \to \mathrm{Conf}(M, k)$$

for which the fibre of the composite

$$Emb(S^1, M) \to \mathrm{Conf}(M, k)$$

is denoted $Emb_k(S^1, M)$. The above maps then restrict to natural maps

$$Emb_k(S^1, M) \to \Omega \mathrm{Conf}(M, k)$$

which will be addressed elsewhere.

Furthermore, there are maps on the level of function spaces

$$\Omega \mathrm{Conf}(M, k) \to Emb(S^1, M \times \mathbb{R})$$

which specialize to the Alexander-Markov closing of a braid in case $M = \mathbb{R}^2$ as given in [9]. In the case of $M = \mathbb{R}^2$, the induced map

$$\amalg_{k\geq 0}\Omega\mathrm{Conf}(M,k) \to Emb(S^1, M \times \mathbb{R})$$

induces a surjection on path-components [9].

A list of sections for this article is as follows.

(1) Introduction
(2) Product decompositions
(3) Loop spaces of configuration spaces for Euclidean spaces
(4) Loop spaces of sphere bundles in tangent bundles
(5) Loop spaces of configuration spaces for punctured manifolds M
(6) Orbit configuration spaces
(7) Arrangements, and redundant arrangements
(8) On configurations modulo the spinor groups
(9) A synopsis

The author would like thank Abbas Bahri, and the other organizers of this conference for providing an extremely stimulating mathematical environment. The author would also like to thank Felix Browder who taught him both in basic functional analysis, and how a mathematician approches problems, as well as Haim Brezis for this opportunity to learn some interesting mathematics.

2. Product decompositions

Let M be a manifold of dimension m with Q_i a subset of M having cardinality i a fixed positive integer. If M is a manifold homeomorphic to $N - \{\mathrm{point}\}$ for a manifold N without boundary, then there is a homotopy equivalence

$$\Omega\mathrm{Conf}(M,k) \to \prod_{0\leq i\leq k-1} \Omega(M - Q_i)$$

as pointed out in [9].

The proof of the previous assertion is a special case of a general result concerning loop spaces of fibrations. That is, a fibration

$$E \to B$$

with inclusion of the fibre $i : F \to E$ for which B, E, and F are path-connected, and with a cross-section σ satisfies the property that there is a homotopy equivalence

$$\Omega(B) \times \Omega(F) \to \Omega(E)$$

given by multiplying $\Omega(\sigma) \times \Omega(i)$. With the hypotheses that M is homeomorphic to $N - \{\mathrm{point}\}$, the natural projection maps

$$\mathrm{Conf}(M,n) \to \mathrm{Conf}(M,n-1)$$

are bundle projections with a section [15]. The asserted product decomposition follows.

However, this product decomposition fails to preserve the multiplicative structure given by concatenation of loops, a failure analogous to a standard group theoretic property of semi-direct products of groups as exemplified by the symmetric group on 3 letters:

(1) The symmetric group on 3 letters is isomorphic as a set to to the product of $\mathbb{Z}/2\mathbb{Z} \times \mathbb{Z}/3\mathbb{Z}$.

(2) The multiplication in the underlying group is different than that in the product decomposition.

The loop space $\Omega\mathrm{Conf}(M, k)$ may be regarded in a similar way as having "building blocks" given by $\Omega(M - Q_i)$. The way in which these blocks are assembled is one of the useful features of the configuration space. In addition, there are analogues of semi-direct products of groups which give the homotopy type of the loop space of many configuration spaces.

One way in which these factors are "twisted" together is reflected homologically by the "infinitesimal braid relations" stated precisely in the next section with the assumption that the underlying manifold is a Euclidean space. For more general manifolds M, there are additional universal relations which arises from topological properties of the tangent bundle of M. This information is described in the following three sections.

3. Loop spaces of configuration spaces for Euclidean spaces

If $n > 2$, the product decomposition listed in Example 1.1 of the Introduction specializes to a homotopy equivalence $\Omega\mathrm{Conf}(\mathbb{R}^n, k) \to \prod_{0 \leq i \leq k-1} \Omega(\mathbb{R}^n - Q_i)$ for which the homology of each factor $\Omega(\mathbb{R}^n - Q_i)$ is isomorphic to a primitively generated tensor algebra. The homology of the loop space $\Omega\mathrm{Conf}(\mathbb{R}^n, k)$ is twisted in a way alluded to above, and described next.

Fix an integer $1 \leq i \leq k - 1$. There is a choice of algebra generators $B_{i,j}$, spherical homology classes of degree $n-2$, $1 \leq j < i$, for the homology of $\Omega(\mathbb{R}^n - Q_i)$, a tensor algebra $T[B_{i,1}, B_{i,2}, \cdots, B_{i,i-1}]$ [**9, 15**]. Commutation relations among the $B_{i,j}$ arise as the (graded) infinitesimal braid relations or "Yang-Baxter Lie relations" described next.

Let $V_n(q)$ denote the graded free abelian group of rank n which is concentrated in degree q and has basis

$$\{x_{n+1,1}, x_{n+1,2}, \ldots, x_{n+1,n}\}.$$

Let $L[V_n(q)]$ denote the free graded Lie algebra generated by $V_n(q)$ and let

$$\lambda : \bigoplus_{n=1}^{k-1} V_n(q) \longrightarrow \mathcal{G}$$

be a homomorphism of graded abelian groups where $\mathcal{G}$ is a graded Lie algebra. Let $B_{i,j} = \lambda(x_{i,j})$. The "(graded) infinitesimal braid relations" or "Yang-Baxter Lie relations" are as follows [**20, 9, 15**]:

(1) $[B_{i,j}, B_{s,t}] = 0$ if $\{i, j\} \cap \{s, t\} = \phi$,
(2) $[B_{i,j}, B_{i,t} + (-1)^q B_{t,j}] = 0$ for $1 \leq j < t < i \leq k$, and
(3) $[B_{t,j}, B_{i,j} + B_{i,t}] = 0$ for $1 \leq j < t < i \leq k$.

Let $\mathcal{L}_k(q)$ denote the graded free Lie algebra generated by the module $\bigoplus_{n=1}^{k-1} V_n(q)$ modulo the graded infinitesimal braid relations.

THEOREM 3.1. *If* $\mathcal{G}$ *is a graded Lie algebra and* $\lambda : \bigoplus_{n=1}^{k-1} V_n(q) \to \mathcal{G}$ *is a homomorphism of graded abelian groups such that* $\lambda(x_{i,j})$ *satisfy the graded infinitesimal braid relations, then there is a unique homomorphism of graded Lie algebras*

$$g : \mathcal{L}_k(q) \longrightarrow \mathcal{G}$$

which factors λ. Furthermore, there are homomorphisms of Lie algebras

$$L[V_n(q)] \to \mathcal{L}_k(q)$$

for $1 \leq n \leq k - 1$ such that the induced homomorphism of graded abelian groups

$$\bigoplus_{1 \leq n \leq k-1} L[V_n(q)] \to \mathcal{L}_k(q)$$

is an isomorphism (of graded abelian groups).

One consequence is the following result proven in [**9, 15**].

THEOREM 3.2. *If $M = \mathbb{R}^n$ for $n > 2$, then the homology of the loop space $\Omega\mathrm{Conf}(\mathbb{R}^n, k)$ is isomorphic to the universal enveloping algebra of $\mathcal{L}_k(n-2)$ as a Hopf algebra . Furthermore, the elements $B_{i,j}$ are given by the Hurewicz image of the fundamental cycle for a sphere via maps $B_{i,j}$.*

T. Kohno has given representations of Vassiliev invariants of pure braids by iterated integrals of Green's forms as well as the Bott-Taubes integral [**22, 24**]. Kohno has also shown that iterated integrals also give the rational cohomology of $\Omega(\mathrm{Conf}(\mathbb{R}^n, k))$.

The homology classes $B_{i,j}$ above arise from elementary maps of products of spheres into configuration spaces after passage to loop spaces. The maps themselves are given by the locus of points represented by "planets in orbit". One choice of the requisite maps is

$$\mu_k : (S^{n-1})^k \to \mathrm{Conf}(\mathbb{R}^n, k+1)$$

obtained be regarding k points in a the unit sphere as giving k particles in orbit around a central object with a specific formula listed next. Let $(z_1, z_2, \cdots, z_k)$ denote an ordered k-tuple of points in S^{n-1}, the points in $\mathbb{R}^n$ of unit norm. Define

$$\mu_k((z_1, z_2, \cdots, z_k)) = (0, z_1, 2 \cdot z_2, 3 \cdot z_3, \cdots, k \cdot z_k).$$

The classes $B_{i,j}$ are in the image induced maps which are analogous to the μ_k. The graded infinitesimal braid relations arise by analyzing the images of the respective fundamental cycles in homology for

$$\Omega(\mu_k) : \Omega((S^{n-1})^k) \to \Omega\mathrm{Conf}(\mathbb{R}^n, k+1).$$

A relationship between the "infinitesimal braid relations" and the classical homotopy groups of the 2-sphere, as well as the unstable Adams spectral sequence, or the Bousfield-Kan spectral sequence are mentioned in Section 9 here, and in [**14**].

4. Loop spaces of sphere bundles in tangent bundles

The main case here arises from the universal sphere bundle with a section. Consider the natural action of $SO(n)$ on the unit sphere S^{n-1}. Fix an axis, restrict to a copy of $SO(n-1)$ which fixes that axis, and consider the bundle with cross-section

$$ESO(n-1) \times_{SO(n-1)} S^{n-1} \to BSO(n-1).$$

Thus the loop space of this bundle is homotopy equivalent to the product

$$SO(n-1) \times \Omega S^{n-1}.$$

However, the multiplicative structure of the loop space is not that of the product as reflected by the following homological property [**9**] in which homology is taken with coefficients in a field $\mathbb{F}$. Additional information required to state the next theorem is that the homology of ΩS^{n-1} is isomorphic to a tensor algebra with fundamental cycle ι_{n-2} in degree $n-2$, and the homology of $SO(n-1)$ is an exterior algebra with prescribed generators x_i in degree i in case $\mathbb{F} = \mathbb{F}_2$. The element y_{n-2} in the next theorem is a non-zero primitive element in degree $n-2$ for n odd with $1/2 \in \mathbb{F}$.

THEOREM 4.1. *If $n > 2$, the following formulas are satisfied in*

$$H_*(\Omega(ESO(n-1) \times_{SO(n-1)} S^{n-1}); \mathbb{F}).$$

(i) *If $\mathbb{F} = \mathbb{F}_2$, then the commutator $[\iota_{n-2}, x_{n-2}] = \iota_{n-2} \otimes x_{n-2} - x_{n-2} \otimes \iota_{n-2}$ satisfies*

$$[\iota_{n-2}, x_{n-2}] = \iota_{n-2}^2 .$$

(ii) *If $1/2 \in \mathbb{F}$ and $n - 1 \equiv 0(2)$, then*

$$[\iota_{n-2}, y_{n-2}] = \lambda \iota_{n-2}^2$$

for a choice of unit λ, and if $n - 1 \equiv 1(2)$, then

$$[\iota_{n-2}, x_{n-2}] = 0.$$

5. Loop spaces of configuration spaces for punctured manifolds M

This section gives information about loop spaces of configuration spaces for more general manifolds M by using the information in the previous two sections. It is sometimes the case that the homology groups of the loop space $\Omega \mathrm{Conf}(M, k)$ are torsion free. If not, restrict to field coefficients, $\mathbb{F}$. In addition, let $Prim H_*(\Omega M; \mathbb{F})$ denote the Lie algebra of primitive elements in the Hopf algebra $H_*(\Omega M; \mathbb{F})$ for which the Lie bracket is given by

$$[a, b] = a \otimes b - (-1)^{|a| \cdot |b|} b \otimes a$$

where $|a|$ denotes the degree of the element a.

The relations in the homology of these loop spaces arise from naive additional topology as follows. Braids correspond to motions of points in $\mathbb{R}^2$ through time with a k-stranded braid regarded as the locus of points of k distinct particles in the Euclidean plane. One could imagine considering analogues of braids by consider subspaces A of a manifold together with motions of k disjoint copies of A through time. The subspace A replaces the role of points in $\mathbb{R}^2$.

A naive idea of "braiding of a subspace A" of M is required from [9] as follows. A manifold M is said to be A-dominated provided the following hold:

(1) There exist embeddings $e_i : A \to M$, i = 1,2, with disjoint images.
(2) The map e_1 is isotopic to e_2.
(3) The induced maps $\Omega(e_i) : \Omega(A) \to \Omega(M)$ (with respect to different basepoints) induce surjections in homology (with $\mathbb{F}$-coefficients).

The homological property that the induced map $\Omega(A) \to \Omega(M)$ gives an epimorphism in homology provides a way to measure motions of disjoint copies of A through time. Furthermore, there is a "braid-like" group attached to these homological measures [12].

In addition, this homological analogue of the notion of braiding subspaces A as above (1) facilitates analysis of the homology of $\Omega \mathrm{Conf}(M, k)$ as a Hopf algebra, and (2) gives a Lie algebraic analogue of the classical wreath product construction on the level of discrete groups. This structure leads to the so-called "extended infinitesimal braid relations" defined below which hold in the homology of loop spaces of certain configuration spaces.

Consider a graded Lie algebra L over a field $\mathbb{F}$ together with the Lie algebra given by the k-fold sum of L with itself, $\oplus_k L$. Consider the coproduct of Lie algebras

$$\mathcal{L}_k(q) \amalg (\oplus_k L),$$

and let x_i denote the element given by an element x of L in the i-th coordinate of $\oplus_k L$. Define

$$\mathcal{L}_k(q) \wr L$$

as the quotient of the coproduct of Lie algebras $\mathcal{L}_k(q)$, and $\oplus_k L$ modulo the "extended infinitesimal braid relations" as follows [9]:

(1) $[B_{i,j}, x_s] = 0$ if $\{i,j\} \cap \{s\} = \phi$,
(2) $[B_{i,j}, x_i + x_j] = 0$.

One choice of graded Lie algebra L is given by the module of primitive elements in the homology of a loop space, $PrimH_*(\Omega X; \mathbb{F})$. The notion of braiding a subspace A as well as the twist which arises in Theorem 4.1 associated to the loop space of the universal sphere bundle with section is used to prove the next result [9].

THEOREM 5.1. *Let M be a simply-connected m-dimensional manifold, $m > 2$, which satisfies*

(1) *$M = N - \{point\}$ for a manifold N without boundary,*
(2) *$w_{m-1}(\tau(M)) = 0$ where $\tau(M)$ is the tangent bundle of M if $\mathbb{F} = Z/2Z$,*
(3) *the euler class of $\tau(M)$ is zero if characteristic$(\mathbb{F}) \neq 2$,*
(4) *the Hopf algebra $H_*(\Omega M; \mathbb{F})$ is a both primitively generated, and is isomorphic to the universal enveloping algebra of the Lie algebra of primitive elements $PrimH_*(\Omega M; \mathbb{F})$, and*
(5) *M is A-dominated for some A.*

Then $H_*(\Omega \mathrm{Conf}(M, k); \mathbb{F})$ is isomorphic to the universal enveloping algebra of

$$\mathcal{L}_k(m-2) \wr \mathcal{G}$$

where $\mathcal{G}$ is the Lie algebra of primitive elements, $PrimH_*(\Omega M; \mathbb{F})$.

In case the underlying manifold has a Euclidean factor, a similar conclusion follows next [9].

THEOREM 5.2. *Let M be a simply-connected m-dimensional manifold, $m > 1$, such that the Hopf algebra $H_*(\Omega M; \mathbb{F})$ is both primitively generated, and isomorphic to the universal enveloping algebra of the Lie algebra of primitive elements $PrimH_*(\Omega M; \mathbb{F})$. Then $H_*(\Omega(\mathrm{Conf}(\mathbb{R} \times M, k)); \mathbb{F})$ is isomorphic to the universal enveloping algebra of $\mathcal{L}_k(m-1) \wr PrimH_*(\Omega M; \mathbb{F})$.*

EXAMPLE 5.3. Let $n > 3$, and $A = \Sigma \mathbb{CP}^{n-1}$. The manifolds $SU(n)$ are A-dominated. Furthermore, the manifolds $SU(3)$, and $Sp(2)$ are not A-dominated for any A [9]. There are analogous computations for the loop space of the configuration space of M given by $M = SU(n+k)/SU(n)$ and $\mathbb{F}$ is given by the rational numbers [9].

Work of Xicoténcatl [27, 28] gives information about configurations of orbits in a manifold with a free action of a group as well as various projective spaces.

A striking result of Félix and Thomas [**18**] is that the natural inclusion

$$M - \{\text{point}\} \to M$$

induces a surjection on loop space homology with rational coefficients in case the following conditions are satisfied:

(1) M is 1-connected.

(2) There exist at least 2 linearly independent algebra generators in $\bar{H}^*(M; \mathbb{Q})$.

Applying [**18**] in this case, the natural map $\Omega\mathrm{Conf}(\mathbb{M}, k) \to \Omega M$ induces a surjection on rational homology whenever M satisfies (1), and (2).

THEOREM 5.4. *If* $\Omega(M - point) \to \Omega(M)$ *induces a surjection on* $H_*(\ ; \mathbb{F})$, *then* $H_*(\Omega\mathrm{Conf}(\mathbb{M}, k); \mathbb{F})$ *is isomorphic to* $H_*(\Omega(M); \mathbb{F}) \otimes H_*(\Omega\mathrm{Conf}(\mathbb{M} - point, k - 1); \mathbb{F})$ *as a* $H_*(\Omega\mathrm{Conf}(\mathbb{M} - point, k - 1); \mathbb{F})$-*module.*

6. Orbit configuration spaces

The loop spaces of configuration spaces with additional natural symmetries are addressed in this section, and is based on work in [**6, 27, 28, 10**]. Let M denote a manifold on which the discrete group G acts freely, and properly discontinuously. Let $\mathrm{Conf}^G(M, k)$ denote the orbit configuration space

$$\mathrm{Conf}^G(M, k) = \{(m_1, \ldots, m_n) \in M^k \mid G \cdot m_i \cap G \cdot m_j = \varnothing \ \text{if} \ i \neq j\}.$$

The natural projection map $M \to M/G$ is the projection map for a covering space. If G is a discrete group, then by [**28**], there are fibrations

$$\mathrm{Conf}^G(M, k) \to \mathrm{Conf}^G(M, i)$$

with fibre over the point $(p_1, p_2, \cdots, p_i)$ in $\mathrm{Conf}^G(M, i)$ given by

$$\mathrm{Conf}^G(M - Q_i^G, k - i)$$

where $G \cdot p$ denotes the G-orbit of p, and

$$Q_i^G = \amalg_{1 \leq j \leq i} G \cdot p_j.$$

Furthermore, there is a natural action of G^k on $\mathrm{Conf}^G(M, k)$ together with a natural covering space

$$G^k \to \mathrm{Conf}^G(M, k) \to \mathrm{Conf}(M/G, k).$$

The groups, and manifolds addressed in this section are as follows.

(1) The space M is either the upper 1/2-plane $\mathbb{H}^2 = SL(2, \mathbb{R})/SO(2)$ or a product $\mathbb{H}^2 \times \mathbb{C}^q$.

(2) The group G is a discrete subgroup of $PSL(2, \mathbb{R})$ where it is assumed that G acts freely, and properly discontinuously on $\mathbb{H}^2$ by fractional linear transformations, trivially on $\mathbb{C}^q$, and diagonally on the product $\mathbb{H}^2 \times \mathbb{C}^q$. In particular, the natural projection

$$\mathbb{H}^2 \to \mathbb{H}^2/G$$

is the projection for a covering space.

Loop spaces of $\mathrm{Conf}^G(M, k)$ considered below have features analogous to the previous section, and give Vassiliev invariants of braids on surfaces [**19, 23, 24, 10**]. Modifications of the Lie algebra obtained from the descending central series for a discrete group G, listed next, are useful for this section.

For each strictly positive integer q, there is a canonical (and trivially defined) graded Lie algebra $E_0^*(G)_q$ attached to the one obtained from the descending central series for G, and which is defined as follows.

(1) Fix a strictly positive integer q.

(2) Let $\Gamma^m(G)$ denote the m-th stage of the descending central series for G.

(3) $E_0^{2mq}(G)_q = \Gamma^m(G)/\Gamma^{m+1}(G)$,

(4) $E_0^i(G)_q = \{0\}$, if $i \not\equiv 0 \mod 2q$, and

(5) the Lie bracket is induced by that for the associated graded for the $\Gamma^m(G)$.

For the rest of this section, it will be assumed that all spaces are path-connected, and homology is either torsion free homology or is taken with field coefficients. Thus the homology of $X \times X$ is isomorphic to $H_*X \otimes H_*X$, and the module of primitive elements in the homology of X, $\operatorname{Prim} H_*(X)$, is the module generated by elements α which have trivial coproduct. In addition, the Lie algebras L below are assumed to be either defined over a field $\mathbb{F}$, or are free modules over $\mathbb{Z}$. Thus the universal enveloping algebra of L exists, and is denoted $U[L]$.

Fix a base-point $\mathbf{x} = (x_1, \cdots, x_k)$ in $\operatorname{Conf}(S_g, k)$. Let $\mathbf{y}$ denote a fixed choice of base-point for $\operatorname{Conf}^G(\mathbb{H}^2, k)$, and x_0 a base-point for S_g. The fundamental group of the surface S_g is denoted $\pi_1(S_g, x_0) = G$, and the fundamental group of the orbit configuration space is denoted $\pi_1(\operatorname{Conf}^G(\mathbb{H}^2, n), \mathbf{y}) = P_n(S_g)^0$.

THEOREM 6.1. [**10**] *Let n, and q be fixed natural numbers. There are isomorphisms of Lie algebras*

$$E_0^*(P_k(S_g)^0)_q \to \operatorname{Prim} H_*\Omega(\operatorname{Conf}^G(\mathbb{H}^2 \times \mathbb{C}^q, k)),$$

and

$$U[E_0^*(P_n(S_g)^0)_q] \to H_*\Omega(\operatorname{Conf}^G(\mathbb{H}^2 \times \mathbb{C}^q, k)).$$

Furthermore, these Lie algebras are prescribed as follows:

(1) *There are sub-Lie algebras of $E_0^*(P_k(S_g)^0)_q$ given by $L[i]$ the free Lie algebra generated by elements $B_{i,j}^\sigma$ for fixed i with $1 \leq j < i \leq k$ of degree $2q$ and where σ runs over the elements of the group G.*

(2) *There is an isomorphism of abelian groups given by the natural additive extension*

$$\bigoplus_{2 \leq i \leq k} L[i] \to E_0^*(P_k(S_g)^0)_q.$$

(3) *A complete set of relations are as follows.*

    (a) *If $\{i, j\} \cap \{s, t\} = \phi$, then $[B_{i,j}^\sigma, B_{s,t}^\tau] = 0$.*

    (b) *If $1 \leq j < s < i \leq k$, then $[B_{i,j}^\tau, B_{i,s}^{\tau\sigma^{-1}} + B_{s,j}^\sigma] = 0$.*

    (c) *If $1 \leq j < s < i \leq k$, then $[B_{s,j}^\sigma, B_{i,j}^\tau + B_{i,s}^{\tau\sigma^{-1}}] = 0$.*

    (d) *The antisymmetry relation, and Jacobi identity for a graded Lie algebra.*

The following additional properties are satisfied for $\operatorname{Conf}^G(\mathbb{H}^2, n)$.

THEOREM 6.2. [**10**] *Let n be fixed natural number. The following properties are satisfied.*

(1) *The orbit configuration space $\operatorname{Conf}^G(\mathbb{H}^2, k)$ is a $K(P_k(S_g)^0, 1)$. The wreath product of the symmetric group Σ_k with G, $\Sigma_k \wr G$, acts properly discontinuously on $\operatorname{Conf}^G(\mathbb{H}^2, k)$. The orbit space*

$$\operatorname{Conf}^G(\mathbb{H}^2, k)/\Sigma_k \wr G$$

is homeomorphic to $\mathrm{Conf}(\mathbb{H}^2/G, k)/\Sigma_k$, *a* $K(\Pi, 1)$ *where* Π *is the k-stranded braid group for the surface* $\mathbb{H}^2/G$.

(2) *The fibration* $\mathrm{Conf}^G(\mathbb{H}^2, k) \to \mathrm{Conf}^G(\mathbb{H}^2, k-1)$ *has (i) trivial local coefficients in homology, and (ii) a cross-section.*

(3) *The Lie algebra* $E_0^*(\pi_1(\mathrm{Conf}^G(\mathbb{H}^2, k)))$ *is isomorphic to that given in Theorem 6.1.*

(4) *The integral homology of* $\mathrm{Conf}^G(\mathbb{H}^2, k)$ *is given additively by*

$$H_* \mathrm{Conf}^G(\mathbb{H}^2, k) \cong H_*(C_1) \otimes H_*(C_2) \otimes \cdots \otimes H_*(C_{k-1})$$

where C_i *is the infinite bouquet of circles* $\bigvee_{|Q_i^G|} S^1$ *where* Q_i^G *as defined as above.*

Horizontal chord diagrams of a closed oriented surface of genus g, S_g, are described in [**19, 10**]. In case $g > 0$, these horizontal diagrams give analogous constructions as those in genus 0 as given in [**22, 23, 24**]. Let $\mathcal{A}_n(S_g)^0$ and $\mathcal{A}_n(S_g)$ denote the algebras of horizontal chord diagrams of S_g as introduced, and studied in [**19**]. The elliptic cases are special, and given in [**23, 13**].

THEOREM 6.3. [**19, 10**] *The algebra*

$$\mathcal{A}_n(S_g)^0$$

is isomorphic to the universal enveloping algebra of $E_0^*(\pi_1 \mathrm{Conf}^G(\mathbb{H}^2, n), \mathbf{y})$, *which is* $E_0^*(P_n(S_g)^0)$ *where* $S_g = \mathbb{H}^2/G$.

Work of T. Kohno [**21, 22, 23**], as well as subsequent work of M. Falk and R. Randell [**17**] gives the structure of the Lie algebra associated to the descending central series for the k-stranded pure braid group, the fundamental group of $\mathrm{Conf}(\mathbb{R}^2, n)$. Work in [**13**] gives the associated Lie algebras for loop spaces of $\mathrm{Conf}^L(\mathbb{C} \times \mathbb{C}^q, n)$ where L is the the standard lattice of integral points in $\mathbb{C}$ acting by translations on $\mathbb{C}$. Work of J. González-Meneses, and L. Paris [**19**] gives the structure of the Lie algebras associated to the descending central series for the fundamental group of $\mathrm{Conf}^G(\mathbb{H}^2, n)$.

These Lie algebras arise in an analysis of Vassiliev invariants for pure braids on a surface. The theorem above gives a comparison between these Lie algebras, and their analogues for certain choices of loop spaces. Analogous groups were shown to be given by morphisms of coalgebras in [**12**]. In addition, recent work of Papadima, and Suciu [**25**] give related structures when the associated Lie algebras are of finite type. The groups which occur here for a surface of genus greater than 0 are not of finite type, and it is unclear whether there is an extension of the work in [**25**] to these choices of Lie algebras.

There is a natural structure of graded Poisson algebra structure for the homology of an iterated loop space [**11**]. This Poisson algebra given by

$$H_* \Omega^k(\mathrm{Conf}^G(\mathbb{H}^2 \times \mathbb{C}^q, n))$$

is described in the next theorem in case $2q + 1 > k > 1$. Loosely speaking, the resulting structure is the free Poisson algebra subject to the "Poisson analogues" of the "infinitesimal braid relations" Theorem 6.1.

THEOREM 6.4. [**10**]

(1) *If $k > 1$, the homology of $\Omega^k(\mathrm{Conf}^G(\mathbb{H}^2 \times \mathbb{C}^q, n))$, with any field coefficients, is a graded Poisson algebra with Poisson bracket given by the Browder operation $\lambda_{k-1}[-,-]$ for the homology of a k-fold loop space.*

(2) *If $2q + 1 > k > 1$, the homology of $\Omega^k \mathrm{Conf}^G(\mathbb{H}^2 \times \mathbb{C}^q, n)$ with coefficients in a field $\mathbb{F}$ of characteristic zero, is the free Poisson algebra generated by elements*

$$B_{i,j}^\sigma$$

of degree $2q+1-k$ for $1 \le j < i \le n$, and σ in G modulo the "infinitesimal Poisson surface braid relations" given as follows:

(a) *If $\{i,j\} \cap \{s,t\} = \phi$, then $\lambda_{k-1}[B_{i,j}^\sigma, B_{s,t}^\tau] = 0$.*

(b) *If $1 \le j < s < i \le n$, then $\lambda_{k-1}[B_{i,j}^\tau, B_{i,s}^{\tau\sigma^{-1}} + B_{s,j}^\sigma] = 0$.*

(c) *If $1 \le j < s < i \le n$, then $\lambda_{k-1}[B_{s,j}^\sigma, B_{i,j}^\tau + B_{i,s}^{\tau\sigma^{-1}}] = 0$.*

(d) *The antisymmetry relation, and Jacobi identity for a graded Poisson algebra.*

(3) *There is a map*

$$E^2 : \mathrm{Conf}^G(\mathbb{H}^2 \times \mathbb{C}^q, n) \to \Omega^2 \mathrm{Conf}^G(\mathbb{H}^2 \times \mathbb{C}^{q+1}, n)$$

which induces a homology isomorphism in degree $2q + 1$. The associated loop map $\Omega(E^2) : \Omega\mathrm{Conf}^G(\mathbb{H}^2 \times \mathbb{C}^q, n) \to \Omega^3 \mathrm{Conf}^G(\mathbb{H}^2 \times \mathbb{C}^{q+1}, n)$ induces an isomorphism on $H_{2q}(-; \mathbb{Z})$. Furthermore, the image of the map $\Omega(E^2)$ in homology is the subalgebra generated by the classes of degree $2q$.

7. Arrangements, and redundant arrangements

Given a hyperplane arrangement in a complex vector space of dimension ℓ, there is a natural associated arrangement of codimension k subspaces in a complex vector space of dimension $k\ell$, a "redundant arrangement" as follows [7]. Let $\mathcal{A}$ be a hyperplane arrangement in $\mathbb{C}^\ell$, a finite collection of codimension one affine subspaces, with complement $M(\mathcal{A}) = \mathbb{C}^\ell \setminus \bigcup_{H \in \mathcal{A}} H$. Given a hyperplane $H \subset \mathbb{C}^\ell$, let H^k be the codimension k affine subspace of $\mathbb{C}^{k\ell} = (\mathbb{C}^\ell)^k$ consisting of all k-tuples of points in $\mathbb{C}^\ell$, each of which lies in H. For each positive integer k, the elements of the hyperplane arrangement $\mathcal{A}$ may be used in this way to obtain a *redundant* arrangement $\mathcal{A}^k$ of complex codimension k subspaces in $\mathbb{C}^{k\ell}$, with complement $M(\mathcal{A}^k) = \mathbb{C}^{k\ell} \setminus \bigcup_{H \in \mathcal{A}} H^k$.

Let $G = \pi_1(M(\mathcal{A}))$ be the fundamental group of the complement of the fiber-type arrangement $\mathcal{A}$ in $\mathbb{C}^\ell$, and let $E_0^*(G)$ be the Lie algebra obtained from the descending central series of G. An important case is given by $M = \mathbb{C}^k \setminus \{0\}$ where $\Gamma = \mathbb{Z}/p\mathbb{Z}$ acts freely on M by rotations. The orbit configuration spaces $\mathrm{Conf}^\Gamma(M, \ell)$ of ordered ℓ tuples of points in distinct orbits were the subject of [6] and [27], the results of which combine to give results analogous to theorems above concerning loop spaces of configuration spaces.

When $\mathcal{A}$ is a fiber-type hyperplane arrangement, the behavior of the family of spaces $\{X_\ell(\mathbb{C}^k) = M(\mathcal{A}^k), k \ge 1\}$ is reminiscent of that of the family $\mathrm{Conf}(\mathbb{R}^n, k)$ of configuration spaces. Many of the results above are inspired by [7] which are analogous to the case where configuration spaces are replaced by certain complements of hyperplane arrangements, and which are described next.

THEOREM 7.1. [7] *If $\mathcal{A}$ denotes a fiber type complex hyperplane arrangement, and $G = \pi_1(M(\mathcal{A}))$ the fundamental group of the complement of $\mathcal{A}$. For $k \ge 1$,* *the homology of $\Omega M(\mathcal{A}^{k+1})$, the loop space of the complement of the redundant subspace arrangement $\mathcal{A}^{k+1}$ in $\mathbb{C}^{(k+1)\ell}$, is isomorphic to the universal enveloping algebra of the graded Lie algebra $E_0^*(G)_k$. Moreover,*

(a) *The image of the Hurewicz homomorphism*

$$\pi_*(\Omega M(\mathcal{A}^{k+1})) \to H_*(\Omega M(\mathcal{A}^{k+1}); \mathbb{Z})$$

is isomorphic to $E_0^*(G)_k$; and

(b) *The Hurewicz homomorphism induces isomorphisms of graded Lie algebras*

$$\pi_*(\Omega M(\mathcal{A}^{k+1}))/\operatorname{Torsion} \to \operatorname{Prim} H_*(\Omega M(\mathcal{A}^{k+1}); \mathbb{Z}) \cong E_0^*(G)_k,$$

where the Lie algebra structure of the source is induced by the Samelson product.

There is an analogous reult concerning the structure of the Poisson algebra given by the rational homology of the iterated loop space of $\Omega^j M(\mathcal{A}^{k+1})$ for $j < 2(k+1)$.

THEOREM 7.2. *Let $\mathcal{A}$ be a fiber-type hyperplane arrangement in $\mathbb{C}^\ell$ with exponents $\{d_1, \ldots, d_\ell\}$. Then, for each $k \geq 1$,*

(a) *There is a map*

$$E^2 : M(\mathcal{A}^{k+1}) \to \Omega^2 M(\mathcal{A}^{k+2})$$

such that the associated loop map

$$\Omega(E^2) : \Omega M(\mathcal{A}^{k+1}) \to \Omega^3 M(\mathcal{A}^{k+2})$$

induces an isomorphism on $H_{2k}(-; \mathbb{Z})$. Furthermore, the image of the map $\Omega(E^2)$ in homology is the subalgebra generated by the classes of degree $2k$.

(b) *If $q > 1$, the homology of $\Omega^q M(\mathcal{A}^{k+1})$, with any field coefficients, is a graded Poisson algebra with Poisson bracket given by the Browder operation $\lambda_{q-1}[-,-]$ for the homology of a q-fold loop space.*

(c) *If $q \geq 1$, then $\Omega^q M(\mathcal{A}^{k+1})$ is homotopy equivalent to $\prod_{j=1}^\ell \Omega^q(\bigvee_{d_j} S^{2k+1})$.*

(d) *If $1 < q < 2k+1$, the homology of $\Omega^q M(\mathcal{A}^{k+1})$, with coefficients in a field $\mathbb{F}$ of characteristic zero, is the free Poisson algebra generated by elements β_H of degree $2k + 1 - q$ for $H \in \mathcal{A}$ modulo the relations*

$$\lambda_{q-1}[\beta_X, \beta_H] = 0,$$

for codimension two flats $X \in \mathsf{L}(\mathcal{A})$ and hyerplanes $H \in \mathcal{A}$ containing X, where $\beta_X = \sum_{X \subset H} \beta_H$.

The relations in this last theorem give a "Poisson analogue" of the infinitesimal braid relations for redundant arrangements obtained from fibre-type hyperplane arrangements.

8. On configurations modulo the spinor groups

The purpose of this section is to record a proof of Example 1.2. The first step arises as a standard alternative description for the classical Hopf fibration. That is consider the bundle

$$ESO(3) \times_{SO(3)} S^2 \to BSO(3)$$

with fibre S^2. It is classical that the homotopy orbit space $ESO(3) \times_{SO(3)} S^2$ is homotopy equivalent to $BSO(2) = \mathbb{C}P^\infty = K(\mathbb{Z}, 2)$. Thus the homotopy theoretic fibre of a degree 1 map $S^2 \to BSO(2)$ is given by $SO(3)$ for which the induced map

$\Delta : SO(3) \to S^2$ also induces an isomorphism of the level of $\pi_3(SO(3)) \to \pi_3(S^2)$. Furthermore, there is an induced map of bundles

$$
\begin{array}{ccccc}
\{point\} & \longrightarrow & S^2 & \xrightarrow{\ id\ } & S^2 \\
\downarrow & & \downarrow & & \downarrow \\
\mathbb{RP}^\infty & \longrightarrow & ESpin(3) \times_{Spin(3)} S^2 & \longrightarrow & ESO(3) \times_{SO(3)} S^2 \\
\downarrow{\scriptstyle id} & & \downarrow & & \downarrow \\
\mathbb{RP}^\infty & \longrightarrow & BSpin(3) & \longrightarrow & BSO(3).
\end{array}
$$

Hence the map $\Delta : Spin(3) \to S^2$ obtained from the fibration $ESpin(3) \times_{Spin(3)} S^2 \to BSpin(3)$ induces the classical Hopf map up to homotopy (up to an ambiguity given by ± 1).

A second feature of the Hopf map $\eta : S^3 \to S^2$ is that there is an induced splitting after looping: The composite

$$
S^1 \times \Omega S^3 \xrightarrow{E \times \Omega \eta} \Omega S^2 \times \Omega S^2 \xrightarrow{multiply} \Omega S^2
$$

is a homotopy equivalence where E denotes the Freudenthal suspension. Furthermore, the map $\Omega(\eta) : \Omega S^3 \to \Omega S^2$ induces an isomorphism on the level of second homology groups, and sends the fundamental cycle in $H_2 \Omega S^3$ to the square of the fundamental cycle in $H_1 \Omega S^2$. This standard fact is recorded in the next lemma.

LEMMA 8.1. *The composite*

$$
S^1 \times \Omega S^3 \xrightarrow{E \times \Omega \eta} \Omega S^2 \times \Omega S^2 \xrightarrow{multiply} \Omega S^2
$$

is a homotopy equivalence where E denotes the Freudenthal suspension, and $\eta : S^3 \to S^2$ is the Hopf map. Furthermore, the map on homology induced by $\Omega(\eta) : \Omega S^3 \to \Omega S^2$ is given by $\Omega(\eta)_* : T[\iota_2] \to T[\iota_1]$ where $T[\iota_j]$ is the tensor algebra generated by a class ι_j of degree j with

$$
\Omega(\eta)_*(\iota_2) = \iota_1^2.
$$

There is an analogous induced map

$$
\Delta_k : Spin(3) \to \mathrm{Conf}(\mathbb{R}^3, k)
$$

which is used to obtain the main result of this section as follows.

THEOREM 8.2. *If $k \geq 2$, then the following statements hold.*

(1) *The space $\Omega\mathrm{Conf}(\mathbb{R}^3, k)$ is homotopy to*

$$
\Omega(ESpin(3) \times_{Spin(3)} \mathrm{Conf}(\mathbb{R}^3, k)) \times \Omega S^3,
$$

and there is a retraction for the induced map

$$
\Omega(\Delta_k) : \Omega Spin(3) \to \Omega\mathrm{Conf}(\mathbb{R}^3, k).
$$

(2) *The induced map in homology $\Omega(\Delta_k)_* : H_*\Omega Spin(3) \to H_*\Omega\mathrm{Conf}(\mathbb{R}^3, k)$ with any field coefficients is a monomorphism of Hopf algebras.*

(3) *The homology Serre spectral sequence with any field coefficients for the multiplicative fibration*

$$
\Omega Spin(3) \xrightarrow{\Omega(\Delta_k)} \Omega\mathrm{Conf}(\mathbb{R}^3, k) \xrightarrow{\Omega(proj)} \Omega(ESpin(3) \times_{Spin(3)} \mathrm{Conf}(\mathbb{R}^3, k))
$$

collapses.

(4) *The Euler-Poincaré series for the homology of* $\Omega(ESpin(3) \times_{Spin(3)} \mathrm{Conf}(\mathbb{R}^3, k))$ *is*
$$(1 + t) \cdot [(1 - 2t) \cdots (1 - (k-1)t)]^{-1}.$$

(5) *The image of the induced map in homology* $\Omega(\Delta_k)_*$ *is the subalgebra generated by the element*
$$\sum_{1 \leq j < i \leq k} B_{i,j}{}^2 + \sum_{1 \leq s < t < i \leq k} [B_{i,s}, B_{i,t}].$$

Thus the homology of $\Omega(ESpin(3) \times_{Spin(3)} \mathrm{Conf}(\mathbb{R}^3, k))$ *is given by the tensor algebra generated by elements* $B_{i,j}$ *of degree 1 for* $1 \leq j < i \leq k$ *modulo the 2-sided ideal generated by the "infinitesimal braid relations" together with the relation* $\sum_{1 \leq j < i \leq k} B_{i,j}{}^2 + \sum_{1 \leq s < t < i \leq k} [B_{i,s}, B_{i,t}] = 0,$

PROOF. The product decomposition in the first part follows from Lemma 8.1 in case $k = 2$. The cases of $k > 2$ follow from the fact that the projection map to any two coordinates $proj : \mathrm{Conf}(\mathbb{R}^3, k) \to \mathrm{Conf}(\mathbb{R}^3, 2)$ is $Spin(3)$-equivariant.

Notice that the natural composite
$$\Omega Spin(3) \xrightarrow{\Omega(\Delta_k)} \Omega \mathrm{Conf}(\mathbb{R}^3, k) \xrightarrow{\Omega(proj)} \Omega \mathrm{Conf}(\mathbb{R}^3, 2).$$

gives a retraction for the map $\Omega(\Delta_k)$ with source $\Omega Spin(3)$. Thus the induced map in homology $\Omega(\Delta_k)_*$ is a split monomorphism of Hopf algebras, and so the homology of $\Omega \mathrm{Conf}(\mathbb{R}^3, k)$ is a free module over the homology of $\Omega Spin(3)$. Hence the Serre spectral sequence for the multiplicative fibration
$$\Omega \mathrm{Conf}(\mathbb{R}^3, 3) \longrightarrow \Omega(ESpin(3) \times_{Spin(3)} \mathrm{Conf}(\mathbb{R}^3, k))$$

collapses, and parts 2, 3, and 4 of the theorem follow.

To compute the image of the homology of $\Omega Spin(3)$ in that of $\Omega \mathrm{Conf}(\mathbb{R}^3, k)$, some additional topology is used. Notice that the homotopy equivalence
$$\zeta : S^2 \to \mathrm{Conf}(\mathbb{R}^3, 2)$$

specified by the formula $\zeta(z) = (z, -z)$ is equivariant with respect to the natural $SO(3)$-action on both source, and target. Thus there is a fibre homotopy equivalence

$$
\begin{array}{ccc}
S^2 & \longrightarrow & \mathrm{Conf}(\mathbb{R}^3, 2) \\
\downarrow & & \downarrow \\
ESpin(3) \times_{Spin(3)} S^2 & \longrightarrow & ESO(3) \times_{SO(3)} \mathrm{Conf}(\mathbb{R}^3, 2) \\
\downarrow & & \downarrow \\
BSpin(3) & \xrightarrow{\mathrm{id}} & BSpin(3).
\end{array}
$$

Thus, $\Omega(\Delta_2)_* : H_* \Omega Spin(3) \to H_* \Omega \mathrm{Conf}(\mathbb{R}^3, 2)$ satisfies the formula
$$\Omega(\Delta)_*(\iota_2) = \iota_1^2$$

by Lemma 8.1. It follows that the image of ι_2 in $H_2 \Omega \mathrm{Conf}(\mathbb{R}^3, k)$ is both primitive, and equal to $\sum_{1 \leq j < i \leq k} B_{i,j}{}^2$ plus possible other terms.

By inspection of $H_2 \Omega \mathrm{Conf}(\mathbb{R}^3, k)$, a basis for the module of primitives in degree 2 is given by
$$\{B_{i,j}{}^2 | 1 \leq j < i \leq k\} \amalg \{[B_{i,s}, B_{i,t}] | 1 \leq s < t < i \leq k\}.$$

Hence it suffices to show that each basis element $[B_{i,s}, B_{i,t}]$ for $1 \le s < t < i \le k$ is a summand of $\Omega(\Delta_k)_*(\iota_2)$. If $k > 1$, it will be checked next that

$$\Omega(\Delta_k)_*(\iota_2) = \sum_{1 \le j < i \le k} B_{i,j}{}^2 + \sum_{1 \le s < t < i \le k} [B_{i,s}, B_{i,t}]$$

from which the last step of Theorem 8.2 follows.

Consider $S^2 \times S^2$ together with the map the map

$$\mu_k : (S^{n-1})^k \to \mathrm{Conf}(\mathbb{R}^n, k+1)$$

for $k = 2$ as defined in Section 3 which is $Spin(3)$-equivariant specialized to $n = 3$. Thus there is a commutative diagram

$$
\begin{array}{ccc}
Spin(3) \times (S^2 \times S^2) & \xrightarrow{1 \times \mu_3} & Spin(3) \times \mathrm{Conf}(\mathbb{R}^3, 3) \\
\downarrow{\scriptstyle \gamma_1} & & \downarrow{\scriptstyle \gamma_2} \\
S^2 \times S^2 & \longrightarrow & \mathrm{Conf}(\mathbb{R}^3, 3)
\end{array}
$$

for which γ_i denotes the natural action of $Spin(3)$. Thus there is a commutative diagram

$$
\begin{array}{ccc}
\Omega(Spin(3)) \times \Omega(S^2 \times S^2) & \xrightarrow{1 \times \Omega(\mu_3)} & \Omega(Spin(3)) \times \Omega(\mathrm{Conf}(\mathbb{R}^3, 3)) \\
\downarrow{\scriptstyle \Omega(\gamma_1)} & & \downarrow{\scriptstyle \Omega(\gamma_2)} \\
\Omega(S^2 \times S^2) & \xrightarrow{\Omega(\mu_3)} & \Omega(\mathrm{Conf}(\mathbb{R}^3, 3)).
\end{array}
$$

A direct computation gives that

$$\Omega(\gamma_1)_*(\iota_2 \otimes 1 \otimes 1) = x_1^2 + x_2^2$$

where x_i denotes the two natural 1-dimensional fundamental cycles for $H_1\Omega(S^2 \times S^2)$. Furthermore,

    (1) $\Omega(\mu_3)_*(x_1) = B_{2,1}$, and
    (2) $\Omega(\mu_3)_*(x_2) = B_{3,1} + B_{3,2}$

by a computation analogous to that in section 4 of [9]. Thus

$$\Omega(\mu_3)_*(x_1^2 + x_2^2) = B_{2,1}{}^2 + (B_{3,1} + B_{3,2})^2.$$

Notice that

$$(B_{3,1} + B_{3,2})^2 = B_{3,1}^2 + B_{3,2}^2 + B_{3,1} \cdot B_{3,2} + B_{3,2} \cdot B_{3,1}.$$

The element $B_{3,1} \cdot B_{3,2} + B_{3,2} \cdot B_{3,1}$ is equal to the graded commutator

$$[B_{3,1}, B_{3,2}]$$

as $B_{i,j}$ has degree 1 in this case. Hence

$$\Omega(\mu_3)_*(x_1^2 + x_2^2) = B_{3,1}^2 + B_{3,2}^2 + [B_{3,1}, B_{3,2}],$$

and the theorem follows by varying choices of projections of the configuration space to 3 coordinates. $\square$

9. A synopsis

The purpose of this section is to list some connections, applications, as well as questions concerning the above structures.

Vassiliev invariants, as well as related structure arise in the context of the homology of the loop space of Euclidean space [20, 21, 22, 23, 24]. In addition, T. Kohno has shown that the rational cohomology of the loop space of $\mathrm{Conf}(\mathbb{R}^n, k)$ for $n > 1$ is always given in terms of integrals associated to Green's forms.

The situation with loop spaces of redundant arrangements may be similar. It is natural to ask about features of iterated integrals giving the deRham cohomology, and what these iterated integrals are measuring in these cases. A related question is whether there is a relationship between the Poisson algebra structure for iterated loop spaces of orbit configuration spaces, and the Goldman bracket as described in [24] as well as analogous more recent structures involving free loop spaces.

A connection to homotopy theory is given next. In [14], there are choices of maps of free groups on k-letters to the $k + 1$-st pure braid group P_{k+1}, the fundamental group of $\mathrm{Conf}(\mathbb{R}^2, k + 1)$ given by

$$\Theta_k : F_k \to P_{k+1}.$$

These homomorphisms are closely tied to the homotopy groups of the 2-sphere. Using the structure of the Lie algebra obtained from the descending central series of P_{k+1}, it is shown that Θ_k is faithful. One application of this point is that the homotopy groups of the 2-sphere are natural sub-quotients of the pure braid groups [14, 3].

The method above of showing that Θ_k is an embedding reduces to a question concerning Lie algebras. Namely, to show that these group homomorphisms are embeddings, it suffices to show that the induced map on the level of Lie algebras obtained from the filtration quotients of the descending cenral series is a monomorphism [14].

This method shows that a non-stable version of the classical Adams spectral sequence is "measured" by Vassiliev invariants of pure braids. That is the method of Bousfield-Kan to construct the unstable Adams spectral sequence (with the modifications obtained by replacing the descending central series by the mod-p descending central series) is precisely filtering certain simplicial groups by their descending central series. The associated graded gives the E^0-term of the spectral sequence. On the other-hand, the associated graded Lie algebra for the pure braid group above gives the Vassiliev invariants of pure braids via [21, 22, 23, 24].

Remarks concerning stable vector bundles over configuration spaces and their loop spaces are listed next. If $n > 2$, then the single suspension of either $\mathrm{Conf}(\mathbb{R}^n, k)$ or $\Omega\mathrm{Conf}(\mathbb{R}^3, k)$ is homotopy equivalent to a bouquet of spheres (as follows directly from the results above). Thus the groups of homotopy classes of pointed maps

$$[X, B^2 G],$$

or

$$[X, BG]$$

follow directly in case X is given by $\mathrm{Conf}(\mathbb{R}^n, k)$ or $\Omega\mathrm{Conf}(\mathbb{R}^n, k)$ with BG equal to BO or BU. Let

$$\Omega_* : [X, B^2 G] \to [\Omega(X), BG]$$

be the homomorphism obtained by looping a map. It seems likely that none of the elements in the image of the map Ω_* restrict to non-trivial ℓ-plane bundles over $\Omega\mathrm{Conf}(\mathbb{R}^n, k)$ for $2 \leq \ell < \infty$. It is natural to ask whether there is an analogue of [1] for which configuration spaces are replaced by their loop spaces.

References

[1] A. P. Balachandran, A. Daughton, Z.-C. Gu, and R. D. Sorkin, *Spin-statistics theorems without relativity or field theory*,International Journal of Modern Physics, v. 8, **17**(1993), 2993–3044.

[2] D. Bar-Natan, *On the Vassiliev knot invariants*, Topology, **34**(1995), 423–472.

[3] J. Berrick, F. R. Cohen, Y. L. Wong, and J. Wu *Configurations, braids, and homotopy groups*, submitted.

[4] V. Chari, A. Pressley, *A Guide to Quantum Groups*, Cambridge University Press, (1994), Cambridge, England.

[5] K. T. Chen, *Iterated path integrals*,**83**(1977), 831–879.

[6] D. C. Cohen, *Monodromy of fiber-type arrangements and orbit configuration spaces*, Forum Math. **13** (2001), 505–530.

[7] D. C. Cohen, F. R. Cohen, M. Xicoténcatl, *Lie Algebras Associated to Fiber-Type Arrangements*, IMRN, **29** (2003), 1591–1621.

[8] F. R. Cohen, *Combinatorial group theory in homotopy theory, I*, preprint on webpage.

[9] F. R. Cohen and S. Gitler, *Loop spaces of Configuration spaces, braid-like groups, and knots*, Proceedings of the 1998 Barcelona Conference in Algebraic Topology, and *On loop spaces of configuration spaces*, Trans. Amer. Math. Soc. **354** (2002), 1705–1748.

[10] F. R. Cohen, T. Kohno, M. Xicoténcatl, *Orbit configuration spaces associated to discrete subgroups of $PSL(2, \mathbb{R})$*, preprint available.

[11] F.R. Cohen, T.J. Lada and J.P. May, *The homology of iterated loop spaces*, Lecture Notes in Math., vol. **533**, Springer-Verlag (1976).

[12] F. R. Cohen and T. Sato, *On groups of morphisms of coalgebras*, preprint available.

[13] F. R. Cohen, M. Xicoténcatl, *On orbit configuration spaces associated to the Gaussian integers: homology and homotopy groups*, Topology Appl. **118** (2002), 17–29.

[14] F. R. Cohen, and J. Wu, *On braid groups, free groups, and the loop space of the 2-sphere*, to appear.

[15] E. Fadell and S. Husseini,
(a) *The space of loops on configuration spaces and the Majer-Terracini index*, Topological Methods in Nonlinear Analysis, Journal of the Julius Schauder Center and
(b) *Geometry and Topology of Configuration Spaces*, Springer-Verlag, Springer Monographs in Mathematics, (2001).

[16] E. Fadell and L. Neuwirth, *Configuration spaces*, Math. Scand., **10** (1962), 111-118.

[17] M. Falk, R. Randell, *The lower central series of a fiber-type arrangement*, Invent. Math., **82** (1985), 77–88.

[18] Y. Félix and J.-C. Thomas,
(a) *Effet d'un attachement cellulaire dans l'homologie de l'espace des lacets*, Ann. Inst. Fourier, Grenoble 39, 1(1989), 207-224 and
(b) *Homologie des espaces de lacets des espaces de configuration*, Ann. Inst. Fourier (Grenoble) 44 (1994), no. 2, 559-568.

[19] J. González-Meneses and L. Paris, *Vassiliev invariants for braids on surfaces*, preprint, 2000, to appear in Trans. Amer. Math. Soc..

[20] T. Kohno, *Série de Poincaré-Koszul associée aux groupes de tresses pures*, Invent. Math., **82**(1985), 57–75.

[21] T. Kohno, *Linear represenations of braid groups and classical Yang-Baxter equations*, Contemp. Math., **78**(1988), 339-363.

[22] T. Kohno, *Vassiliev invariants and de Rham complex on the space of knots*, Contemp. Math. **179**(1994), 123-138.

[23] T. Kohno, *Elliptic KZ system, braid groups of the torus and Vassiliev invariants*, Topology and its Applications, **78**(1997), 79-94.

[24] T. Kohno, *Conformal Field Theory and Topology*, Iwanami Series in Modern Mathematics, Translations of Mathematical Monographs, A.M.S., **210**(2002).

[25] S. Papadima, and A. Suciu *Homotopy Lie algberas, lower central series, and the Koszul property*, preprint, 2001.

[26] D. Quillen, *On the associated graded ring of a group ring*, J. Algebra **10**(1968), 411-418.

[27] M.A. Xicoténcatl, *Orbit Configuration spaces, infinitesimal braid relations in homology and equivariant loop spaces*, Ph.D Thesis, The University of Rochester (1997).

[28] M.A. Xicoténcatl, *The Lie algebra of the pure braid group* , Bol. Soc. Mat. Mexicana **3**, Vol. 6 (2000), 55-62.

DEPARTMENT OF MATHEMATICS, UNIVERSITY OF ROCHESTER, ROCHESTER, NY 14627 U.S.A.
E-mail address: cohf@math.rochester.edu

Global Energy Minimizers for Free Boundary Problems and Full Regularity in Three Dimensions

Luis A. Caffarelli, David Jerison, and Carlos E. Kenig

This paper is dedicated to Felix Browder and Haim Brezis

ABSTRACT. We show that in three-dimensional space, the only global minimizer for a well-known one phase free boundary problem is the half-space solution. As a consequence, we deduce full regularity of energy-minimizing solutions of this free boundary problem both in the one-phase and two-phase case, namely, the free boundary is a smooth surface at every interior point. The method also shows that the simplest nonplanar critical point for the energy functional is not a minimizer in dimensions 4 through 6, indicating that the first nonsmooth example should occur in dimension 7.

§1. Introduction

An energy minimizer for the functional

$$J(v, B) = \int_B |\nabla v|^2 + \chi_{v>0}$$

is a continuous function $u \in H^1_{loc}(\mathbf{R}^n)$ such that for any ball B and any continuous function v defined on $\overline{B}$ such that $v = u$ on ∂B, $J(v, B) \geq J(u, B)$. Our main result is as follows.

THEOREM. *Suppose that $u \geq 0$ is a nonzero energy minimizer for $J(\cdot, B)$ for every ball $B \subset \mathbf{R}^3$ and that u is homogeneous of degree 1: $u(rx) = ru(x)$. Then after rotation, $u(x) = x_3^+ \equiv \max(x_3, 0)$.*

The significance of this theorem is that in conjunction with well-known theorems concerning the regularity of free boundaries, it implies full, classical regularity in variational free boundary problems in 3 dimensions, both in the one and the two-phase case. We formulate this corollary as follows. Consider smooth functions $Q_\pm$ on $\mathbf{R}^3$ satisfying

---

1991 *Mathematics Subject Classification*. Primary 35J25, 35B65; Secondary 35J05.
Key words and phrases. two-phase, free boundary, one-phase, energy minimizers.
The three authors were supported in part by NSF grants.

$Q_+(x) - Q_-(x) \geq c > 0$, a bounded domain $\Omega \subset \mathbf{R}^3$, and a function u that minimizes

$$I(v) = \int_\Omega (|\nabla v|^2 + Q_+(x)\chi_{\{v>0\}} + Q_-(x)\chi_{\{v\leq 0\}})dx$$

among all functions in $H^1(\Omega) \cap C(\Omega)$ satisfying $v = f$ on some portion of $\partial\Omega$. Denote $\Omega^+(u) = \{x : u(x) > 0\} \cap \Omega$ and $\Omega^-(u) = \{x \in \Omega : u(x) \leq 0\}^\circ$. The free boundary is defined as $F(u) = (\partial\Omega^+(u)) \cap \Omega$. Denote $u^\pm(x) = \max(\pm u(x), 0)$. It is a routine matter to confirm that $\Delta u^\pm = 0$ on $\Omega^\pm(u)$.

COROLLARY. *Let u be the minimizer of I as above for a domain $\Omega \subset \mathbf{R}^3$. Then $u^\pm$ is a smooth function in $\overline{\Omega^\pm(u)} \cap \Omega$ and the free boundary $F(u)$ is a smooth surface. Moreover, the free boundary condition*

$$(u_\nu^+)^2 - (u_\nu^-)^2 = Q_+ - Q_- \quad on\ F(u)$$

is satisfied in the classical sense, where u_ν denotes the normal derivative.

A brief history of this problem is as follows. In [**AC**] in the one-phase case ($u \geq 0$) and [**ACF**] in the two-phase case, Alt, Caffarelli and Friedman developed the full regularity theory for a class of free boundary problems in two dimensions. They proved partial regularity results in higher dimensions. G. S. Weiss [**W1**] improved these partial regularity results to show that free boundary surfaces have singularities of codimension at least 3. In particular, his theorem says that in $\mathbf{R}^3$ the singularities are points. Our theorem shows that there are no singularities in dimension 3.

The regularity proved here is valid only for local energy minimizers, not higher critical points. Other critical points of the functional can be singular even in $\mathbf{R}^3$ as we now show.[1] Following [**AC**], for $n \geq 3$ choose $t_n > 0$ to be the unique constant such that the positive harmonic function U (unique up to a scalar multiple) in the cone $\Gamma = \{x \in \mathbf{R}^n : |x_n| < t_n\sqrt{x_1^2 + \cdots + x_{n-1}^2}\}$ such that $U = 0$ on $\partial\Gamma$ is homogeneous of degree 1: $U(rx) = rU(x)$. (The constant t_n and the elementary function U are computed explicitly in [**AC**] in the case $n = 3$. See also the higher-dimensional calculations below.) Symmetry with respect to rotation and dilation implies that U_ν is constant on $\partial\Gamma\backslash\{0\}$. Therefore, one can choose the scalar multiple so that $U_\nu = -1$. Extend U to all of $\mathbf{R}^n$ by defining $U = 0$ on $\mathbf{R}^n\backslash\Gamma$. Then U is a critical point for the functional $J(v, B)$ for any ball B, but the free boundary $\partial\Gamma$ has a singular point at the origin.

In the case $n = 3$, Alt and Caffarelli showed in [**AC**] that the function U is *not* an energy minimizer. They prove this by computing the asymptotic behavior of the functional for a perturbation of the solution obtained by subtracting a small multiple of the fundamental solution $|x|^{-1}$ in $\mathbf{R}^3$. Their computation already suggests that our theorem should be true. The problem is to find a perturbation that works in general, not just for the specific function U.

The perturbation that works in general is obtained by subtracting a small multiple of the harmonic function with boundary values $|x|^{-1/2}$ instead of $|x|^{-1}$. Actually, we begin with perturbations that are compactly supported away from the origin, and compute the

---

[1]In [**W2**], G. S. Weiss showed that the free boundary of a stationary solution to the variational problem for I in dimension 3 is singular only at a discrete set of points. The example mentioned here shows that Weiss's theorem is best possible for nonminimizing solutions.

second variation. Choosing variations that vanish near the origin is counterintuitive: one has the impression that one must modify the singularity in order to decrease the energy. But it turns out that one can decrease the energy while leaving the singularity unchanged. The key point is that positivity of the second variation is a dilation-invariant inequality, from which one can easily infer the best candidate for a perturbation with negative second variation. The dilation invariance also suggests that in higher dimensions one should subtract a small multiple of the harmonic function with boundary values $|x|^{-(n-2)/2}$. Using this perturbation we obtain a partial result in higher dimensions as follows.

PROPOSITION. *The spherically symmetric function U defined above is not an energy minimizer in dimensions $n \leq 6$.*

In dimensions $n \geq 7$, on the other hand, the second variation in the direction $-|x|^{-(n-2)/2}$ is positive, which suggests that U is a global minimizer when $n \geq 7$. From the work of G. S. Weiss [**W1**], which draws on a strong analogy with minimal surfaces, one knows that there is a critical dimension k^* ($3 \leq k^* \leq \infty$) such that energy minimizing free boundaries are smooth for $n < k^*$, that they can have singular points in dimension $n = k^*$, and that in higher dimensions the singularities form a closed set of dimension at most $n - k^*$. The present paper shows that $k^* \geq 4$. The Proposition and its proof suggest that $k^* = 7$.

This paper resolves the regularity problem fully only in dimension 3. The reason our proof only works in dimension 3 is that it relies on the fact that in $\mathbf{R}^3$ connected cones with positive mean curvature are convex. It also makes use of the Gauss-Bonnet formula on the 2-sphere.

§2. Proof of the Theorem and its Corollary.

Let u be as in the Theorem. Denote $\Gamma = \{x \in \mathbf{R}^3 : u(x) > 0\}$. Then $\partial\Gamma$ is smooth except possibly at the origin, and u is smooth in $\overline{\Gamma}$ except at the origin. Indeed if $\partial\Gamma$ were nonsmooth at a point other than the origin, then by homogeneity it would be nonsmooth on a whole ray. But according to the theorem of G. S. Weiss [**W1**], the singular set in 3-space is at most a discrete set of points.

At any smooth point of the free boundary u is infinitely differentiable and the variational equations are satisfied in the classical sense:

$$(1) \qquad \Delta u = 0 \quad \text{on } \Gamma; \quad u_\nu = -1 \text{ on } \partial\Gamma\backslash\{0\}$$

The first lemma gives a necessary condition for a global minimizer that is valid in all dimensions. We begin by specifying the sign conventions for the normal ν and the mean curvature H of $\partial\Gamma$. Consider any point $P \in \partial\Gamma$ except the origin. Rotate the coordinate system around P so that $\partial\Gamma$ is the locally represented as the graph $x_n = \phi(x')$ ($x' = (x_1, \cdots, x_{n-1})$) of a function satisfying $\nabla\phi = 0$ at P. Here and in the rest of the paper we use the convention that the outer normal to Γ is $\nu = e_n$. Thus $u > 0$ in the region $x_n < \phi(x')$ in a neighborhood of P, and $\nabla u(P) = -e_n$. The mean curvature H at P is the sum of the principal curvatures, or, equivalently, at P,

$$(2) \qquad H = \sum_{i=1}^{n-1} \phi_{ii}$$

where the subscript i represents partial derivative with respect to x_i.

LEMMA 1. *(Main lemma)* Let $n \geq 3$. *Suppose that u is an energy minimizer for J in $\mathbf{R}^n$, that u is homogeneous of degree 1, and that $\Gamma = \{x \in \mathbf{R}^n : u(x) > 0\}$ has a smooth boundary except possibly at the origin. Then for every $f \in C_0^\infty(\mathbf{R}^n \backslash \{0\})$, $f \geq 0$,*

$$(3) \qquad \int_{\partial\Gamma} H f^2 \, d\sigma \leq \int_{\Gamma} |\nabla f|^2 \, dx$$

It follows that

$$(4) \qquad \int_{\partial\Gamma} H f^2 \, d\sigma \leq \int_{\Gamma} |\nabla F_\infty|^2 \, dx$$

where F_∞ is the harmonic function on Γ that tends to zero at infinity in Γ and has boundary values f on $\partial\Gamma$.

PROOF. We prove the lemma by showing that (3) is equivalent to the nonnegativity of the second variation of J. Fix a ball $B = B_R$ centered at the origin of radius R such that $\text{supp } f \subset B$. (At the end of the proof we will let the radius of the ball tend to infinity.) Define $\Omega = \Gamma \cap B$ and define F (depending on R) equal to the harmonic function in Ω with boundary values f on $\partial\Omega$. F and u are both harmonic functions that vanish on the boundary in a neighborhood of the origin. The region Ω is a nontangentially accessible domain (see [**JK**]) and it follows from the comparison theorem of [**JK**] that $F \leq Cu$ in a neighborhood of the origin and also in a neighborhood of $(\partial B) \cap \overline{\Gamma}$ in $\overline{\Omega}$. Let

$$\Omega_\epsilon = \{x \in \Omega : u(x) > \epsilon F(x)\}$$

For $\epsilon > 0$ sufficiently small, $\partial\Omega_\epsilon$ agrees with $\partial\Omega$ near the origin and near $(\partial B) \cap \overline{\Gamma}$, and is smoothly bounded otherwise. Let $v_\epsilon = u - \epsilon F$ on $\overline{\Omega}_\epsilon$ and $v_\epsilon = 0$ on $B \backslash \Omega_\epsilon$, then $v_\epsilon = u$ on ∂B.

Integrating by parts on Ω_ϵ,

$$\int_{B \cap \{v_\epsilon > 0\}} |\nabla v_\epsilon|^2 = \int_{(\partial B) \cap \Gamma} u(u_\nu - \epsilon F_\nu) \, d\sigma$$

Hence

$$J(u, B) - J(v_\epsilon, B) = \epsilon \int_{(\partial B) \cap \Gamma} u F_\nu \, d\sigma + \text{vol} \, (0 < u < \epsilon F)$$

Integrating by parts on Ω,

$$\int_{(\partial B) \cap \Gamma} u F_\nu \, d\sigma = \int_{(\partial B) \cap \Gamma} (u F_\nu - u_\nu F) \, d\sigma$$

$$= \int_{\partial(B \cap \Gamma)} (u F_\nu - u_\nu F) \, d\sigma - \int_{(\partial\Gamma) \cap B} F \, d\sigma$$

$$= - \int_{(\partial\Gamma) \cap B} F \, d\sigma$$

because $F = 0$ on $(\partial B) \cap \Gamma$, $u = 0$ on $\partial\Gamma$, and $u_\nu = -1$ on $\partial\Gamma \backslash \{0\}$. Combining the formulas above,

$$(5) \qquad J(u, B) - J(v_\epsilon, B) = -\epsilon \int_{(\partial\Gamma) \cap B} f \, d\sigma + \text{vol} \, (0 < u < \epsilon F)$$

To describe the asymptotic behavior of $J(v_\epsilon, B)$ in more detail we need several remarks.

REMARK 1. $u_{\nu\nu} = -H$ on $\partial\Gamma$ except at the origin.

PROOF. With the notations above, $u(x', \phi(x')) = 0$ near P. Differentiating with respect to x_i and x_j,

$$u_i + \phi_i u_n = 0; \quad u_{ij} + u_{in}\phi_j + \phi_{ij}u_n + \phi_i u_{nj} + \phi_i\phi_j u_{nn} = 0$$

for $i, j = 1, \ldots n - 1$. Evaluate at P using $\phi_i = 0$ and $u_n = -1$ to obtain

$$u_{ij} = \phi_{ij}, \quad i, j = 1, \ldots n - 1$$

Therefore, at P,

$$(6) \qquad \sum_{i=1}^{n-1} u_{ii} = H$$

Moreover, since u is harmonic,

$$(7) \qquad u_{nn} = -\sum_{i=1}^{n-1} u_{ii} = -H,$$

which proves Remark 1.

REMARK 2. $H \geq 0$. In particular, in dimension 3, $\mathbf{R}^3\backslash\Gamma$ is a finite union of convex cones.

PROOF. Since u is homogeneous of degree 1, $|\nabla u|$ is bounded in $\Gamma = \{u > 0\}$. Furthermore, $|\nabla u|^2$ is subharmonic: $\Delta|\nabla u|^2 = 2|\nabla^2 u|^2 \geq 0$. By the maximum principle $|\nabla u|^2$ achieves its maximum on $\partial\Gamma$. But $|\nabla u|$ achieves its maximum ($|\nabla u| \equiv 1$) at every point of $\partial\Gamma$ (except the origin), so at every boundary point except the origin, $\nu \cdot \nabla|\nabla u|^2 \geq 0$. Consider the coordinate system as above in which $\nu = e_n$, then at P,

$$(u_1^2 + \cdots + u_n^2)_n \geq 0 \implies 2\sum_{i=1}^{n} u_i u_{in} \geq 0 \implies 2u_n u_{nn} \geq 0$$

since $u_i = 0$, $i = 1, \ldots, n - 1$. Finally, $u_n = -1$, so $u_{nn} \leq 0$, and it follows from (7) that $H \geq 0$.

The cone $\partial\Gamma$ has one principal curvature equal to zero. In dimension 3, $H \geq 0$ implies that the other principal curvature is also nonnegative. Thus $\mathbf{R}^3\backslash\Gamma$ is a finite union of convex cones.

REMARK 3. Let $z : U \to \mathbf{R}^n$ defined on an open subset U of $\mathbf{R}^{n-1}$ be a local parametrization of the surface $\partial\Gamma$ and let $\nu = \nu(s)$ be the unit normal to $\partial\Gamma$ at $z(s)$ pointing away from Γ. In the coordinate system $x(s,t) = z(s) - t\nu(s)$, the volume element $dV = (1 + tH + O(t^2))d\sigma(s)dt$, where $d\sigma(s)$ is the $n - 1$ area element on the surface.

PROOF. Note $\nu(s) = -\nabla u(z(s))$. Therefore, the Jacobian matrix $A(s,t) = \partial x/\partial(s,t)$ has entries

$$a_{ij} = \frac{\partial z_i}{\partial s_j}(s) + t u_{ik}(z(s))\frac{\partial z_k}{\partial s_j}(s)$$

$i = 1,\ldots,n$ and $j = 1,\ldots,n-1$, and $a_{in} = u_i(z(s))$. To compute $dV = |\det(A(s,t))|dsdt$ in terms of $d\sigma(s) = |\det(A(s,0))|ds$, choose a linear change of variable in s (with origin at s_0) and a rotation in x centered at $z(s_0)$ so that at the fixed point s_0, $\partial x_i/\partial s_j = \delta_{ij}$ for $i,j < n$, $\partial x_n/\partial s_j = 0$ for $j < n$. At $z(s_0)$, $u_i = 0$ ($i < n$) and $u_n = -1$. It follows that

$$\det(A(s_0,t)) = -1 - t\sum_{i=1}^{n-1} u_{ii} + O(t^2) = (1 + tH + O(t^2))\det(A(s_0,0))$$

since $\det(A(s_0,0)) = -1$ and $\sum_{i=1}^{n-1} u_{ii} = H$. This proves the formula for dV.

With the notations of Remark 3, the equation for $\partial\Omega_\epsilon$ is

$$u(z(s) - t\nu(s)) = \epsilon F(z(s) - t\nu(s))$$

Taking Taylor expansions at $t = 0$, this can be written

$$t - H\frac{t^2}{2} + O(t^3) = \epsilon(F - tF_\nu + O(t^2))$$

with H, F and F_ν evaluated at $z(s)$ for each fixed s. (The functions u and F are smooth and the error terms $O(t^2)$ and $O(t^3)$ are uniformly bounded because the region $(\partial\Omega)\backslash(\partial\Omega_\epsilon)$ is bounded away from ∂B and the origin.) For t sufficiently small, $t = O(\epsilon)$. Solving for t as a function of s, the equation for $\partial\Omega_\epsilon$ can be written as

$$t = \epsilon F + \epsilon^2 F^2\frac{H}{2} - \epsilon^2 FF_\nu + O(\epsilon^3)$$

with all functions evaluated at $z(s)$. This asymptotic description of $\partial\Omega_\epsilon$ is valid on a finite collection of coordinate neighborhoods covering the support of f on the boundary. With the help of a partion of unity one obtains

$$\mathrm{vol}\,(0 < u < \epsilon F) = \int_{(\partial\Gamma)\cap B}\int_0^{\epsilon F + \epsilon^2 F^2\frac{H}{2} - \epsilon^2 FF_\nu + O(\epsilon^3)} (1 + tH + O(t^2))dtd\sigma$$

$$= \int_{(\partial\Gamma)\cap B}(\epsilon F + \epsilon^2 F^2\frac{H}{2} - \epsilon^2 FF_\nu)d\sigma$$

$$+ \int_{(\partial\Gamma)\cap B}\frac{1}{2}(\epsilon F + \epsilon^2 F^2\frac{H}{2} - \epsilon^2 FF_\nu)^2 Hd\sigma + O(\epsilon^3)$$

$$= \int_{(\partial\Gamma)\cap B}(\epsilon F + \epsilon^2 F^2 H - \epsilon^2 FF_\nu)d\sigma + O(\epsilon^3)$$

Putting together the calculations above,

$$J(u,B) - J(v_\epsilon,B) = \epsilon^2\int_{(\partial\Gamma)\cap B}(F^2 H - FF_\nu)d\sigma + O(\epsilon^3)$$

Since u is a minimizer, and $v_\epsilon = u$ on ∂B, $J(u, B) - J(v_\epsilon, B) \leq 0$ and hence

$$\int_{(\partial \Gamma) \cap B} (F^2 H - FF_\nu) d\sigma \leq 0$$

Integrating by parts, $\int_{(\partial \Gamma) \cap B} FF_\nu d\sigma = \int_\Omega |\nabla F|^2 dx$. Thus the inequality can be written

$$(8) \qquad \int_{(\partial \Gamma) \cap B} f^2 H d\sigma \leq \int_\Omega |\nabla F_R|^2 dx$$

(Here we have restored the dependence on R. Recall that $F = F_R$ is the harmonic function in $B_R \cap \Gamma$ with the same boundary values as f.) Since F_R minimizes the Dirichlet integral,

$$\int_\Omega |\nabla F_R|^2 dx \leq \int_\Omega |\nabla f|^2 dx.$$

This concludes the proof of (3).

Before taking the limit as $R \to \infty$, we point out that the limiting inequality (4) is not needed for the proof of the Theorem. It will be used in the explicit computations in higher dimensions.

Fix a function $f \in C_0^\infty(\mathbf{R}^n \backslash \{0\})$ supported in a ball B_{R_0}. For each $R > R_0$, we revise the notation to restore the dependence on R by considering F_R, the harmonic function on $\Omega_R = B_R \cap \Gamma$ such that $F_R = f$ on $\partial \Omega_R$.

Denote by F the minimal nonnegative solution to the Dirichlet problem $\Delta F = 0$ in Γ and $F = f$ on $\partial \Gamma$, that is the infimum of all nonnegative supersolutions. Since f is bounded, C and $C|x|^{2-n}$ are supersolutions for sufficiently large C, and hence

$$0 \leq F \leq C(1 + |x|)^{2-n}$$

It follows that $F = F_\infty$ is the unique solution described in (4). We will continue to use the notation F for this function.

Consider $\phi \in C_0^\infty(B_{3R} \backslash B_{R/2})$ such that $|\nabla \phi| \leq C/R$ and $\phi = 1$ on $B_{2R} \backslash B_R$. If $R > 2(R_0 + 1)$, then

$$\int_\Gamma \phi^2 |\nabla F|^2 dx = -2 \int_\Gamma F\phi \nabla \phi \cdot \nabla F dx$$

$$\leq 2 \int_\Gamma |\phi \nabla F| |F \nabla \phi| dx$$

$$\leq \frac{1}{2} \int_\Gamma |\phi \nabla F|^2 dx + 2 \int_\Gamma |F \nabla \phi|^2 dx$$

(The boundary terms in the integration by parts above are zero because $f = 0$ on $\mathbf{R}^n \backslash B_{R/2}$.) Hence,

$$\int_\Gamma \phi^2 |\nabla F|^2 dx \leq 4 \int_\Gamma |F \nabla \phi|^2 dx$$

In particular,

$$(9) \qquad \int_{\Gamma \cap B_{2R} \backslash B_R} |\nabla F|^2 dx \leq CR^{2-n}$$

The same maximum principle argument shows $F_{R_1} \leq C(1+|x|)^{2-n}$ and the same integration by parts shows that for $R > 2(R_0 + 1)$

$$(10) \qquad \int_{\Gamma \cap B_{2R} \setminus B_R} |\nabla F_{R_1}|^2 dx \leq CR^{2-n}$$

where one extends F_{R_1} to all of Γ by $F_{R_1} = 0$ on $\Gamma \setminus B_{R_1}$. Here C is a constant depending on f and n, but not on R_1 or R. Next apply the same integration by parts argument but with a cut-off function $\psi \in C_0^\infty(B_{R_1})$ satisfying $\psi = 1$ in $|x| < R_1/2$ and $|\nabla \psi| \leq C/R_1$. The boundary terms are zero because $F - F_{R_1} = 0$ on the entire boundary $\partial \Gamma$. One obtains

$$\int_\Gamma \psi^2 |\nabla(F - F_{R_1})|^2 dx \leq 4 \int_\Gamma |(F - F_{R_1})\nabla \psi|^2 dx$$

and hence,

$$\int_{\Gamma \cap B_{R_1/2}} |\nabla(F - F_{R_1})|^2 dx \leq CR_1^{2-n}$$

In all, using (9) and (10),

$$\int_\Gamma |\nabla(F - F_{R_1})|^2 dx \leq CR_1^{2-n} + \sum_{j=0}^\infty \int_{\Gamma \cap (B_{2^j R_1} \setminus B_{2^{j-1}R_1})} (2|\nabla F|^2 + 2|\nabla F_{R_1})|^2) dx \leq CR_1^{2-n}$$

Since R_1^{2-n} tends to zero as $R_1 \to \infty$,

$$\lim_{R_1 \to \infty} \int_\Gamma |\nabla F_{R_1}|^2 dx = \int_\Gamma |\nabla F|^2 dx$$

Finally, note that for $R > R_0$, F_R minimizes the Dirichlet integral in the sense that

$$\int_\Gamma |\nabla F_R|^2 dx = \inf_v \int_\Gamma |\nabla v|^2 dx$$

where the infimum is over all $v \in C_0^\infty(B_R)$ such that $v = f$ on $\partial \Gamma$. It follows that the Dirichlet integral of F_R decreases as R increases and the limit satisfies

$$\int_\Gamma |\nabla F|^2 dx = \inf_v \int_\Gamma |\nabla v|^2 dx$$

where the infimum is over all $v \in C_0^\infty(\mathbf{R}^n)$ such that $v = f$ on $\partial \Gamma$. In other words, F minimizes the Dirichlet integral on the whole cone Γ and (4) holds. This concludes the proof of Lemma 1.

LEMMA 2. *Suppose that Γ is an open cone in $\mathbf{R}^3$ and $\gamma = (\partial \Gamma) \cap S^2$ (the intersection of the cone with the unit sphere S^2) is a finite union of smooth curves. Then the mean curvature H of $\partial \Gamma$ can be written as*

$$H = \frac{1}{r} \kappa(x/r), \quad r = |x|$$

where κ is the curvature of the curve γ in the unit sphere S^2.

PROOF. The mean curvature H is homogeneous of degree -1 so it suffices to confirm that $H = \kappa$ on the unit sphere. (Note that in our case $H \geq 0$ and $\mathbf{R}^3 \backslash \Gamma$ is a collection of convex cones. The intersection of each of these convex cones with the unit sphere S^2 is a geodesically convex subset of the sphere. Thus one sees that the boundary has nonnegative curvature, $\kappa \geq 0$, even without knowing that $H = \kappa$ on the sphere.)

Parametrize the curve γ locally by a function $\alpha(t)$. The formula for the geodesic curvature of γ at $\alpha(t)$ is $\kappa = J(\alpha'(t)) \cdot \alpha''(t)/|\alpha'(t)|^3$, where J is rotation by $90°$ counterclockwise in the sphere, when viewed from outside the sphere (see [**Exercise 9,ON p. 337**]). Without loss of generality we may assume that $t = 0$, $\alpha(0) = (1, 0, 0)$, $\alpha'(0) = (0, 1, 0)$, and hence $J(\alpha'(0)) = (0, 0, 1)$. Then at $(1, 0, 0)$, $\kappa = \alpha_3''(0)$. The surface $\partial\Gamma$ is described near $(1, 0, 0)$ as the graph $x_3 = \phi(x_1, x_2)$ of a smooth ϕ satisfying

$$\phi(sx_1, sx_2) = s\phi(x_1, x_2), \quad \alpha_3(t) = \phi(\alpha_1(t), \alpha_2(t))$$

for s near 1 and t near 0. Differentiating with respect to s at $s = 1$, $(x_1, x_2) = (1, 0)$, one finds that

$$\phi_1(1, 0) = \phi(1, 0) = 0, \quad \phi_{11}(1, 0) = 0$$

Differentiating once with respect to t at $t = 0$, one has

$$\alpha_3' = \phi_1 \alpha_1' + \phi_2 \alpha_2'$$

Evaluating at $t = 0$ gives $\phi_2(1, 0) = 0$. Differentiating a second time,

$$\kappa = \alpha_3'' = \phi_{11}(\alpha_1')^2 + 2\phi_{12}\alpha_1'\alpha_2' + \phi_{22}(\alpha_2')^2 + \phi_1\alpha_1'' + \phi_2\alpha_2'' = \phi_{22}$$

since at $t = 0$, $\alpha_1' = 0$, $\alpha_2' = 1$, $\phi_2 = 0$ and $\phi_1 = 0$. Finally, $H = \phi_{11} + \phi_{22} = \phi_{22}$ since $\phi_{11} = 0$. In all, $\kappa = H$ on S^2 and Lemma 2 is proved.

We can now conclude the proof of the Theorem. If u is an energy minimizer, then inequality (3) applied to any radial function $f(r)$, $r = |x|$, $f \in C_0^\infty(0, \infty)$ says

$$\int_{\partial\Gamma} H f^2 d\sigma \leq \int_\Gamma |\nabla f|^2 dx$$

On $\partial\Gamma$, the surface area element can be written $d\sigma = r\,dr\,ds$ where ds is arclength measure on $\gamma = S^2 \cap \partial\Gamma$. Let $\Omega = S^2 \cap \Gamma$, then using $H = \kappa/r$, the inequality can be written

$$\int_\gamma \int_0^\infty \kappa(\alpha(s)) f(r)^2 dr\,ds \leq \int_0^\infty f'(r)^2 r^2 dr |\Omega|$$

where $|\Omega|$ is the area of Ω in the unit sphere, S^2, and $\alpha(s)$ is an arclength parametrization of γ. Let $g, h \in C^\infty([0, \infty))$ be such that $g \geq 0$, $h \geq 0$, and $g(r) = 1$ for $0 \leq r \leq 1/2$, $g(r) = 0$ for $r \geq 3/4$, $h(r) = 0$ for $0 \leq r \leq 1$, $h(r) = 1$ for $r \geq 2$. For $\epsilon > 0$, let $f_\epsilon(r) = h(r/\epsilon)g(r)r^{-1/2}$. Then as $\epsilon \to 0$,

$$\int_0^\infty f_\epsilon'(r)^2 r^2 dr = \int_0^\infty \left(-\frac{1}{2} r^{-3/2} h(r/\epsilon)g(r) \right)^2 r^2 dr + O(1)$$

$$= \frac{1}{4} \int_0^\infty \left(r^{-1/2} h(r/\epsilon)g(r) \right)^2 dr + O(1)$$

$$= \frac{1}{4} \int_0^\infty f_\epsilon(r)^2 dr + O(1)$$

Moreover, $\int_0^\infty f_\epsilon(r)^2 dr = c \log(1/\epsilon) + O(1)$ for some $c > 0$, so that

$$\lim_{\epsilon \to 0} \frac{\int_0^\infty f'_\epsilon(r)^2 r^2 dr}{\int_0^\infty f_\epsilon(r)^2 dr} = \frac{1}{4}$$

Consequently,

$$\int_\gamma \kappa \, ds \leq \frac{1}{4}|\Omega|$$

Denote by V_j, $j = 1, \ldots, m$ the connected components of $S^2 \backslash \Omega$, and let γ_j by the boundary curve of V_j. Recall that by Remark 2, $H \geq 0$ and therefore $\kappa \geq 0$. In other words, the regions V_j are convex subsets of S^2. The Gauss-Bonnet formula [**ON, p.375**] says

$$|V_j| + \int_{\gamma_j} \kappa \, ds = 2\pi,$$

and therefore,

$$2\pi m - \sum_{j=1}^m |V_j| = \sum_{j=1}^m \int_{\gamma_j} \kappa \, ds = \int_\gamma \kappa \, ds \leq \frac{1}{4}|\Omega|.$$

Since $\sum_{j=1}^m |V_j| = 4\pi - |\Omega|$, one has

$$0 < \frac{3}{4}|\Omega| \leq 4\pi - 2m\pi$$

It follows that $m = 1$, and the complement of Ω is a single connected convex component of S^2. Therefore, Ω contains a hemisphere. But homogeneity rules out any larger domain than a hemisphere. Indeed, for a general domain $\Omega \subset S^2$ the associated homogeneous harmonic function has the form $\phi(x/r)r^\beta$ in which ϕ is lowest Dirichlet eigenfunction for Ω. The correponding eigenvalue of ϕ is $-\beta(\beta+1) = -\lambda$ (for general n, $-\beta(\beta+n-2) = -\lambda$). As the domain increases, the value of λ decreases, so the value of β decreases as well. In the hemisphere the homogeneous solution has degree 1, so any larger domain has an associated harmonic function of homogeneity $\beta < 1$. Therefore Ω is a hemisphere and u is the positive part of a linear function. This concludes the proof of the Theorem.

PROOF OF THE COROLLARY. First, the theorem of [**ACF**] says that u is Lipschitz continuous in Ω. Suppose that $u(x_0) = 0$, and denote $u_r(x) = \frac{1}{r}u(x_0 + rx)$. Any limit function of a subsequence u_{r_j}, $r_j \to 0$ is a minimizer for the functional analogous to I with $Q_\pm$ replaced by the constants $Q_\pm(x_0)$, or, equivalently, the functional

$$J(v, B) = \int_B |\nabla v|^2 + (Q_+(x_0) - Q_-(x_0))\chi_{v>0}$$

for any ball $B \subset \mathbf{R}^3$. (After a change of variable, we may assume without loss of generality that $Q_+(x_0) - Q_-(x_0) = 1$.) It follows from Lipschitz continuity and a theorem of G. S. Weiss [**Theorem 4.1,W2**] that such limit functions are homogeneous of degree 1. The key step is to characterize the limit functions. If the rescaled limit has two phases, then it was already characterized in [**ACF**]. It is a so-called two-plane solution, namely, u such that $u > 0$ in a half-space, $u < 0$ in the complementary half-space, and u is linear on each half-space.[2] Thus, the only case that remains is the case in which the limit of u_{r_j} is a homogeneous global minimizing solution with only one phase ($u \geq 0$). The scale-invariant nondegeneracy estimate of [**Theorem 3.1, ACF**] implies that the rescaled limits centered at free boundary points are not identically zero. Therefore, the Theorem implies that the rescaled limit is a one-plane solution, linear in one half-space and zero in the complementary half-space. In sum, we have established in all cases that there exists subsequence u_{r_j} that tends uniformly on compact subsets to a half-space solution. This property and a routine argument using nondegeneracy imply that the free boundary is flat near x_0 in the sense of [**AC**], and hence the theorems of [**C2,C1**] imply that u is a classical solution near x_0. This concludes the proof of the Corollary.

Note that even in the two-phase case, the rescaled (blow up) limit may still have only one phase. Thus, one needed to characterize all homogeneous one-phase energy minimizers in order to obtain regularity in the two-phase case.

§3. Higher Dimensional Results

We now prove the Proposition. To compute t_n, Γ and U, consider, more generally, a harmonic function u depending only on r and θ where $x_n = r\cos\theta$ and $r = |x|$. Then u satisfies

$$\left(\frac{\partial^2}{\partial r^2} + \frac{n-1}{r}\frac{\partial}{\partial r} + \frac{n-2}{r^2}(\cot\theta)\frac{\partial}{\partial\theta} + \frac{1}{r^2}\frac{\partial^2}{\partial\theta^2}\right)u = 0$$

If $u = r^z v_z(\theta)$, then $v = v_z$ satisfies

$$(11) \qquad v'' + (n-2)(\cot\theta)v' + z(z+n-2)v = 0$$

Consider the solution to (11) with $z = 1$ and the symmetry $V(\theta) = V(\pi-\theta)$. This function is unique up to a multiple. Choose θ_0, $0 < \theta_0 < \pi/2$ so that $V(\theta_0) = V(\pi - \theta_0) = 0$ and $V > 0$ in $\theta_0 < \theta < \pi - \theta_0$. Then $U(x) = rV(\theta)$ and Γ is defined by $\theta_0 < \theta < \pi - \theta_0$, and, in the notation of the introduction, $t_n = \cos\theta_0$.

Next, consider the unique (up to a multiple) positive solution W to (11) with $z = -(n-2)/2$ and the symmetry $W(\theta) = W(\pi - \theta)$. Then $G(x) = r^{-(n-2)/2}W(\theta)$ is the homogeneous harmonic function whose boundary values on $\partial\Gamma$ are a multiple of $r^{-(n-2)/2}$.

Let H be the mean curvature of $\partial\Gamma$. Then $H = H_1/r$ where H_1 is the mean curvature of Γ at $r = 1$. H_1 is a dimensional constant determined by θ_0. We claim that if U is a minimizer, then (4) implies

$$(12) \qquad H_1 \leq |W'(\theta_0)/W(\theta_0)|$$

We will then evaluate both sides of this inequality numerically. This is a routine matter because V and W are explicit well-known special functions.

To prove the claim, recall that if $F = F_\infty$ is the harmonic function in Γ with boundary values f, then (4) can be written

$$\int_{\partial\Gamma} Hf^2 d\sigma \leq \int_\Gamma |\nabla F_\infty|^2 dx = \int_{\partial\Gamma} FF_\nu d\sigma$$

---

[2] As explained in [**ACF**], Lemma 6.6 and Remark 6.1, the proof is valid only in dimension $n = 2$, but the same method works in higher dimensions provided the case of equality in a certain isoperimetric inequality on the sphere holds only for rotationally symmetric caps. That fact was proved subsequently by Brothers and Ziemer in [**BZ**].

Denote $\Lambda f = F_\nu$ and dilation $T_r f(x) = f(rx)$, then Λ is an operator of degree -1: $T_r \Lambda = r^{-1}\Lambda T_r$. Consider radial functions $f = f(r)$. Denote $h(r) = r^{(n-2)/2}f(r)$, then since $d\sigma = r^{n-2}drds$ where ds is the measure on $\partial\Gamma \cap \{r = 1\}$ (two $(n-2)$-spheres),

$$\int_{\partial\Gamma} Hf^2 d\sigma = H_1 \int_{\partial\Gamma\cap\{r=1\}} \int_0^\infty h(r)^2 \frac{dr}{r} ds$$

and

$$\int_{\partial\Gamma} f\Lambda f d\sigma = \int_{\partial\Gamma\cap\{r=1\}} \int_0^\infty h(r)r^{n/2}\Lambda(r^{-(n-2)/2}h)\frac{dr}{r} ds$$

The operator $h \to r^{n/2}\Lambda(r^{-(n-2)/2}h)$ commutes with dilation T_r and hence is represented as a Fourier-Mellin multiplier operator on $L^2(\mathbf{R}^+, dr/r)$ with symbol

$$m(\xi) = r^{-i\xi}r^{n/2}\Lambda(r^{-(n-2)/2}r^{i\xi})$$

In other words, defining the Mellin transform by $\tilde{h}(\xi) = \displaystyle\int_{-\infty}^\infty h(r)r^{-i\xi}\frac{dr}{r}$,

$$\int_{-\infty}^\infty r^{n/2}\Lambda(r^{-(n-2)/2}h)r^{-i\xi}\frac{dr}{r} = m(\xi)\tilde{h}(\xi)$$

Using the Plancherel theorem, (4) implies

$$H_1 \int_{-\infty}^\infty |\tilde{h}(\xi)|^2 d\xi \leq \int_{-\infty}^\infty |m(\xi)||\tilde{h}(\xi)|^2 d\xi$$

for all $\tilde{h}$ such that $h \in C_0^\infty(0, \infty)$. But this class of h is dense in $L^2(\mathbf{R}^+, dr/r)$ and hence the inequality is valid for all $\tilde{h} \in L^2(\mathbf{R})$. Thus,

$$H_1 \leq \min_\xi |m(\xi)|$$

We only make use of this inequality at $\xi = 0$:

(13) $$H_1 \leq |m(0)|$$

(We expect $|m|$ to take its smallest value at $\xi = 0$; note that $m(\xi) \approx c|\xi|$ as $|\xi| \to \infty$.) Since $G = W(\theta_0)r^{-(n-2)/2}$ on $\partial\Gamma$, $r^{n/2}G_\nu = m(0)W(\theta_0)$. On $r = 1$, the normal derivative G_ν equals the derivative with respect to θ. In other words,

(14) $$|m(0)| = |W'(\theta_0)/W(\theta_0)|$$

Combining (13) and (14), gives (12).

It remains to compute both sides of (12) numerically in each dimension. Change variables to $t = \cos\theta$. Then $-1 \leq t \leq 1$ and we seek solutions that are even in t to the differential equation for $w(t) = v(\theta)$,

$$(1 - t^2)w'' + (1 - n)tw' + z(z + n - 2)w = 0$$

(This is, of course, a Legendre equation — the n-dimensional equation for spherical harmonics used to find the zonal harmonic. So the solutions are well-known special functions.) The two cases that interest us are the functions $f_n(t) = V(\theta)$ with $z = 1$ and $g_n(t) = W(\theta)$ with $z = -(n - 2)/2$. Thus f_n is the (unique up to scalar multiple) even function of t satisfying

$$(1 - t^2)f_n'' - (n - 1)tf_n' + (n - 1)f_n = 0$$

For example,

$$f_3(t) = 1 - \frac{1}{2}t \ln\left(\frac{1+t}{1-t}\right); \quad f_4(t) = \left(1 - t^2\right)^{-1/2}\left(1 - 2t^2\right)$$

Recall that t_n is defined so that $f_n(t_n) = 0$ (smallest $t_n > 0$). Then $H_1 = (n-2)\kappa$ where $\kappa = \cot\theta_0 = t_n/\sqrt{1 - t_n^2}$ is one of the $n - 2$ nonzero principal curvatures of $\partial\Gamma$ at $r = 1$.

Next, $W(\theta) = g_n(t)$ where g_n is the even solution of

$$(1 - t^2)g_n'' - (n-1)tg_n' - \frac{1}{4}(n-2)^2 g_n = 0$$

Moreover,

$$\frac{d}{d\theta} = -\sin\theta\frac{d}{dt} = -\sqrt{1 - t^2}\frac{d}{dt}$$

Therefore, the inequality (12) is equivalent to

(12′)
$$(n-2)\frac{t_n}{\sqrt{1 - t_n^2}} \leq \sqrt{1 - t^2}\frac{g_n'(t_n)}{g_n(t_n)}$$

Define

$$c_n \equiv \frac{g_n'(t_n)(1 - t_n^2)}{t_n g(t_n)}$$

(12′) is violated if $c_n < n - 2$.

For the purposes of numerical calculation we used Mathematica. The even solutions can be expressed in terms of Legendre P and Q functions as follows.

$$f_n(t) = (1 - t^2)^{-(n-3)/4}P[(n-1)/2, (n-3)/2, t], \quad n \text{ even}$$

$$f_n(t) = (1 - t^2)^{-(n-3)/4}Q[(n-1)/2, (n-3)/2, t], \quad n \text{ odd}$$

For $n \equiv 0, 1, 3 \bmod 4$, g_n is the even part of

$$(1 - t^2)^{-(n-3)/4}P[-1/2, (n-3)/2, t]$$

For $n \equiv 2 \bmod 4$, g_n is the even part of

$$(1 - t^2)^{-(n-3)/4}P[-1/2, -(n-3)/2, t]$$

One can compute numerically,

$$
\begin{aligned}
t_3 &= 0.833557 & c_3 &= 0.23 \\
t_4 &= 0.707107 & c_4 &= 1.0 \\
t_5 &= 0.623175 & c_5 &= 2.2 \\
t_6 &= 0.563016 & c_6 &= 3.6 \\
t_7 &= 0.517331 & c_7 &= 5.3 \\
t_8 &= 0.481158 & c_8 &= 7.1 \\
t_9 &= 0.451615 & c_9 &= 9.1 \\
t_{10} &= 0.426905 & c_{10} &= 11.2 \\
t_{11} &= 0.405841 & c_{11} &= 13.4 \\
t_{12} &= 0.387611 & c_{12} &= 15.7 \\
t_{13} &= 0.371631 & c_{13} &= 18.1
\end{aligned}
$$

Note that $c_n < n - 2$ for $n = 3, 4, 5, 6$. Hence the corresponding cones are not minimizing. (We already knew that for $n = 3$.) This concludes the proof of the Proposition.

On the other hand, $c_n > n - 2$ for $n \geq 7$. (For $n \geq 13$ one can prove analytically using differential inequalities that $c_n > n - 2$ without recourse to numerical approximations. The proof is omitted.) Therefore, it can be expected that these cones are energy minimizers in dimension $n \geq 7$.

We mention in closing that asymptotics for perturbation by r^{-1} as in [**AC**] can be carried out for any cone $\Gamma \subset \mathbf{R}^3$. The result can be expressed in terms of a global inequality concerning the spherical cross-section $\Omega = \Gamma \cap \partial B_1$. The calculation is in some respects more complicated than the one used here. The first term in asymptotics in ϵ after the first power is a term of order $\epsilon^{3/2}$ instead of ϵ^2. The coefficient on $\epsilon^{3/2}$, whose sign determines whether the energy increases or decreases, is a dimensional constant times

$$\int_\Omega V(\eta)^{-3/2} d\eta$$

where where V is the Dirichlet eigenfunction on the portion of the sphere Ω. This integral is divergent, but the correct value is the one taken in the sense of analytic continuation in the exponent. The sign of this expression seems very hard to determine except for explicit domains Ω, such as a rotation-invariant one. In any case, the computations carried out here suggest that perturbations with homogeneity $r^{-(n-2)/2}$, not r^{2-n} are the most likely to decrease the energy.

References

[**AC**]   H. W. Alt and L. A. Caffarelli, *Existence and regularity for a minimum problem with free boundary*, J. Reine Angew. Math. vol 325 (1981), 105–144.

[**ACF**]   H. W. Alt, L. A. Caffarelli, and A. Friedman, *Variational problems with two phases and their free boundaries*, Transactions A. M. S. vol 282 (1984), 431–461.

[**BZ**]   J. Brothers and W. Ziemer, *Minimal rearrangements of Sobolev functions.*, J. Reine Angew. Math. vol 384 (1988), 153–179.

[**C1**]   L. A. Caffarelli, *A Harnack inequality approach to the regularity of free boundaries. Part I: Lipschitz free boundaries are $C^{1,\alpha}$*, Rev. Mat. Iberoamericana vol 3 (1987), 139–162.

[**C2**]   ______, *A Harnack inequality approach to the regularity of free boundaries. Part II: Flat free boundaries are Lipschitz*, Comm. Pure Appl. Math. vol 42 (1989), 55–78.

[**ON**]   B. O'Neill, *Elementary Differential Geometry*, Academic Press, New York, 1966.

[**JK**]   D. S. Jerison and C. E. Kenig, *Boundary behavior of harmonic functions in nontangentially accessible domains*, Advances in Math. vol 46 (1982), 80–147.

[**W1**]   G. S. Weiss, *Partial regularity for a mimimum problem with free boundary*, Journal Geom. Anal. vol 9 (1999), 317–326.

[**W2**]   ______, *Partial regularity for weak solutions of an elliptic free boundary problem*, Comm. P. D. E. vol 23 (1998), 439–457.

LUIS A. CAFFARELLI, DEPARTMENT OF MATHEMATICS, UNIVERSITY OF TEXAS, AUSTIN, TX 78712

E-mail address: caffarel@math.utexas.edu

DAVID JERISON, DEPARTMENT OF MATHEMATICS, MASSACHUSETTS INSTITUTE OF TECHNOLOGY, CAMBRIDGE, MA 02139

E-mail address: jerison@math.mit.edu

Carlos E. Kenig, Department of Mathematics, University of Chicago, Chicago, IL 60637
E-mail address: cek@math.uchicago.edu

Contemporary Mathematics
Volume **350**, 2004

Stability and Instability of the Reissner-Nordström Cauchy Horizon and the Problem of Uniqueness in General Relativity

Mihalis Dafermos

This paper is dedicated to F. Browder and H. Brezis

ABSTRACT. This talk will describe some recent results [**16**] regarding the problem of uniqueness in the large (also known as *strong cosmic censorship*) for the initial value problem in general relativity. The interest in the issue of uniqueness in this context stems from its relation to the validity of the principle of determinism in classical physics. As will be clear from below, this problem does not really have an analogue in other equations of evolution typically studied. Moreover, in order to isolate the essential analytic features of the problem from the complicated setting of gravitational collapse in which it arises, some familiarity with the conformal properties of certain celebrated special solutions of the theory of relativity will have to be developed. This talk is an attempt to present precisely these features to an audience of non-specialists, in a way which hopefully will fully motivate a certain characteristic initial value problem for the spherically-symmetric Einstein-Maxwell-Scalar Field system. The considerations outlined here leading to this particular initial value problem are well known in the physics relativity community, where the problem of uniqueness has been studied heuristically [**1**, **22**] and numerically [**2**, **3**]. In [**16**], the global behavior of generic solutions to this IVP, and in particular, the issue of uniqueness, is completely understood. Only a sketch of the ideas of the proof is provided here, but the reader may refer to [**16**] for details.

1. General Relativity and its Initial Value Problem

The general theory of relativity is thought to provide the correct classical description for the interaction of gravity with matter. This description is embodied in a system of partial differential equations on a four dimensional manifold M, the so-called *Einstein equations*, which relate the Ricci curvature $R_{\mu\nu}$ of an unknown metric $g_{\mu\nu}$ to the energy-momentum tensor $T_{\mu\nu}$ of matter:

$$(1) \qquad R_{\mu\nu} - \frac{1}{2}Rg_{\mu\nu} = 2T_{\mu\nu}.$$

1991 *Mathematics Subject Classification.* Primary: 83C75, 83C57; Secondary 35L70, 35Q75.
Key words and phrases. Strong cosmic censorship, Einstein equations.

To complete the classical picture of a physics based on a collection of fields satisfying a closed system of equations, one must also consider the laws which govern the evolution of the matter fields generating the energy-momentum tensor on the right hand side of (1). (One important special case is when there is no matter, the so-called vacuum. Then (1) with vanishing right hand side is a closed system of quasilinear hyperbolic equations.) In general, one arrives at quite complicated systems of equations. However, from the perspective of classical physics, all phenomena are in principle described by the solutions of such a system. Moreover, for these systems, the initial value problem is natural, just as in classical dynamics.

Thus, from one point of view, the general theory of relativity is a classical physical theory that can be studied mathematically in parallel with other field theories of nineteenth-century classical physics. Indeed, the equations of general relativity exhibit similar local behavior with other equations of evolution as regards, for instance, the issues of local existence and uniqueness of solutions to the initial value problem. When one turns, however, to the initial value problem in the large, the Einstein equations present features that have no analogue in other typical equations of mathematical physics. The subject of this talk will be what appears, at least at first sight, as the most pathological of these features, namely the possibility of loss of uniqueness of the solution of the initial value problem *without loss of regularity*. This possibility is at the center of what is known as the *strong cosmic censorship conjecture* formulated by Penrose [21].

The reason why the theory of the initial value problem in the large for the Einstein equations is richer than for other non-linear wave equations is that the global geometry of the characteristics is not constrained *a priori* by any other structure. This geometry, which corresponds precisely to the conformal geometry for the vacuum equations, is *a priori* unknown. It turns out that many features of the initial value problem for hyperbolic equations that one takes for granted actually depend on certain global properties of the geometry of the characteristics; the question of uniqueness indicated above is one of these.

The best way to gain some intuition for what kind of conformal geometric structure develops in the course of evolution in general relativity–and what are the implications of this structure–is to carefully examine the special solutions of the theory. In fact, almost all conjectures and intuition regarding the theory in the end derives from simple properties of such solutions. Moreover, since our focus of interest is global *geometric* structure, there is no substitute in building intuition than a good pictorial representation. This talk will rely very much on such "pictures". It should be noted, however, that in the spherically symmetric context in which we shall be working, these "pictures", besides conveying intuition, also carry complete and precise information and can be treated on the same level as symbols or formulas.

The assumption of spherical symmetry and associated pictorial representations will be carefully discussed in the next section. We will then proceed to examine a series of special solutions which will lead to a particular initial value problem. Finally, theorems describing the solutions of the initial value problem will be formulated and their proofs will be discussed.

In regard to uniqueness, it turns out that there is always a spacetime which can be uniquely associated to initial data[1]. This is the so-called *maximal domain of development* [5]. It is the "biggest" spacetime which admits the given initial hypersurface as initial data and is at the same time *globally hyperbolic*, i.e. all inextendible causal curves intersect the initial hypersurface precisely once. This latter property ensures that the domain of dependence property holds. The question of uniqueness in general relativity is thus the issue of the *extendiblity* of this maximal domain of development. If it is extendible, then the solution is not unique. Since, as noted earlier, there is really no substitute for a pictorial representation, we defer further discussion of this till later on.

2. Spherical Symmetry

The current state of affairs in the theory of quasilinear hyperbolic partial differential equations in several space variables is such that global, large data problems appear beyond reach. For there to be any hope of making headway, it seems that some sort of reduction must be made to a problem where the number of the independent variables is no more than two. For hyperbolic equations of evolution, such reductions in general are accomplished by considering symmetric solutions, or equivalently, symmetric initial data. In general relativity, symmetry assumptions are formulated in terms of a group which acts by isometry on the spacetime and preserves all matter.

The only 2-dimensional symmetry group that is compatible with the notion of an isolated gravitating system, i.e. that can act on asymptotically flat spacetimes, is $SO(3)$. Solutions invariant under such an action are called *spherically symmetric*. As we shall see below, most of the expected phenomena of gravitational collapse of isolated gravitating systems, and the fundamental questions that these phenomena pose, can be suggested by the spherically symmetric solutions of various Einstein-matter systems. Moreover, the conformal structure of these solutions, which is the essential ingredient for the phenomena we wish to discuss, can be completely represented on the blackboard. (Or on paper!) The reason for this is simple: The space of group orbits

$$Q = M/SO(3)$$

can be given the structure of a 2-dimensional Lorentzian manifold. Restricting to Q which are *maximal domains of development* of initial data, it follows that these can be globally conformally represented as bounded domains in $1 + 1$-dimensional Minkowski space. The images of such representations are called *Penrose diagrams*; from these, the conformal geometry can be immediately read off as the characteristics are just the lines at $\pi/4$ or $-\pi/4$ radians from the horizontal:

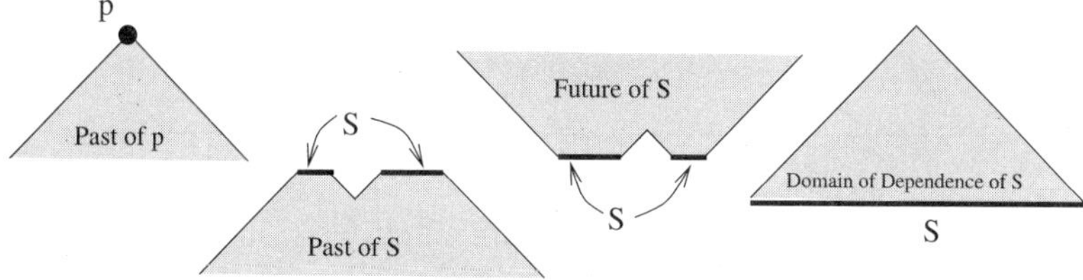

---

[1]For the vacuum, initial data is a Riemannian 3-manifold $(M, \tilde{g})$, along with a symmetric 2-tensor K satisfying the constraint equations that would arise if K were to be the second fundamental form of M realized as a hypersurface in a Ricci flat 4-manifold.

3. Minkowski space

To gain some familiarity with these diagrams, it is perhaps best to begin with 4-dimensional Minkowski space from this point of view, i.e. with the Penrose diagram of the maximal domain of development of Minkowski initial data. Here the diagram is as follows:

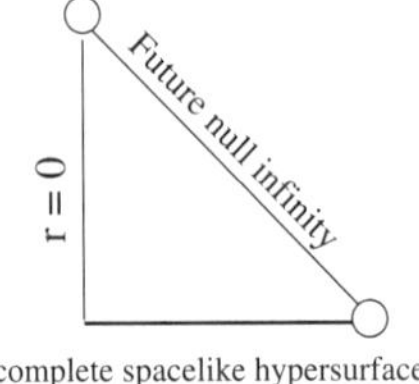

The r referred to above is a function on Q defined to be a multiple of the square root of the area of the group orbit corresponding to the points of Q. The line labelled $r = 0$ is thus the axis of symmetry. The line labelled "future null infinity" is not part of the spacetime but should be thought of as a "boundary" at infinity. The same applies to its two endpoints, "spacelike infinity" and "future timelike infinity". The latter corresponds to the "endpoint" of all inextendible future timelike geodesics.

The above Minkowski space is of course future causally geodesically complete, i.e. all causal curves can be extended to infinite affine parameter. Thus we have the analogue of global existence and uniqueness. That these properties of Minkowski space are stable to small perturbations (*without* any symmetry assumptions) is a deep theorem of Christodoulou and Klainerman [**13**].

4. Schwarzschild

Having understood the conformal diagram of Minkowski space, we turn to a more interesting solution: the Schwarzschild solution. This is actually a one-parameter family of solutions (the parameter is called mass and denoted by m) which contains Minkowski space (the case where $m = 0$). As it is a spherically symmetric vacuum solution, its non-triviality in the case $m \neq 0$ must be generated by topology. Any Cauchy hypersurface has two asymptotically flat ends and topology $S^2 \times R$. "Downstairs", this corresponds to a line with two $r = \infty$ endpoints, not intersecting an axis of symmetry. For convenience, we will choose a time-symmetric initial hypersurface. The maximal development of this "Schwarzschild" initial data then looks like:

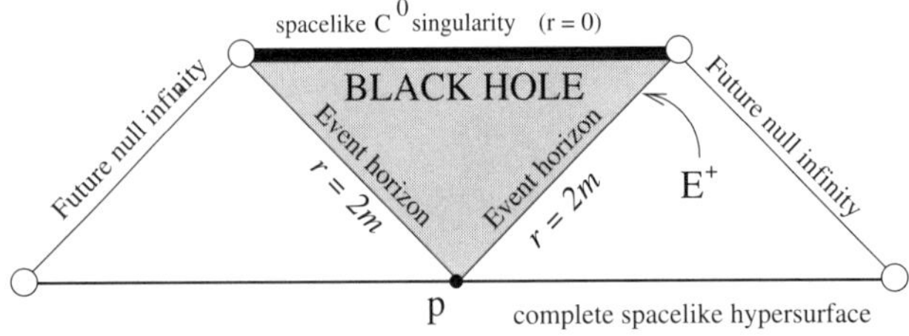

The point p depicted where $r = 2m$ is a minimal surface "upstairs" in the initial hypersurface. As the initial hypersurface was chosen to be time symmetric, this minimal surface is also what is called marginally trapped. The "ingoing" and "outgoing" null cones emanating from this surface, which correspond to the two null rays through p "downstairs", can thus not reach "future null infinity". (This is related to the so-called "singularity theorems" of Penrose.) Moreover, all timelike geodesics emanating from the point p reach in finite time the curve $r = 0$. This curve is not an axis of symmetry but a C^0 singularity! That is to say, there is no extension of the spacetime above through $r = 0$ with a continuous Lorentzian metric.

This singular behavior of the Schwarzschild solution may at first appear to be an undesirable feature. Indeed, historically, it was first considered exactly as such. But it turns out in fact that the behavior outlined above would provide the "ideal" scenario for the end-state of gravitational collapse. This has to do with two specific features of this solution:

(1) Labelling the null rays emanating from the minimal surface as E^+, it turns out that there is a class of timelike observers, namely those who do not cross E^+, who can observe for infinite time and whose causal past is completely regular. The singularity is hidden inside a black hole, and E^+ is called the event horizon. To be more precise about the "completeness" property of the region outside the black hole, fix some outgoing null geodesic which intersects future null infinity and parallel translate a conjugate null vector (i.e. a null vector in the other direction). The affine length of the null curves joining this null geodesic and E^+, where the affine parameter is determined by the aforementioned vector, goes to ∞. In particular, future null infinity can be thought to have infinite affine length. One says that the solution possesses a "past complete future null infinity" or a complete domain of outer communications.

(2) The above spacetime is future inextendible as a C^0 metric. Thus, according to the discussion in the Introduction, this means that the Schwarzschild solution is unique even in the class of very low regularity solutions. The significance of this fact will become clear later.

5. Christodoulou's solutions

At this point it should be noted that while the Schwarzschild solution indeed provides intuition about black holes, it cannot give insight as to whether these can occur in evolution of data where no trapped surfaces are present initially, i.e. whether the kind of behavior outlined above is related in any way to the endstate of gravitational collapse. That Properties 1 and 2 above are indeed general properties of solutions was proven by Christodoulou [6] for the spherically-symmetric Einstein scalar field equations:

$$R_{\mu\nu} - \frac{1}{2} R g_{\mu\nu} = 2 T_{\mu\nu},$$

$$g^{\mu\nu} (\partial_\mu \phi)_{;\nu} = 0,$$

$$T_{\mu\nu} = \partial_\mu \phi \partial_\nu \phi - \frac{1}{2} g_{\mu\nu} g^{\rho\sigma} \partial_\rho \phi \partial_\sigma \phi.$$

For generic solutions of the initial value problem, the Penrose diagram obtained by Christodoulou is as follows:

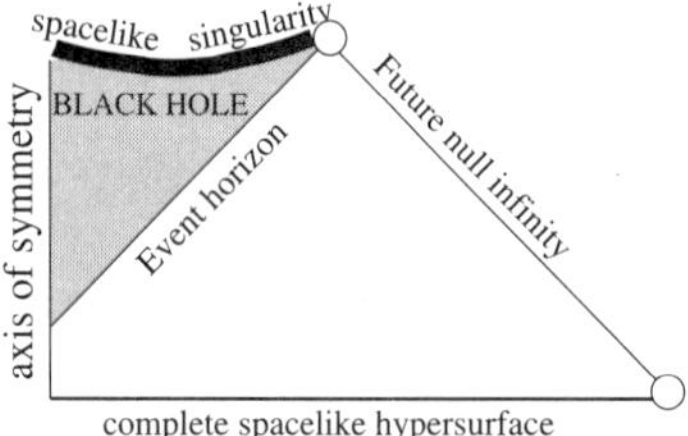

In [**12**], however, Christodoulou explicitly constructs solutions with conformal diagram:

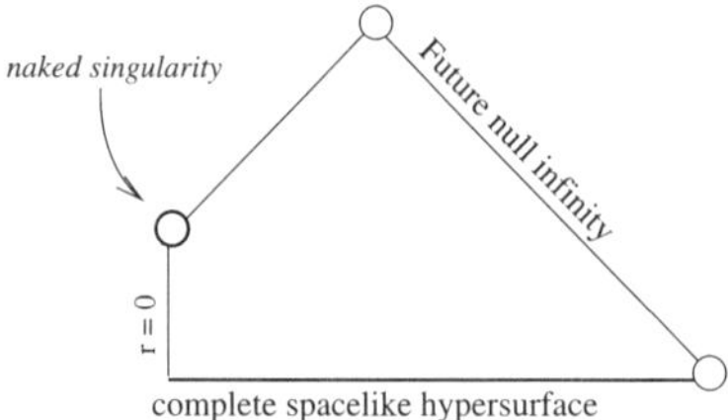

These are so-called "naked singularities".

One might ask why should one consider the coupling with the scalar field. The natural case to consider first, it would seem, is the vacuum. A classical theorem of Birkhoff, however, states that the only spherically symmetric vacuum solutions are Schwarzschild. Thus, matter *must* be included to give the problem enough dynamical degrees of freedom in spherical symmetry. The scalar field is in some sense the simplest, most natural choice.[2]

As far as Property 1 is concerned, Christodoulou's results are the best evidence yet that this is indeed a property of "realistic" gravitational collapse.[3] It turns out however that there is another "competing" set of evidence that indicates that the behavior of Christodoulou's solutions related to Property 2 does not represent "realistic collapse". (Remember that Property 2 is the central question for us, as this is what determines the notion of uniqueness.) This evidence is provided again by the intuition given by special solutions.

6. The Kerr and Reissner-Nordström families

One may consider the Schwarzschild family of solutions as embedded in a larger, 2-parameter family of solutions called the *Kerr solutions*. Here the parameters are called mass and angular momentum, and Schwarzschild corresponds to vanishing angular momentum. For all non-vanishing values of angular momentum, the internal structure of the black hole is completely different, and, as we shall see momentarily, much more "problematic", as Property 2 will fail. Thus the introduction of even an arbitrarily small amount of angular momentum–a phenomenon that cannot be "seen" by spherically-symmetric models–seems to change everything and cast

---

[2]It satisfies a linear equation and does not form singularities in the absense of coupling, it is hyperbolic so does not change the hyperbolic character of the equations, and moreover, its characteristics coincide with those of the metric g, etc. . .

[3]The statement that for generic initial data, the domain of outer communications possesses a complete null infinity is known as "weak cosmic censorship".

doubt on the conclusions derived from the spherically symmetric Einstein-Scalar Field model.

To summarize our "unhappy" situation, it seems that the phenomenon which plays a fundamental role in the issue we want to study is incompatible with the assumptions we have to make in order to render it mathematically tractable. It would seem that understanding the black hole region of realistic gravitational collapse using a spherically symmetric model is a lost cause.

Fortunately, there is a "solution" to this problem! The effect of angular momentum on gravity turns out to be similar to the effect of charge. (As John Wheeler puts it, charge is a poor man's angular momentum.) Indeed, there is a very close similarity between the conformal structure of the Kerr family and a 2-parameter *spherically symmetric* family of solutions to the Einstein-Maxwell equations:

$$R_{\mu\nu} - \frac{1}{2}g_{\mu\nu}R = 2T_{\mu\nu}$$

$$F^{\mu\nu}_{;\nu} = 0,$$

$$F_{[\mu\nu,\rho]} = 0,$$

$$T\mu\nu = F_{\mu\lambda}F_{\nu\rho}g^{\lambda\rho} - \frac{1}{4}g_{\mu\nu}F_{\lambda\rho}F_{\sigma\tau}g^{\lambda\sigma}g^{\rho\tau},$$

the so-called *Reissner-Nordström* solution. Here the parameters are mass m and charge e. For $e = 0$ one retrieves the Schwarzschild family, while for $0 < e < m$ one obtains:

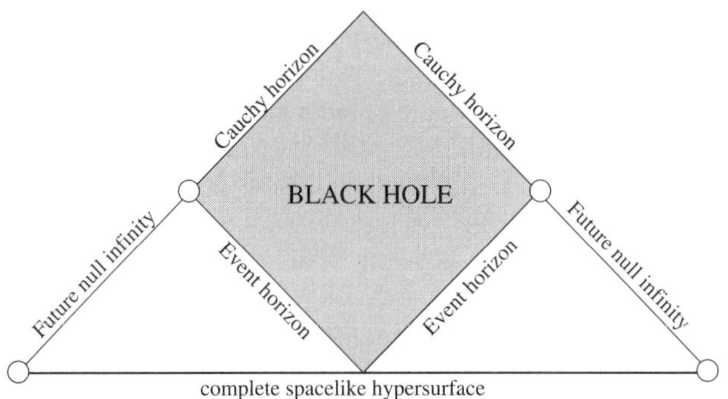

The $r = 0$ singularity of the Schwarzschild solution has disappeared! The above spacetime is completely regular up to the edges. These new edges that "complete" the triangle, however, are at a *finite* distance from the initial data, in the sense that all timelike geodesics joining those edges with the initial hypersurface have finite length. This solution thus has a regular future boundary and is extendible (in C^∞!) beyond it.

What fails at the boundary of this maximal domain of development of initial data is thus not the regularity of the solution, but rather, global hyperbolicity. Any extension of Q will contain past inextendible causal geodesics not intersecting the initial hypersurface:

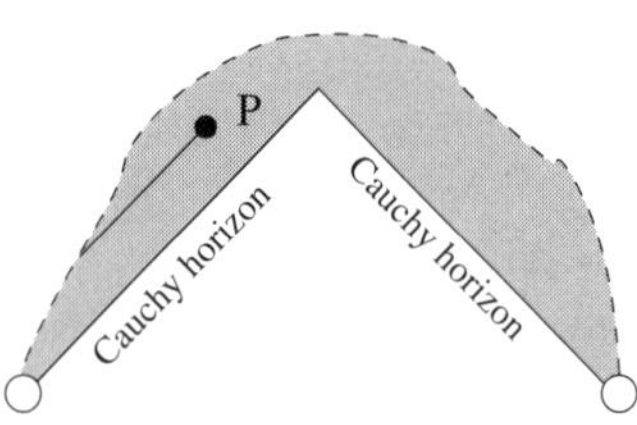

Points in such an extension but not in Q itself cannot be determined by initial data, in the exact same way that the solution to a linear wave equation $\Box\Psi = 0$ at the point P depicted below, cannot be uniquely determined by its values in the shaded set S:

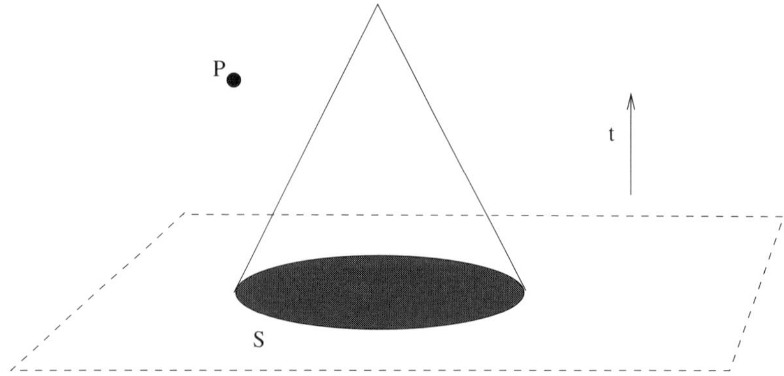

What has happened in the Reissner-Nordström solution is that the situation depicted above developed, but from an S that was complete.

7. Strong Cosmic Censorship

The physical interpretation of the above situation is that the classical principle of determinism fails, but without any sort of loss in regularity that would indicate that the domain of the classical theory has been exited. It is in that sense that this kind of behavior is widely considered by physicists to be problematic[4]. On the other hand, numerical calculations (Penrose and Simpson [23]) on the behavior of linear equations on a Reissner-Nordström background indicate that a naturally defined derivative blows up at the Cauchy horizon. This was termed the blue-shift effect. Thus, Penrose argued, the pathological behavior of the Reissner-Nordström solution might be unstable to perturbation. This led him to conjecture, more generally,

Strong Cosmic Censorship For generic initial data, in an appropriate class, the maximal domain of development is inextendible.

In view of our discussion in the introduction, this can be thought of as the conjecture that, for generic initial data, the solution is unique wherever it can be defined. Of course, the context in which this should be applied to (i.e. what

---

[4]Of course, this failure of determinism only applies to observers who enter into black holes. In particular, Property 1 ensures that determinism holds in the domain of outer communications (hence the term "weak" cosmic censorship). On the other hand, it seems that fundamental principles of physics should be valid everywhere, including the interiors of black holes.

equations, what class of initial data should be considered, etc...) and the notion of extendibility are left open. We will comment more on that later.

8. The Einstein-Maxwell-Scalar Field system

It might seem at first that the proper setting for discussing the problem of whether Cauchy horizons arise in evolution, for generic data in spherical symmetry, is the Einstein-Maxwell equations. Unfortunately, these suffer in fact from the same drawback as the Einstein vacuum equations, namely they do not possess the required dynamical degrees of freedom. One necessarily has to include more matter, and again the simplest choice, as in the work of Christodoulou, is a scalar field. Thus one is easily led to the coupled Einstein-Maxwell-Scalar Field system:

$$R_{\mu\nu} - \frac{1}{2}g_{\mu\nu}R = 2T_{\mu\nu} = 2(T^{em}_{\mu\nu} + T^{sf}_{\mu\nu})$$

$$F^{\mu\nu}_{;\nu} = 0,$$

$$F_{[\mu\nu,\rho]} = 0,$$

$$g^{\mu\nu}(\partial_\mu\phi)_{;\nu} = 0,$$

$$T^{em}_{\mu\nu} = F_{\mu\lambda}F_{\nu\rho}g^{\lambda\rho} - \frac{1}{4}g_{\mu\nu}F_{\lambda\rho}F_{\sigma\tau}g^{\lambda\sigma}g^{\rho\tau},$$

$$T^{sf}_{\mu\nu} = \partial_\mu\phi\partial_\nu\phi - \frac{1}{2}g_{\mu\nu}g^{\rho\sigma}\partial_\rho\phi\partial_\sigma\phi.$$

It turns out that in spherical symmetry the Maxwell part of the equation decouples. Since the scalar field ϕ carries no charge, a non-trivial Maxwell field can only be present if an initial complete spacelike hypersurface has non-trivial topology. In particular, this model is not suitable for considering the formation of black holes, as in the work of Christodoulou. Thus we will consider the problem where there is already a black hole present initially. To take the simplest possible formulation that captures the essense of the problem at hand, one can prescribe initial data for the system on two null rays, such that one corresponds to the event horizon of a Reissner-Nordström solution, and the other carries "arbitrary" matching data:

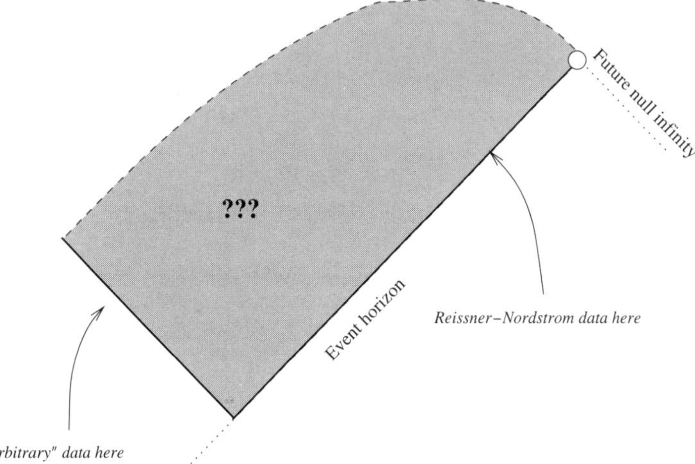

To make the most of the method of characteristics, we introduce null coordinates, i.e. coordinates (u, v) such that the metric on Q takes the form $-\Omega^2 dudv$. Here we select the v-axis to be the event horizon, and the u axis to be the conjugate ray on which we presribe our data. The unknowns are then just r, Ω and ϕ, and the electromagnetic part contributes a constant $e \neq 0$ which is computed from the initial data. To write the equations as a first order system, we define $\partial_u r = \nu$, $\partial_v r = \lambda$, $r\partial_u \phi = \zeta$, $r\partial_v r = \zeta$, and also ϖ, what one can call the "renormalized" Hawking mass[5], defined by

$$1 - \frac{2\varpi}{r} + \frac{e^2}{r^2} = |\nabla r|^2 = -4\Omega^{-2}\lambda\nu.$$

We then have:

$$(2) \qquad\qquad \partial_u r = \nu,$$

$$(3) \qquad\qquad \partial_v r = \lambda,$$

$$(4) \qquad\qquad \partial_v \nu = \nu\left(-\frac{2\lambda}{1-\mu}\frac{1}{r^2}\left(\frac{e^2}{r}-\varpi\right)\right),$$

$$(5) \qquad\qquad \partial_u \varpi = \frac{1}{2}(1-\mu)\left(\frac{\zeta}{\nu}\right)^2 \nu,$$

$$(6) \qquad\qquad \partial_v \varpi = \frac{1}{2}(1-\mu)\left(\frac{\theta}{\lambda}\right)^2 \lambda,$$

$$(7) \qquad\qquad \partial_u \theta = -\frac{\zeta\lambda}{r},$$

$$(8) \qquad\qquad \partial_v \zeta = -\frac{\theta\nu}{r}.$$

9. Statement of the theorems

We can now state the theorems of [**16**]. On the one hand we have:

THEOREM 1. *After restricting the range of the u coordinate, the Penrose diagram of the solution of the I.V.P. described above is as follows:*

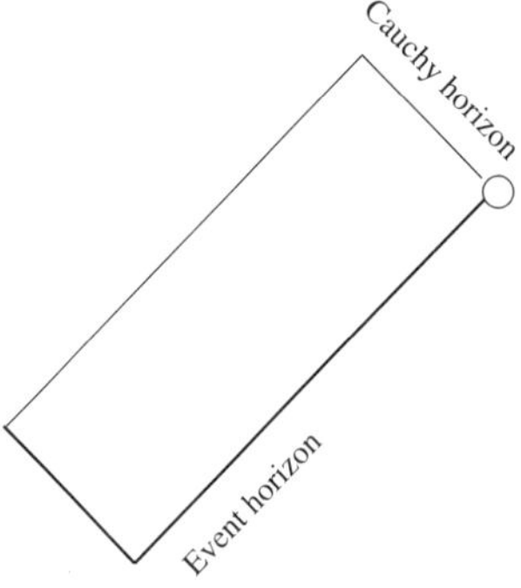

Moreover, r extends to a function on the Cauchy horizon with the property $r \to \varpi_{init} - \sqrt{\varpi_{init}^2 - e^2}$ as $u \to 0$ (where ϖ_{init} is the constant value of ϖ on the initial

---

[5]The Hawking mass m is defined to be $m = \frac{r}{2}(1 - |\nabla r|^2)$. The renormalized version has the property that it is constant in the Reissner-Nordsröm solution and coincides with m at Future Null Infinity.

event *horizon), and the metric can be continuously extended globally across the Cauchy horizon.*

On the other hand, we have:

THEOREM 2. *For "generic" initial data in the class of allowed initial data, ϖ blows up identically on the Cauchy horizon. In particular, the solution is inextendible across the Cauchy horizon as a C^1 metric.*

Thus, strong cosmic censorship is true, according to Theorem 2, if formulated with respect to extendibility in C^1 or higher, but false, according to Theorem 1, if formulated with respect to extendibility in C^0 (See [7] for reasons why one might want to require C^0.) In any case, the formation of a null "weak" singularity indicates a qualitatively different picture of the internal structure of the black hole from any of the previous models described above, and from the original expectations of Penrose.

The scenario of Theorems 1 and 2 was first suggested by Israel and Poisson [22] who put forth some heuristic arguments. Subsequently, a large class of numerics was done for precisely the equations considered here (see [3] for a survey). Because of the blow-up in the mass, this phenomenon was termed "mass inflation".

It should be noted that the imposition of Reissner-Nordström data on the event horizon is somewhat unnatural if the data is viewed as having arisen from generic data for a characteristic value problem where the v-axis is in the domain of outer communications[6]. Similar results to the above theorems, however, can in fact be proven for a wide class of data which includes the kind conjectured to arise from the aforementioned problem. These results will appear in [17]. The relevance of such an extension will only become clear, however, if the problem of determining the correct "generic" decay on the event horizon is mathematically resolved.

10. Some ideas from the proofs

Our initial data are trapped, i.e. ν and λ are both nonpositive. These signs are then preserved in evolution. Note that from the equation

$$(9) \qquad \partial_u\left(\frac{\lambda}{1-\mu}\right) = \left(\frac{\lambda}{1-\mu}\right)\frac{1}{r}\left(\frac{\zeta}{\nu}\right)^2\nu$$

it follows that $\frac{\lambda}{1-\mu}$ is non-increasing in u[7]. What gives the analysis of our equations *in the black hole region* its characteristic flavor is the fact that $\int\frac{\lambda}{1-\mu}dv$ is potentially infinite when integrated for constant u along the whole range of v (it is indeed infinite in initial data, i.e. $u = 0$), and that this infinity can appear in the equations (for instance in (4)) with either a positive or negative sign, depending on the sign of $\frac{e^2}{r} - \varpi$. Note that, by contrast, in the domain of outer communications, this infinity is killed by the $\frac{1}{r^2}$ term since $r \to \infty$ on outgoing rays. In the domain of development of our initial data, r is bounded above by its initial constant value on the event horizon $\varpi_{init} + \sqrt{\varpi_{init}^2 - e^2}$, in view of the signs of λ and ν.

---

[6]It should be emphasized that our rationale in choosing the I.V.P. here was to separate completely the issue of the dynamics of the interior of the black hole from the precise understanding of the set of data that arises in collapse, which is a problem of very different analytical flavor.

[7]That this quantity is non-increasing is in fact a general feature of spherical symmetry, i.e. it depends only on the dominant energy condition, not on the particular choice of matter.

In the Reissner-Nordström solution, the sign of $\frac{e^2}{r} - \varpi$ goes from negative near the event horizon to positive near the Cauchy horizon, while $\int \frac{\lambda}{1-\mu} dv$ remains constant in u and thus infinite, accounting for both what is called the infinite red shift near the event horizon (this makes objects crossing the event horizon slowly disappear to outside observers as they are shifted to the red) and the infinite blue shift near the Cauchy horizon (this accounts for the instability of the Cauchy horizon to *linear* perturbations).

For general solutions of the initial value problem, some of these features of the sign of $\frac{e^2}{r} - \varpi$ turn out to be stable, while others do not. In particular, Theorems 1 and 2 together imply that the sign must become negative near the Cauchy horizon, and not positive! To attack this initial value problem, it is clear that the behavior of this sign is the first thing that must be understood. It turns out that before the effects of the linear instability start to play a role, three geometrically distinct regions develop in evolution, a red-shift, no-shift, and stable blue-shift region, characterized by

$$\frac{e^2}{r} - \varpi < -\epsilon \, , \, \frac{e^2}{r} - \varpi \sim 0 \, , \, \frac{e^2}{r} - \varpi > \epsilon,$$

respectively:

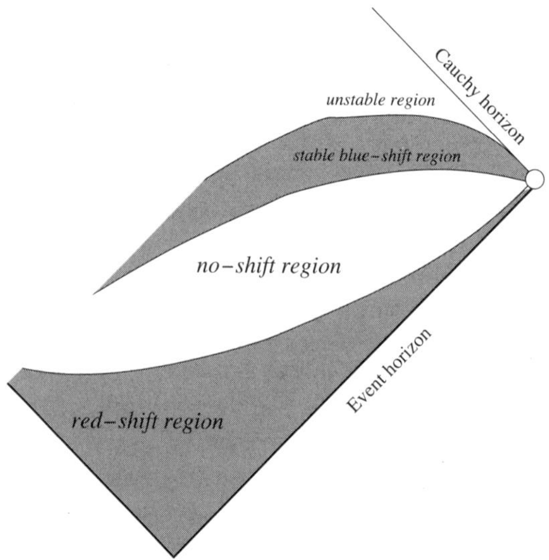

In the red-shift region, $\int \frac{\lambda}{1-\mu} dv$ is unbounded as $u \to 0$, but it appears with a "favorable" sign (favorable as far as controlling ϖ is concerned[8]), in the no-shift region, $\int \frac{\lambda}{1-\mu}$ is uniformly bounded, and in the stable blue-shift region, $\int \frac{\lambda}{1-\mu} dv$ grows with u but at a rate "less" than the growth of certain natural derivative of ϕ. These facts allow us to control all quantities reasonably well up until the future boundary of the stable blue-shift region, though completely different arguments must be applied to each subsequent region as it develops in evolution from the previous one.

---

[8]This sign tends to make ν bigger, and there is an extra ν on the denominator of (5).

Of course, all this work seems only to have pushed forward the problem from the original initial segments to the future boundary of the "stable blue-shift region". But in fact our new "initial" conditions on the future boundary of the stable blue-shift region are much more favorable. The stable blue-shift region that has preceded it ensures that ν has a sufficiently fast decay rate in u. (Remember that blue-shift regions tend to make $|\nu|$ smaller, so they are favorable for controlling r, but unfavorable for controlling ϖ.) Once this rate can be shown to be preserved, it follows by integrating u that one can bound r *a priori* away from 0 in its future, and thus prove the existence of the solution up to the Cauchy horizon. It is clear from what we have said above that if the unstable region remains a blue-shift region (see left diagram below), there is no problem. (This is of course what happens for the Reissner-Nordsröm solution itself.)

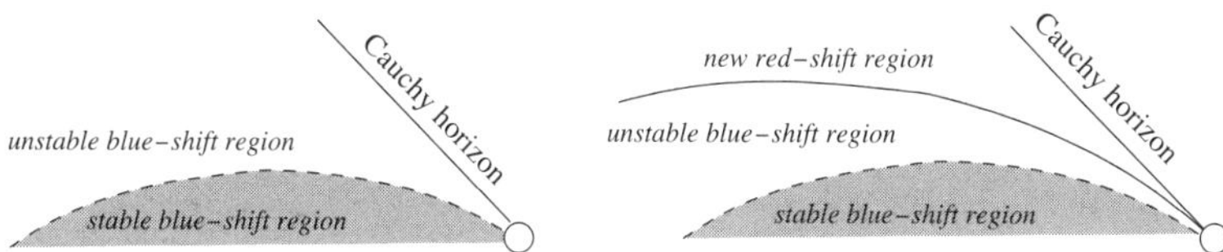

The danger is if a new red-shift region develops (see right diagram above). It turns out, however, that a new *a priori* estimate

$$\int \partial_v \log(\nu) dv = -\int \frac{2\lambda}{1-\mu} \frac{1}{r^2} \left(\frac{e^2}{r} - \varpi \right) dv < C$$

is available in this case, which is independent of the size of ϖ. This makes use of the fact that there is a ϖ hidden also in the denominator in $1 - \mu$. (The estimate depends, however, on our knowledge of the new initial condition of r and a particular bootstrap assumption on its future behavior; in particular this estimate does not hold in the original "red-shift" region.) This is the last element in the proof of Theorem 1.

We now discuss the proof of Theorem 2. As has been noted, at the level of perturbation theory [4], one does see an instability, caused by the blue-shift region. (Requiring some derivative of the scalar field to be positive initially, ζ will decay in u, and thus θ will decay in v, at a slower rate than the decay of λ in v, so that the natural derivative $\frac{\theta}{\lambda} \to \infty$.) On the other hand, the non-linearity tends to diminish this effect, since if the mass indeed increases, in view of Theorem 1, we have a reappearance of a red-shift region (see diagram on the right above). As remarked earlier, this tends to make $|\nu|$ (and also $|\lambda|$) bigger, and thus $\left|\frac{\theta}{\lambda}\right|$ smaller. The proof must thus encorporate something beyond "linear theory arguments".

This extra ingredient is supplied by a very powerful monotonicity peculiar to black hole interiors, or more specifically, to "trapped regions". Integration of (7) and (8) then implies that if θ and ζ are initially of the same signs[9], let's say non-negative, then they remain non-negative, and in fact $\partial_v \zeta \geq 0$ and $\partial_u \theta \geq 0$. Now integration of (6) and using that $\frac{1-\mu}{\lambda}$ is also non-increasing in u, yields that for

---

[9]In view of the fact that the data vanish on the event horizon, this can be considered a generic condition after restricting the domain of the u coordinate. This condition and the non-vanishing of a particular derivative of ϕ at the origin together define the "generic" class of initial data to which Theorem 2 applies.

$$v_2 > v_1, \; u_2 > u_1,$$

$$(10) \qquad \varpi(u_2, v_2) - \varpi(u_2, v_1) \geq \varpi(u_1, v_2) - \varpi(u_1, v_1).$$

It should be mentioned that in view of the sign of λ, both sides of the above inequality are positive.

The broad outline of the proof of Theorem 2 is as follows: First assume that the spacetime looks very much like Reissner-Nordström. Then the linear theory more or less applies, and applying the bounds for θ in the equation (6), one obtains $\varpi \to \infty$, which is a contradiction. Thus one is reduced to proving that any spacetime "quantitatively different" from Reissner-Nordström must have $\varpi \to \infty$.

It is not possible here to explain precisely what "quantitatively different" has to mean. To give a taste of the kind of arguments involved in this "non-linear part" of the proof, we will be content to show that assuming only that ϖ is bounded below by a positive number plus its Reissner-Nordström value (this is indeed a quantitative difference)

$$(11) \qquad \varpi_{Cauchy\ horizon} > \varpi_{init} + 2\epsilon,$$

it follows from (10) that ϖ must in fact blow up identically on the Cauchy horizon. If γ is the future boundary of the stable blue-shift region, the fact that $\varpi_\gamma(u) \to \varpi_{init}$ implies that given any u_0, a sequence of points (u_i, v_i) can be constructed so that the mass differences $\varpi(v_{i+1}, u_i) - \varpi(v_i, u_i)$ are all greater than ϵ:

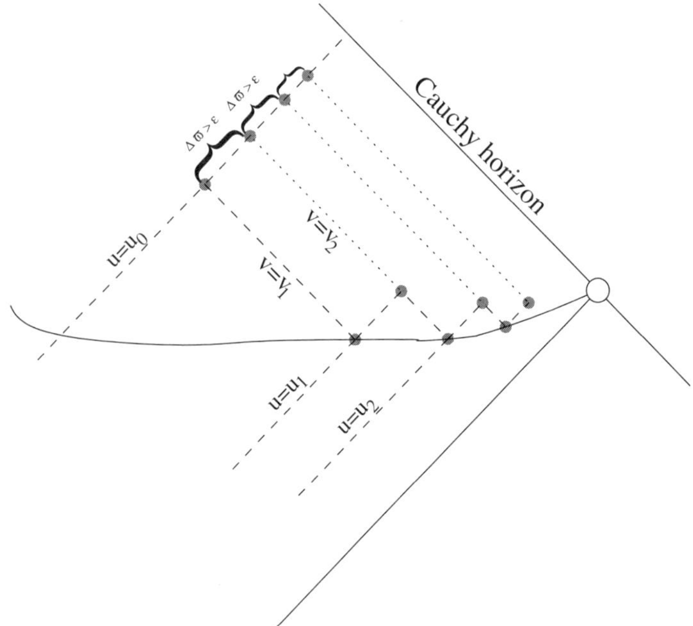

But in view of (10), this mass difference can be added on $u = u_0$ to yield infinite mass at the point where $u = u_0$ meets the Cauchy horizon.

Suffice it to say that the non-linear analysis of black hole regions is quite different than the analysis we are used to. Which parts of this spherically-symmetric picture generalize and which do not remains to be seen.

References

[1] A. Bonanno, S. Droz, W. Israel and S. M. Morsink *Structure of the charged black hole interior* Proc. Roy. Soc. London Ser. A **450** (1995), no. 1940, 553–567

[2] Patrick Brady and John Smith *Black hole singularities: a numerical approach* Phys. Rev. Lett. **75** (1995), no. 7, 1256–1259

[3] Lior M. Burko and Amos Ori, ed. *Internal structure of black holes and spacetime singularities* Israel Institute of Technology, Haifa June 29–July 3, 1997 Ann. of the Isr. Phys. Soc., 13.

[4] S. Chandrasekhar and J. B. Hartle *On crossing the Cauchy horizon of a Reissner-Nordström black-hole* Proc. Roy. Soc. London Ser. A **384** (1982), no. 1787, 301–315

[5] Yvonne Choquet-Bruhat and Robert Geroch *Global aspects of the Cauchy problem in general relativity* Comm. Math. Phys. **14** 1969, 329–335

[6] Demetrios Christodoulou *The instability of naked singularities in the gravitational collapse of a scalar field* Ann. of Math. **149** (1999), no 1, 183–217

[7] Demetrios Christodoulou *On the global initial value problem and the issue of singularities* Classical Quantum Gravity **16** (1999), no. 12A, A23–A35

[8] Demetrios Christodoulou *Self-gravitating relativistic fluids: a two-phase model* Arch. Rational Mech. Anal. **130** (1995), no. 4, 343–400

[9] Demetrios Christodoulou *Bounded variation solutions of the spherically symmetric Einstein-scalar field equations* Comm. Pure Appl. Math **46** (1992), no. 8, 1131–1220

[10] Demetrios Christodoulou *The formation of black holes and singularities in spherically symmetric gravitational collapse* Comm. Pure Appl. Math. **44** (1991), no. 3, 339–373

[11] Demetrios Christodoulou *The problem of a self-gravitating scalar field* Comm. Math. Phys. **105** (1986), no. 3, 337–361

[12] Demetrios Christodoulou *Examples of naked singularity formation in the gravitational collapse of a scalar field* Ann. of Math. **140** (1994), no. 3, 607–653

[13] D. Christodoulou and S. Klainerman *The global nonlinear stability of the Minkowski space* Princeton University Press, Princeton, 1993

[14] Piotr Chruściel *On the uniqueness in the large of solutions of the Einstein's equations ("strong cosmic censorship")* Australian National University, Centre for Mathematics and its Applications, Canberra, 1991

[15] C. J. S. Clarke *The analysis of space-time singularities* Cambridge University Press, Cambridge, 1993.

[16] Mihalis Dafermos *Stability and Instability of the Cauchy Horizon for the Spherically-Symmetric Einstein-Maxwell-Scalar Field equations* to appear, Ann. of Math.

[17] Mihalis Dafermos *The interior of charged black holes and the problem of uniqueness in general relativity* preprint, 2003

[18] S. W. Hawking and G. F. R. Ellis *The large scale structure of space-time* Cambridge Monographs on Mathematical Physics, No. 1. Cambridge University Press, London-New York, 1973

[19] S. Klainerman and I. Rodnianski *Rough solutions for the Einstein vacuum equations* preprint, 2001

[20] Roger Penrose *Gravitational collapse and space-time singularities* Phys. Rev. Lett. **14** (1965), 57–59

[21] Roger Penrose *Gravitational collapse-the role of general relativity* Riv. del Nuovo Cim. **1** (numero speziale) (1969), 252–276

[22] Eric Poisson and Werner Israel *Internal structure of black holes* Phys. Rev. D (3) **41** (1990), no. 6, 1796–1809

[23] Michael Simpson and Roger Penrose *Internal instability in a Reissner-Nordström Black Hole* Int. Journ. Theor. Phys. **7** (1973), no. 3, 183–197

[24] S. T. Yau *Problem Section* in Seminar on Differential Geometry, edited by S. T. Yau, Annals of Math. Studies, Princeton, N. J., 1982

DEPARTMENT OF MATHEMATICS, MASSACHUSETTS INSTITUTE OF TECHNOLOGY
E-mail address: dafermos@math.mit.edu

Contemporary Mathematics
Volume **350**, 2004

Nonlinear Elliptic Equations of Critical Sobolev Growth From a Dynamical Viewpoint

Emmanuel Hebey

Dedicated to Haïm Brézis and Felix Browder

ABSTRACT. Let (M, g) be a smooth compact Riemannian manifold of dimension $n \geq 3$, and $\alpha > 0$. Equations like

$$\Delta_g u + \alpha u = u^{2^\star - 1} \qquad (E_\alpha)$$

where $2^\star$ is the critical Sobolev exponent, have been the target of investigation for decades. However, almost no studies are concerned with the equation as part of a whole family or, equivalently, with these equations when α is considered as a free parameter. The very general question we ask here, to be made precise, is to understand the structure of (E_α) when α varies.

In what follows, we let (M, g) be a smooth compact Riemannian manifold of dimension $n \geq 3$, and let $2^\star = \frac{2n}{n-2}$ be the critical Sobolev exponent for the embedding of the H_1^2-Sobolev space into L^p-Lebesgue spaces. Concerning notation, H_1^2 consists of functions in L^2 whose gradient (more precisely, the pointwise norm of their gradient) is also in L^2. Given $\alpha > 0$, we consider the equation

$$\Delta_g u + \alpha u = u^{2^\star - 1} \qquad (E_\alpha)$$

where u is required to be positive, and $\Delta_g u = -div_g(\nabla u)$ is the Laplacian with the minus sign convention. As a very basic remark, it is easily seen that the constant function $\overline{u}_\alpha = \alpha^{(n-2)/4}$ satisfies (E_α). An intensive literature was dedicated to the study of equation (E_α), and to the study of its model, the Yamabe equation. The question authors usually considered was whether or not such nonlinear elliptic equations of critical Sobolev growth possess one or several solutions. Thanks to a beautiful contribution by Schoen [16], mathematicians also considered the question of the compactness of the solutions to such equations. These studies are concerned with equation (E_α) when the parameter α is fixed. Almost no studies are concerned with the equation as part of a whole family or, equivalently, with these equations when α is considered as a free parameter. Emphasizing the opposition between what we could refer to as a statical approach, and what we could refer to as a

1991 *Mathematics Subject Classification.* Primary: 35J60; Secondary: 58G03.
Key words and phrases. Elliptic equations, Critical exponent, Energy function.

dynamical approach (with respect to the parameter α), the very general question we ask here, to be made precise below, is to understand the structure of the set of solutions of (E_α) when α varies.

Given a function $u \in H_1^2$, we define the energy $E(u)$ of u as the $L^{2^\star}$-norm of u. In other words,

$$E(u) = \left(\int_M |u|^{2^\star} \, dv_g \right)^{\frac{1}{2^\star}}.$$

If u is a solution of (E_α) we recover the classical energy associated to such types of equations. The minimum energy Λ_{min} is then given by $\Lambda_{min} = K_n^{-(n-2)/2}$ where K_n is the sharp constant K in the Euclidean Sobolev inequality $\|u\|_{2^\star} \le K\|\nabla u\|_2$. The explicit value of K_n can be found, for instance, in Hebey [**10**]. The minimum energy is characterized by the property that blowing up sequences of solutions of (E_α) have an energy which is greater than or equal to Λ_{min}, as described in Struwe's decomposition [**17**]. Another energy we define is the following:

DEFINITION 0.1. *Let $\alpha > 0$. We define the energy function E_m associated to equation (E_α) by*

$$E_m(\alpha) = \inf_{u \in S_\alpha} E(u)$$

where S_α consists of the solutions of (E_α).

Basic arguments show that for any $\alpha > 0$, $E_m(\alpha) > 0$. Given $\Lambda > 0$ we let E_Λ be the set of functions u in H_1^2 with $E(u) \le \Lambda$, and

$$S_\Lambda = \left\{ \alpha \text{ s.t. } S_\alpha \cap E_\Lambda \ne \emptyset \right\}$$

so that S_Λ consists of the α's for which (E_α) has a solution of energy less than or equal to Λ.

As already mentioned, the question of understanding the structure of equation (E_α) can be interpreted in two ways. On the one hand, we may want to understand the structure for a given linear term—in other words, we may want to understand the structure of S_α when α is given. On the other hand, we may want to understand the structure as α varies —in other words, we may want to understand the structure of S_Λ or the structure of the union of the S_α's that we intersect with some E_Λ. The problem of describing the structure of S_α for a given α is more classical than the problem of describing the structure of S_Λ, or the problem of understanding the structure of $(\bigcup S_\alpha) \cap E_\Lambda$. In particular, we find in the literature very interesting results describing S_α, especially when (E_α) coincides with the conformal Yamabe equation. For the sake of completeness we quote some of these results.

(1) (Obata [**14**]) Let (S^n, h) be the unit n-sphere, and $\alpha = n(n-2)/4$ (the value of the linear term for which (E_α) is the Yamabe equation). Then

$$S_\alpha = \left\{ \alpha^{\frac{n-2}{4}} \right\} \bigcup \left\{ C_n(\beta)^{\frac{n-2}{4}} (\beta - \cos r)^{1-\frac{n}{2}}, \beta > 1, x_0 \in S^n \right\}$$

where $C_n(\beta) = \frac{n(n-2)}{4}(\beta^2 - 1)$, and $r = d_h(x_0, .)$. In particular, for all $u \in S_\alpha$, $E(u) = \Lambda_{min}$, and S_α is not compact.

(2) (Gidas-Spruck [**8**], Bidaut-Veron [**1**]) Let (M, g) be compact of dimension n. Assume that the Ricci curvature Rc_g of g is such that

$$Rc_g > \frac{4(n-1)}{n(n-2)}\alpha_0 g$$

in the sense of bilinear forms, where α_0 is a positive real number. Then $S_\alpha = \left\{\alpha^{(n-2)/4}\right\}$ if $\alpha \le \alpha_0$.

(3) (Schoen [**16**]) Let (M, g) be compact of dimension n and constant scalar curvature (an assumption we always achieve up to a conformal change of the metric). We assume that (M, g) is not conformally diffeomorphic to the unit sphere, and let $\alpha = \frac{n-2}{4(n-1)} S_g$ where S_g is the scalar curvature of g (so that (E_α) is the Yamabe equation). Then S_α is compact in the C^2-topology. In particular, there exists $\Lambda > 0$ such that $S_\alpha \subset E_\Lambda$.

In addition to studying S_α when α is fixed, the study of S_Λ, and more generally the study of the solutions of (E_α) when α is free, has to be carried out. We propose here a guideline and several questions which, as a whole, constitute a program for the study of (E_α) when α is free. Concerning S_Λ, natural questions to ask are the following ones:

(Q1) Given $\Lambda > 0$, is S_Λ bounded ?

(Q2) Given $\Lambda > 0$, is S_Λ closed in $(0, +\infty)$?

(Q3) Given $\Lambda > 0$, is S_Λ connected ?

As a remark, the particular case of question (Q1) where $\Lambda = \Lambda_{min}$ is equivalent to the long standing conjecture on sharp Sobolev inequalities that was recently proved by Hebey-Vaugon [**11**]. Question (Q2) when $\Lambda = \Lambda_{min}$ is equivalent to the question of whether or not there exist extremal functions for these sharp inequalities. General references for the study of sharp Sobolev inequalities are Druet-Hebey [**4**] and Hebey [**10**].

Concerning the more general question of the understanding of the solutions of (E_α) when α is free, it is reasonable to ask first for a good understanding of the energy function E_m. Natural questions are the following ones:

(Q4) Do we have that $E_m(\alpha) \to +\infty$ as $\alpha \to +\infty$?

(Q5) What is the asymptotic behavior of $E_m(\alpha)$ as $\alpha \to +\infty$?

(Q6) Is E_m a continuous function ?

(Q7) Is E_m a nondecreasing function ?

Basic remarks on these questions are as follows. First, it is easily seen that (Q1) and (Q4) are the same question in the sense that S_Λ is bounded for any Λ if and only if $E_m(\alpha) \to +\infty$ as $\alpha \to +\infty$. This holds also for (Q3) and (Q7) in the sense that E_m is nondecreasing if and only if S_Λ is connected for any Λ. Independently, it is easily checked that for any $\Lambda > 0$,

$$E_m\left([-\infty, \Lambda]\right) = \bigcap_{\varepsilon > 0} S_{\Lambda + \varepsilon} \ .$$

It follows that a positive answer to (Q2) for any Λ implies that E_m is lower semicontinuous. At last, it is easily seen that the energy of the constant function $\overline{u}_\alpha = \alpha^{(n-2)/4}$ is precisely $E(\overline{u}_\alpha) = \alpha^{(n-2)/4} V_g^{1/2^*}$, where V_g is the volume of M with respect to g. A possible refinement of question (Q5) is whether or not $E_m(\alpha)$ has the behavior of $E(\overline{u}_\alpha)$ as $\alpha \to +\infty$. Or again, whether or not $\alpha^{-(n-2)/4} E_m(\alpha)$ has a positive limit as $\alpha \to +\infty$. Of course, it follows from the definition of $E_m(\alpha)$

that we always have that $E_m(\alpha) \leq E(\overline{u}_\alpha)$. We can also ask whether or not there exists k, and for which values of k, we do have that

$$\int_{\alpha_0}^{+\infty} \frac{d\alpha}{E_m(\alpha)^k} < +\infty$$

where $\alpha_0 > 0$ is given.

Many other interesting questions can be asked about the energy function. Among them, as suggested to us by Brézis, we can ask the following:

(Q8) Is E_m a regular function ?

(Q9) Is E_m a convex or a concave function ?

Assuming that the manifold has positive Ricci curvature, we have seen in point (2) above that $S_\alpha = \{\alpha^{(n-2)/4}\}$, and thus that $E_m(\alpha) = \alpha^{(n-2)/4} V_g^{1/2^*}$, if α is small, where V_g is the volume of M with respect to g. It follows that E_m is concave if $n \leq 6$, and convex if $n \geq 6$ when α is small.

Still concerning the problem of understanding the solutions of (E_α) when α is free, and with respect to what was proved by Schoen [16] for the Yamabe equation (see point (1) above), it is also reasonable to ask the following question:

(Q10) Given $\Lambda > 0$, is $\left(\bigcup_{\alpha \geq 0} S_\alpha \right) \bigcap E_\Lambda$ compact in the C^2-topology ?

By convention, $S_0 = \{0\}$ consists of the trivial solution.

The program we just described was started in Druet-Hebey-Vaugon [7]. It is continued in Druet [2, 3], and Druet-Hebey-Robert [5, 6]. Section 1 of this note is devoted to the existence of nonconstant solutions of equation (E_α). In section 2 we give a description of the results obtained in [7]. Section 3 is devoted to comments on questions (Q1)-(Q10). We give a sketch of the proof of the theorems in section 4. We refer to the original article Druet-Hebey-Vaugon [7] for more details.

The author whishes to express his deep thanks to the math department at Rutgers university for its warm hospitality, and to Olivier Druet, Matthew Gursky and Michel Vaugon for very interesting and helpful comments on this work.

1. Nonconstant solutions

A preliminary question to (Q1)-(Q10), which could be motivated by the well-known Pohozaev obstructions [15], is whether or not there exists $\alpha_0 > 0$ such that $S_\alpha = \{\overline{u}_\alpha\}$ if $\alpha \geq \alpha_0$. A positive answer to this question would imply that the above questions are trivial. Fortunately, as shown in Druet-Hebey-Vaugon [7], there are several manifolds for which (E_α) has a nonconstant solution for a sequence (α_k) such that $\alpha_k \to +\infty$ as $k \to +\infty$. Typical examples of such manifolds are the unit n-sphere when n is odd, or the product of a circle with any $(n-1)$-dimensional compact manifold.

We prove the above claim for the product of a circle with any $(n-1)$-dimensional compact manifold. Let (M, g) be a smooth compact $(n-1)$-dimensional manifold, and $(S^1(t), h_t)$ be the circle in $\mathbb{R}^2$ of center 0 and radius t with its standard metric. We let $M_t = S^1(t) \times M$ and $g_t = h_t + g$ be the product metric on M_t. Thanks to

Hebey-Vaugon [**11**] we know that there exists a positive constant B such that for any $u \in H_1^2(M_t)$,

$$\|u\|_{2^*}^2 \le K_n^2 \|\nabla u\|_2^2 + B\|u\|_2^2 \tag{1.1}$$

Let $B_0(g_t)$ be the smallest B in (1.1). It is easily seen, see Hebey [**10**], that there exists $t_0 > 0$ such that for $t \in (0, t_0)$, $B_0(g_t) > V_{g_t}^{-2/n}$, where V_{g_t} is the volume of M_t with respect to g_t. Given k integer, we let G_k be the subgroup of $O(2)$ generated by $G_k(z) = e^{2i\pi/k}z$. Clearly, G_k acts freely on M_t and $M_t/G_k = M_{t/k}$. We fix $t = 1$. For k sufficiently large so that $1/k < t_0$, we let α_k be such that $V_{g_{1/k}}^{-2/n} < \alpha_k < B_0(g_{1/k})$ and consider the minimization problem

$$\lambda_k = \inf_{u \in H_1^2 \backslash \{0\}} \frac{\|\nabla u\|_2^2 + \hat\alpha_k \|u\|_2^2}{\|u\|_{2^*}^2}$$

where $\hat\alpha_k = \alpha_k K_n^{-2}$. Then, $\lambda_k < K_n^{-2}$, and it follows from basic arguments that there exists a minimizer u_k for λ_k. Moreover, this minimizer can be chosen to be positive, smooth, and such that

$$\Delta_{g_{1/k}} u_k + \hat\alpha_k u_k = \lambda_k u_k^{2^*-1}$$

and $\|u_k\|_{2^*} = 1$. Since $\alpha_k > V_{g_{1/k}}^{-2/n}$, u_k is not constant. Let now $\tilde u_k$ be the smooth positive function in $M_1 = S^1(1) \times M$ given by the relation $\tilde u_k/G_k = u_k$. Then,

$$\Delta_{g_1} \tilde u_k + \hat\alpha_k \tilde u_k = \lambda_k \tilde u_k^{2^*-1}$$

and $\|\tilde u_k\|_{2^*} = k^{1/2^*}$. Let $\hat u_k = \lambda_k^{1/(2^*-2)} \tilde u_k$. Noting that $V_{g_{1/k}}^{-2/n} \to +\infty$ as $k \to +\infty$, we get that there exists a sequence $(\hat\alpha_k)$ such that $\hat\alpha_k \to +\infty$ as $k \to +\infty$, and a sequence $(\hat u_k)$ of smooth nonconstant positive solutions of equations $(E_{\hat\alpha_k})$ on (M_1, g_1). This proves the above claim. We also get that $E(\hat u_k) < E(\overline u_{\hat\alpha_k})$ so that $\overline u_{\hat\alpha_k}$ is not a minimizing solution for $(E_{\hat\alpha_k})$.

2. Partial results

As already mentioned, the program we described above was started by Druet-Hebey-Vaugon [**7**]. It is continued in Druet [**2, 3**] and Druet-Hebey-Robert [**5, 6**] where the complete C^0-theory for blow-up and the non conformally flat case in Theorems 2.1 and 2.2 below are discussed. The first result obtained in Druet-Hebey-Vaugon [**7**] can be stated as follows.

THEOREM 2.1 (Druet-Hebey-Vaugon [**7**]). *Let (M, g) be a smooth compact conformally flat Riemannian n-manifold, $n \ge 3$. Given $\Lambda > 0$, there exists α_0 such that (E_α) has no solution of energy less than or equal to Λ if $\alpha \ge \alpha_0$. In particular, S_Λ is bounded for any $\Lambda > 0$, and $E_m(\alpha) \to +\infty$ as $\alpha \to +\infty$.*

Given $\Lambda > 0$, let $\alpha_1(\Lambda)$ be the maximal α for which $(0, \alpha) \subset S_\Lambda$. Let $\alpha_2(\Lambda)$ be the upper bound of S_Λ. Thanks to Theorem 2.1, $\alpha_2(\Lambda) < +\infty$. Independently, it is easily seen that

$$\Lambda^{\frac{n-2}{4}} V_g^{-\frac{2}{n}} \le \alpha_1(\Lambda) \le \alpha_2(\Lambda) \,.$$

The second of these two inequalities is obvious. The first inequality just expresses the fact that for α less than or equal to the left hand side in the above equation, the energy of the constant solution $\overline u_\alpha$ is less than or equal to Λ. Let us now assume that $n \ge 4$. The second result obtained in Druet-Hebey-Vaugon [**7**] can be stated as follows.

THEOREM 2.2 (Druet-Hebey-Vaugon [**7**]). *Let (M, g) be a smooth compact conformally flat Riemannian n-manifold, $n \geq 4$. If Λ is such that*

$$\alpha_1(\Lambda) > \frac{n-2}{4(n-1)} \max_M S_g$$

then S_Λ is closed in $(0, +\infty)$. Moreover, if α is greater than the right hand side of the above equation, then E_m is lower semicontinuous at α.

As a remark, it easily follows from works on the Yamabe problem that we always have that

$$\alpha_1(\Lambda) \geq \frac{n-2}{4(n-1)} \max_M S_g$$

when $\Lambda \geq \Lambda_{min}$. As another remark, since $\alpha_1(\Lambda) \geq \Lambda^{(n-2)/4} V_g^{-2/n}$, the condition of the theorem is automatically satisfied if Λ is sufficiently large. The more general statement we actually prove, see Druet-Hebey-Vaugon [**7**], is that for any

$$\alpha_1 > \frac{n-2}{4(n-1)} \max_M S_g$$

and any $\Lambda > 0$, $S_\Lambda \bigcap [\alpha_1, +\infty)$ is closed. As an easy remark, the conditions on Λ and α in Theorem 2.2 disappear on manifolds of nonpositive scalar curvature. Finally, it should be noted that the condition on Λ is sharp in the following sense: if g is a Riemannian metric on S^n which is conformal to h, but not isometric to h, then

$$\alpha_1(\Lambda_{min}) = \frac{n-2}{4(n-1)} \max_M S_g$$

and $S_{\Lambda_{min}} = \big(0, \alpha_1(\Lambda_{min})\big)$. This easily follows from conformal arguments and was proved in Hebey [**9**].

Theorems 2.1 and 2.2 answer questions (Q1), (Q2), and (Q4) for conformally flat manifolds. They also provide a partial answer to question (Q6), still for conformally flat manifolds.

3. Various comments

An elementary remark on question (Q3) is that S_Λ is connected if and only if $\alpha_1(\Lambda) = \alpha_2(\Lambda)$. Another further remark is that S_Λ is indeed connected if $\Lambda \leq \Lambda_{min}$. This easily follows from basic minimization arguments. For the sake of completeness we give a sketch of the proof of this claim. Given $\alpha > 0$ we let I_α be the functional defined on the nonzero functions of H_1^2 by

$$I_\alpha(u) = \frac{\|\nabla u\|_2^2 + \alpha \|u\|_2^2}{\|u\|_{2^*}^2}.$$

We let also $\Lambda \leq \Lambda_{min}$ and $\alpha \in S_\Lambda$. By the definition of S_Λ, there exists $u_\alpha \in S_\alpha$ such that $E(u_\alpha) \leq \Lambda$. Moreover, $I_\alpha(u_\alpha) \leq \Lambda^{4/(n-2)}$. Noting that for any $\alpha' < \alpha$, $I_{\alpha'}(u_\alpha) < I_\alpha(u_\alpha)$, and that $\Lambda_{min}^{4/(n-2)} = K_n^{-2}$, we get that

$$\inf_{u \in H_1^2 \backslash \{0\}} I_{\alpha'}(u) < \frac{1}{K_n^2}.$$

It follows from this inequality and very basic variational arguments (expressing the fact that the above infimum is below the minimum energy for blow up) that there exists a smooth positive function $u_{\alpha'}$, and $\lambda_{\alpha'} \in (0, \Lambda^{4/(n-2)})$ such that

$$\Delta_g u_{\alpha'} + \alpha' u_{\alpha'} = \lambda_{\alpha'} u_{\alpha'}^{2^*-1}$$

and $\|u_{\alpha'}\|_{2^*} = 1$. In other words, the infimum above is attained. Setting $v_{\alpha'} = \lambda_{\alpha'}^{(n-2)/4} u_{\alpha'}$, we get that $v_{\alpha'}$ is a solution of $(E_{\alpha'})$ such that $E(v_{\alpha'}) \leq \Lambda$. Hence, $\alpha' \in S_\Lambda$ and we proved that when $\Lambda \leq \Lambda_{min}$, then $(0, \alpha] \subset S_\Lambda$ if $\alpha \in S_\Lambda$. In particular, S_Λ is connected. On the other hand, the question is left open for $\Lambda > \Lambda_{min}$.

An elementary remark on question (Q7), which also follows from basic minimization arguments, is that E_m is nondecreasing when α is small. Arguments similar to the ones we just discussed give indeed that if α is such that $E_m(\alpha) < \Lambda_{min}$, then for any $\alpha' \leq \alpha$, $E_m(\alpha') \leq E_m(\alpha)$. On the other hand, the question is left open when α is large.

An elementary remark on question (Q10) is that $\left(\bigcup S_\alpha\right) \cap E_\Lambda$ is indeed compact if $\Lambda < \Lambda_{min}$. This easily follows from the property of Λ_{min} we mentioned above: namely, that blow up does not occur when the energy is less than the minimum energy. Another elementary remark, which follows from the Obata result [14] cited above, is that $\left(\bigcup S_\alpha\right) \cap E_\Lambda$ is not compact on the conformal unit sphere when $\Lambda \geq \Lambda_{min}$. Examples of compact manifolds (M, g), not conformally diffeomorphic to the unit sphere, for which

$$\alpha_1(\Lambda_{min}) = \frac{n-2}{4(n-1)} \max_M S_g$$

and $\left(\bigcup S_\alpha\right) \cap E_\Lambda$ is not compact for $\Lambda \geq \Lambda_{min}$ are in Hebey and Vaugon [12]. A possible reformulation of question (Q10) appears below.

Another remark on question (Q10) is as follows. Define S_α^{min} to be the set of the u's in S_α which are such that $E(u) = E_m(\alpha)$, namely

$$S_\alpha^{min} = \left\{ u \in S_\alpha \text{ s.t. } E(u) = E_m(\alpha) \right\}$$

and define

$$\alpha_0(g) = \frac{n-2}{4(n-1)} \max_M S_g \ .$$

We claim that it follows from the proof of Theorem 2.2 that for $\alpha > \alpha_0(g)$, $S_\alpha^{min} \neq \emptyset$, and that for any $\alpha_1 > \alpha_0(g)$ and any $\Lambda > 0$, $(\bigcup_{\alpha \geq \alpha_1} S_\alpha^{min}) \cap E_\Lambda$ is compact in the C^2-topology, provided that E_m is continuous on $[\alpha_1, \alpha_2]$, where α_2 is such that $E_m(\alpha) > \Lambda$ for $\alpha \geq \alpha_2$. This is easily explained in terms of the Struwe decomposition [17]. We give a sketch of the proof that for any $\alpha_1 > \alpha_0(g)$ and any $\Lambda > 0$, $(\bigcup_{\alpha \geq \alpha_1} S_\alpha^{min}) \cap E_\Lambda$ is compact in the C^2-topology provided that E_m is continuous on $[\alpha_1, \alpha_2]$. We let (α) be a sequence converging to some $\alpha_0 > \alpha_0(g)$, and (u_α) be a sequence such that for any α, $u_\alpha \in S_\alpha^{min} \cap E_\Lambda$. Up to a subsequence, there exists $u_0 \in H_1^2$ such that $u_\alpha \rightharpoonup u_0$ in H_1^2 as $\alpha \to \alpha_0$. Thanks to the Struwe decomposition [17], we can write that

$$u_\alpha = u_0 + \sum_{i=1}^{N} B_i^\alpha + o(1) \quad \text{and} \quad E(u_\alpha) = E(u_0) + \sum_{i=1}^{N} E(B_i^\alpha) + o(1)$$

where the B_i^α's are bubbles with the property that $E(B_i^\alpha) = \Lambda_{min}$, N is an integer, u_0 is a nonnegative solution of (E_{α_0}), and we adopt the convention that there are no sums in the above expressions if $N = 0$. What we get from the proof of Theorem 2.2 (see Druet-Hebey-Vaugon [7] for details) is that u_0 has to be positive if $\alpha_0 > \alpha_0(g)$. Thanks to the maximum principle we actually have only two choices: either $u_0 = 0$ everywhere, or $u_0 > 0$ everywhere. It follows from the definition of $E_m(\alpha_0)$ that $E(u_0) \geq E_m(\alpha_0)$. Assuming that E_m is continuous at α_0, $E(u_\alpha) \to E_m(\alpha_0)$ as $\alpha \to \alpha_0$, and it follows from what we just said that we must have that $E(u_0) = E_m(\alpha_0)$ and $N = 0$. Hence, $u_\alpha \to u_0$ in H_1^2 as $\alpha \to \alpha_0$. We then get the C^2-convergence with standard arguments from regularity theory.

The counterexamples to question (Q10) we mentioned above, in view of what we just said, suggest that a possibly more reasonable formulation of question (Q10) is the following: Given $\alpha_1 > \alpha_0(g)$, and $\Lambda > 0$, do we have that $\left(\bigcup_{\alpha \geq \alpha_1} S_\alpha\right) \bigcap E_\Lambda$ is compact in the C^2-topology ?

For the sake of completeness, we mention that Li and Zhu [13], taking inspiration from previous works by Schoen [16], proved that for 3-dimensional compact Riemannian manifolds (M, g), of positive scalar curvature, the union of the S_α's is compact in the C^2-topology provided that $\alpha \leq \frac{1}{8} S_g$ and (M, g) is not conformally diffeomorphic to the unit 3-sphere. Improvements are in the recent Druet [3].

A final remark concerns question (Q8). We let (M, g) be compact, conformally flat, of dimension $n \geq 4$. As already mentioned, it follows from the proof of Theorem 2.2 that $S_\alpha^{min} \neq \emptyset$ when $\alpha > \alpha_0(g)$. Given $\alpha_0 > \alpha_0(g)$, let (α) be a sequence such that $\alpha \to \alpha_0$, and let $u_\alpha \in S_\alpha^{min}$. Let us also assume that E_m is continuous at α_0. Still thanks to the proof of Theorem 2.2, see the second remark above concerning question (Q10), there exists $u_0 \in S_{\alpha_0}^{min}$ such that, up to a subsequence, $u_\alpha \to u_0$ in the C^2-topology. We write that $u_\alpha = u_0 + \tilde{u}_\alpha$. In particular, $\|\tilde{u}_\alpha\|_{C^0} \to 0$ as $\alpha \to \alpha_0$. Then,

$$E_m(\alpha)^{2^\star} - E_m(\alpha_0)^{2^\star} = 2^\star \int_M u_0^{2^\star - 1} \tilde{u}_\alpha dv_g + o(\|\tilde{u}_\alpha\|_{C^0}) .$$

Thanks to the obvious equation satisfied by $\tilde{u}_\alpha$, we can write that

$$\int_M (\Delta_g \tilde{u}_\alpha + \alpha \tilde{u}_\alpha) u_0 dv_g + (\alpha - \alpha_0) \int_M u_0^2 dv_g$$

$$= \int_M u_\alpha^{2^\star - 1} u_0 dv_g - \int_M u_0^{2^\star} dv_g$$

$$= (2^\star - 1) \int_M u_0^{2^\star - 1} \tilde{u}_\alpha dv_g + o(\|\tilde{u}_\alpha\|_{C^0}) .$$

Independently, integrating by parts and using the equation satisfied by u_0, we can write that

$$\int_M (\Delta_g \tilde{u}_\alpha + \alpha \tilde{u}_\alpha) u_0 dv_g = \int_M u_0^{2^\star - 1} \tilde{u}_\alpha dv_g + o(\|\tilde{u}_\alpha\|_{C^0}) .$$

Combining these different equations, it follows that

$$E_m(\alpha)^{2^\star} - E_m(\alpha_0)^{2^\star} = \frac{n}{2}(\alpha - \alpha_0) \int_M u_0^2 dv_g + o(\|\tilde{u}_\alpha\|_{C^0}) .$$

Summarizing, if we assume that E_m is continuous at α_0, and that $\|\tilde{u}_\alpha\|_{C^0} = O(\alpha - \alpha_0)$ as $\alpha \to \alpha_0$, then $E_m^{2^\star}$ has a derivative at α_0 which is given by

$$\frac{d}{d\alpha} E_m^{2^\star}(\alpha_0) = \frac{n}{2} \int_M u_0^2 dv_g .$$

In particular, if the above assumptions are satisfied for all $\alpha_0 > \alpha_0(g)$, then E_m is increasing.

4. Proof of the theorems

We sketch the proof of Theorem 2.1. The proof of Theorem 2.2 basically follows the same lines. We let (M, g) be compact of dimension $n \geq 3$, and assume that (M, g) is conformally flat. In other words, we assume that up to conformal changes of the metric, we do get local isometries with the Euclidean space. The round sphere and real projective space are examples of conformally flat manifolds. The assumption that (M, g) is conformally flat makes the argument simple. We refer to Druet [2, 3], and Druet-Hebey-Robert [5, 6], for the deep analysis involved in the non conformally flat case.

Following Druet-Hebey-Vaugon [7], we proceed by contradiction. We let $\Lambda > 0$ be given and assume that for any α, equation (E_α) has a solution of energy less than or equal to Λ. By rescaling, we may assume that for any α, there exists a smooth positive function u_α, and there exists $\lambda_\alpha \in (0, \Lambda^{2^*-2}]$, such that

$$\Delta_g u_\alpha + \alpha u_\alpha = \lambda_\alpha u_\alpha^{2^*-1} \qquad (E'_\alpha)$$

and such that $\int_M u_\alpha^{2^*} dv_g = 1$. Multiplying (E'_α) by u_α, and integrating over M, we very easily get that

$$\lim_{\alpha \to +\infty} \int_M u_\alpha^2 dv_g = 0 .$$

Thus, blow up occurs as $\alpha \to +\infty$. Given some point x_0 in M, we say that x_0 is a concentration point for the u_α's if for any $\delta > 0$,

$$\limsup_{\alpha \to +\infty} \int_{B_{x_0}(\delta)} u_\alpha^{2^*} dv_g > 0$$

where $B_x(\delta)$ is the geodesic ball of center x and radius δ. In other words, x_0 is a concentration point for the u_α's if the L^{2^*}-norm of the u_α's does not vanish around x_0 as $\alpha \to +\infty$. Let S be the set consisting of the concentration points for the u_α's. We very easily get through naive concentration point theory, in other words through regularity theory and the De Giorgi-Nash-Moser iterative scheme, that, up to a subsequence, the following holds:

(1) S is a finite set,
(2) $u_\alpha \to 0$ in $C^0_{loc}(M \backslash S)$ as $\alpha \to +\infty$.

Thus, there is a finite number of points such that the u_α's go to zero outside these points, and such that they concentrate at these points like bubbles do. With respect to sharp constant problems in Sobolev type inequalitites, where the energy is always minimal, a main difference here is that we have to deal with multiple bubbles. To some extent, the number of bubbles increases as the bound Λ on the energy increases.

The defining property of a concentration point is that the L^{2^*}-norm of the u_α's concentrate at these points. An important notion now, which has proved to be powerful in the study of sharp constant problems, is the notion of L^2-concentration. We claim here that global L^2-concentration holds for the u_α's. More precisely, we let δ be any positive real number. We let B_δ be the union over the $x \in S$ of the geodesic balls of center x and radius δ. We form the ratio

$$R_\delta(\alpha) = \frac{\int_{M \setminus B_\delta} u_\alpha^2 dv_g}{\int_M u_\alpha^2 dv_g} \, .$$

When α goes to $+\infty$, the two quantities in this ratio go to 0. What we claim here is that the ratio itself also goes to 0 as α goes to $+\infty$, so that

$$\lim_{\alpha \to +\infty} R_\delta(\alpha) = 0$$

and global L^2-concentration holds for the u_α's. This can be proved using very simple arguments.

Now we very briefly indicate how to proceed for the final argument. We let x be any point in S. We let η be a cut-off function centered at x with small support around x. Then we consider the standard Pohozaev identity [15]. Given a smooth bounded domain Ω of the Euclidean space $\mathbb{R}^n$, and a smooth function u with compact support in Ω, this identity, which holds on the Euclidean space, reads as

$$\int_\Omega (x^k \partial_k u) \Delta u \, dx + \frac{n-2}{2} \int_\Omega u \Delta u \, dx = -\frac{1}{2} \int_{\partial \Omega} (x, \nu)(\partial_\nu u)^2 d\sigma \, .$$

We plug ηu_α into this identity. Using the equation satisfied by the u_α's, conformal invariance, and the fact that the manifold is conformally flat, we conclude after some fairly lengthy calculations, that for $\delta > 0$ small, there exists some positive constant C, independent of α, such that

$$\alpha \int_{B_x(\delta)} u_\alpha^2 dv_g \leq C \int_{M \setminus B_x(\delta)} u_\alpha^2 dv_g \, .$$

We sum these different equations over x in S. Thanks to global L^2-concentration, we get a contradiction as α goes to $+\infty$. This proves Theorem 2.1. The proof of Theorem 2.2 basically follows the same lines.

References

[1] M.F.Bidaut-Veron and L.Veron, Nonlinear elliptic equations on compact Riemannian manifolds and asymptotics of Emden equations, *Inventiones Mathematicae*, 106, 1991, 489-539.

[2] O.Druet, Asymptotiques géométriques, phénomènes de concentration et inégalités optimales, *Thèse de l'université de Cergy-Pontoise*, Novembre 2001.

[3] O.Druet, From one bubble to several bubbles. The low-dimensional case, *Preprint*, 2002.

[4] O.Druet and E.Hebey, *The AB program in geometric analysis. Sharp Sobolev inequalities and related problems*, Memoirs of the American Mathematical Society, Vol. 160, No761, 2002.

[5] O.Druet, E.Hebey, and F.Robert, Blow-up theory for elliptic PDEs in Riemannian geometry, *Preprint*, 2002.

[6] O.Druet, E.Hebey, and F.Robert, A C^0-theory for the blow-up of second order elliptic equations of critical Sobolev growth, *Elect. Res. Ann. A.M.S*, 9, 2003.

[7] O.Druet, E.Hebey, and M.Vaugon, Pohozaev type obstructions and solutions of bounded energy for quasilinear elliptic equations with critical Sobolev growth - The conformally flat case, *Nonlinear Analysis, Theory, Methods, and Applications*, 51, 2002, 79-94.

[8] B.Gidas and J.Spruck, Global and local behavior of positive solutions of nonlinear elliptic equations, *Communications on Pure and Applied Mathematics*, 34, 1981, 525-598.

[9] E.Hebey, Fonctions extrémales pour une inégalité de Sobolev optimale dans la classe conforme de la sphère, *Journal de Mathématiques Pures et Appliquées*, 77, 721-733, 1998.

[10] E.Hebey, *Nonlinear Analysis on Manifolds: Sobolev Spaces and Inequalities*, Courant Institute of Mathematical Sciences, Lecture Notes in Mathematics, 5, 1999. Second edition published jointly with the American Mathematical Society.

[11] E.Hebey and M.Vaugon, The best constant problem in the Sobolev embedding theorem for complete Riemannian manifolds, *Duke Mathematical Journal*, 79, 1995, 235-279.

[12] E.Hebey and M.Vaugon, From best constants to critical functions, *Mathematische Zeitschrift*, 237, 2001, 737-767.

[13] Y.Y.Li and M.Zhu, Yamabe type equations on three dimensional Riemannian manifolds, *Communications in Contemporary Mathematics*, 1, 1999, 1-50.

[14] M.Obata, The conjectures on conformal transformations of Riemannian manifolds, *Journal of Differential Geometry*, 6, 1971, 247-258.

[15] S.I.Pohozaev, On the eigenfunctions of the equation $\Delta u + \lambda f(u) = 0$, *Soviet Math. Dokl.*, 6, 1965, 1408-1411.

[16] R.Schoen, On the number of constant scalar curvature metrics in a conformal class, Differential Geometry: A symposium in honor of Manfredo Do Carmo (H.B.Lawson et K.Tenenblat eds.), Wiley, 1991, 311-320.

[17] M. Struwe, A global compactness result for elliptic boundary value problems involving limiting nonlinearities, *Mathematische Zeitschrift*, 187, 1984, 511-517.

EMMANUEL HEBEY, UNIVERSITÉ DE CERGY-PONTOISE, DÉPARTEMENT DE MATHÉMATIQUES, SITE DE SAINT-MARTIN, 2 AVENUE ADOLPHE CHAUVIN, 95302 CERGY-PONTOISE CEDEX, FRANCE
E-mail address: `Emmanuel.Hebey@math.u-cergy.fr`

Contemporary Mathematics
Volume **350**, 2004

Hamiltonian Formalisms for Multidimensional Calculus of Variations and Perturbation Theory

Frédéric Hélein

This paper dedicated to Haim Brezis and Felix Browder

ABSTRACT. In a first part we propose an introduction to multisymplectic formalisms, which are generalisations of Hamilton's formulation of Mechanics to the calculus of variations with several variables: we give some physical motivations, related to the quantum field theory, and expound the simplest example, based on a theory due to T. de Donder and H. Weyl. In a second part we explain quickly a work in collaboration with J. Kouneiher on generalizations of the de Donder–Weyl theory (known as Lepage theories). Lastly we show that in this framework a perturbative classical field theory (analog of the perturbative quantum field theory) can be constructed.

1. Introduction

The main question investigated in this text concerns the construction of a Hamiltonian description of classical fields theory compatible with the principles of special and general relativity, or more generally with any effort towards understanding gravitation like string theory, supergravity or Ashtekar's theory: since space-time should merge out from the dynamics we need a description which does not assume any space-time/field splitting a priori. This means that there is no space-time structure given *a priori*, but space-time coordinates should instead merge out from the analysis of what are the observable quantities and from the dynamics. From this point of view, as we will see, the Lepage–Dedecker theory seems to be much more appropriate than the de Donder–Weyl one. This is the philosophy that we have followed in [**21**]. Here a caveat is in order, in the classical one-dimensional Hamiltonian formalism: we start with a Lagrangian action functional

$$\mathcal{L}[c] := \int_{t_0}^{t_1} L(t, c(t), \dot{c}(t))\, dt,$$

defined on a set of smooth[1] paths $\{t \longmapsto c(t) \in \mathcal{Y}\}$. Here $\mathcal{Y}$ is a smooth k-dimensional manifold and L is a smooth function on $[t_0, t_1] \times T\mathcal{Y}$ ($T\mathcal{Y}$ is the tangent

---

1991 *Mathematics Subject Classification.* Primary 58E30; 53D99; 53C80; 81T99.
Key words and phrases. Mathematical Physics; Analysis of PDEs.
[1]here we assume for instance that c is $\mathcal{C}^2$

bundle to $\mathcal{Y}$: we denote by q a point in $\mathcal{Y}$ and by $v \in T_q\mathcal{Y}$ a vector tangent to $\mathcal{Y}$ at q). The critical points of $\mathcal{L}$ satisfy the Euler–Lagrange equation

$$\frac{d}{dt}\left(\frac{\partial L}{\partial v^i}(t, c(t), \dot{c}(t))\right) = \frac{\partial L}{\partial q^i}(t, c(t), \dot{c}(t)).$$

For all fixed time $t \in [t_0, t_1]$, the Legendre transform is the mapping

$$\begin{array}{ccc} T\mathcal{Y} & \longrightarrow & T^*\mathcal{Y} \\ (q, v) & \longmapsto & (q, p) = (q, \partial L(t, q, v)/\partial v), \end{array}$$

where $q \in \mathcal{Y}$, $v \in T_q\mathcal{Y}$ and $p \in T_q^*\mathcal{Y}$. In cases where, for all time t, this mapping is a diffeomorphism, we define the Hamiltonian function $H : [t_0, t_1] \times T^*\mathcal{Y} \longrightarrow \mathbb{R}$ by

$$H(t, q, p) := \langle p, V(t, q, p)\rangle - L(t, q, V(t, q, p)),$$

where $(q, p) \longmapsto (q, V(t, q, p))$ is the inverse mapping of the Legendre mapping. Then it is well-known that $t \longmapsto c(t)$ is a solution of the Euler–Lagrange equations if and only if $t \longmapsto (c(t), \partial L(t, c(t), \dot{c}(t))/\partial v) =: (c(t), \pi(t))$ is a solution of the *Hamilton equations*

$$\frac{dc^i}{dt}(t) = \frac{\partial H}{\partial p_i}(t, c(t), \pi(t)), \quad \text{and} \quad \frac{d\pi_i}{dt}(t) = -\frac{\partial H}{\partial q^i}(t, c(t), \pi(t)).$$

Thus this converts the second order Euler–Lagrange equations into the flow equation of the non autonomous vector field $X_{H,t}$ defined over $T^*\mathcal{Y}$ by

$$(1) \qquad\qquad X_{H,t} \lrcorner \, \Omega + dH_t = 0.$$

where $\Omega := \sum_{i=1}^{k} dp_i \wedge dq^i$ is the symplectic form over $T^*\mathcal{Y}$, $H_t(q, p) := H(t, q, p)$ and "$\lrcorner$" denote the interior product, i.e. for any tangent vector $\xi \in T_{(q,p)}(T^*\mathcal{Y})$, $\xi \lrcorner \, \Omega$ is the 1-form such that $\xi \lrcorner \, \Omega(V) = \Omega(\xi, V)$, $\forall V \in T_{(q,p)}(T^*\mathcal{Y})$

Instead of viewing the dynamics as the motion of a point in some space, like for instance the phase space $T^*\mathcal{Y}$, we can use another approache which consists in determining how an observable quantity, such as the position or the momentum of a particle, evolves. This is achieved by using the Poisson bracket operation

$$\begin{array}{ccc} \mathcal{C}^\infty(T^*\mathcal{Y}) \times \mathcal{C}^\infty(T^*\mathcal{Y}) & \longrightarrow & \mathcal{C}^\infty(T^*\mathcal{Y}) \\ (F, G) & \longmapsto & \{F, G\}, \end{array}$$

where

$$\{F, G\} := \sum_{i=1}^{k}\left(\frac{\partial F}{\partial p_i}\frac{\partial G}{\partial q^i} - \frac{\partial F}{\partial q^i}\frac{\partial G}{\partial p_i}\right).$$

Then, for all Hamiltonian trajectory $t \longmapsto (c(t), \pi(t))$, and for all $F \in \mathcal{C}^\infty(T^*\mathcal{Y})$, we have

$$\frac{dF(c(t), \pi(t))}{dt} = \{H, F\}(c(t), \pi(t)).$$

For example in the particular case where the variational problem is autonomous (i.e. L does not depend on t) then H does not depend on time and we deduce from the skewsymmetry of the Poisson bracket that the energy is conserved along the trajectories, a special case of Noether theorem when the problem is invariant by time translation.

Eventually this formulation of the dynamics is a good preliminary for modelling the evolution of the quantum version of our problem: for instance by replacing the functions in $\mathcal{C}^\infty(T^*\mathcal{Y})$ by Hermitian self-adjoint operators and the Poisson bracket by the commutator $[\cdot,\cdot]$ we "guess" the Heisenberg evolution equation

$$i\hbar\frac{d\widehat{F}}{dt} = [\widehat{H},\widehat{F}],$$

consequently the commutation relations $[\widehat{p}_i,\widehat{q}^j] = i\hbar\delta_i^j$, is nothing but a formalisation of Heisenberg incertainty principle.

All that leads to a now "well understood" strategy of building a mathematical description of a quantum particle governed by a Hamiltonian functions H (with the restriction that, among other things, the correspondence $H \longmapsto \widehat{H}$ is far from being uniquely defined). Starting from a variational formulation of Newton's law of mechanics, it is thus possible to formally derive the Schrödinger equation more or less by following the steps discussed above.

Now the more challenging question is to produce a similar analysis for fields theories.

Quantum fields theory[2] results from the efforts of physicists in order to cure the shortcoming of the Schrödinger equation. Indeed, this latter equation is not invariant by the group of special relativity, the Poincaré group, but by the (projective) Galilean group. This was one of the motivations which led Dirac to its famous equation. Another concern was the interactions of charged particles and a relativistic field, namely the electro-magnetic field governed by Maxwell equations. A second reason for fields theory is that, unlike the Schrödinger equation, they allow interactions of a variable number of particles. The simplest exemple is the Klein–Gordon equation for scalar fields $\varphi : \mathbb{R}^{1,3} \longrightarrow \mathbb{R}$ (or $\mathbb{C}$):

$$\frac{1}{c^2}\frac{\partial^2\varphi}{\partial t^2} - \Delta\varphi + m^2\varphi = 0.$$

---

[2]Starting from the framework of classical physics, the concept of a field at first might evoke ideas about macroscopic systems, for example velocity fields or temperature fields in fluids and gases, etc. Fields of this kind will not concern us, however; they can be viewed as derived quantities which arise from an averaging of microscopic particle densities. Our subjects are the fundamental fields that describe matter on a microscopic level: it is the quantum-mechanical wave function $\psi(x,t)$ of a system which can be viewed as a field from which the observable quantities can be deduced. In quantum mechanics the wave function is introduced as an ordinary complex-valued function of space and time. In Dirac's terminology it has the character of a "c number". Quantum field theory goes one step further and treats the wave function itself as an object which has to undergo quantization. In this way the wave function $\psi(x,t)$ is transmuted into a field operator $\hat{\psi}(x,t)$, which is an operator-valued quantity (a "q number") satisfying certain commutation relations. This process, often called "second quantization", is quite analogous to the route that in ordinary quantum mechanics leads from a set of classical coordinates q_i to a set of quantum operators $\hat{q}_i$. There is one important technical difference, though, since $\hat{\psi}(x,t)$ is a field, i.e. an object which depends on the coordinate x. The latter plays the role of a "continuous-valued" index, in contrast to the discrete index i, which labels the set q_i. Field theory therefore is concerned with systems having an infinite number of degrees of freedom. The concept of field quantization has far-reaching consequences and is one of the cornerstones of modern physics. Field quantization provides an elegant language to describe particle systems. Moreover the theory naturally leads to the insight that there are field quanta which can be created and annihilated. These field quanta come in many guises and are found in virtually all areas of physics.

This is the Euler–Lagrange equation for the variational problem

$$\mathcal{L}[\varphi] := \int_{\mathbb{R}^{1,3}} \frac{1}{2} \left(\frac{1}{c^2} \left| \frac{\partial \varphi}{\partial t} \right|^2 - |\nabla \varphi|^2 - m^2 |\varphi|^2 \right) dt dx^1 dx^2 dx^3.$$

Here $\mathbb{R}^{1,3}$ is the Minkowski space. Note that the integral $\mathcal{L}[\varphi]$ may not be defined, but if φ smooth or in H^1_{loc} then, for any compact subset $K \subset \mathbb{R}^{1,3}$, the integral $\mathcal{L}_K[\varphi]$ of the Lagrangian density over K is defined and we say that φ is a critical point of $\mathcal{L}$ if and only if, for any K the restriction of φ to K is a critical point of $\mathcal{L}_K$. At this level we can address the following questions: is it possible to follow the same lines as for a 1-dimensional variational problem and build a Hamiltonian formulation of the Klein–Gordon equation ? And can we deduce a quantization procedure for that equation ?

The answer is positive using an approache developed by physicists and is known as the canonical quantization of fields: one chooses a time coordinate t over the Minkowski space-time and looks, for any value of t_0, at the instantaneous state of the field, i.e. the restriction of φ on the Cauchy hypersurface $\{t = t_0\}$. Then we picture the whole history of the field as an evolution, parametrized by t, of a point in the infinite dimensional space $\{(x^1, x^2, x^3) \longmapsto \varphi(x^1, x^2, x^3)\}$. One associates to each path in this infinite dimensional space an action, which is just the one obtained by using Fubini theorem through the splitting $\mathbb{R}^{1,3} \simeq \mathbb{R} \times \mathbb{R}^3$. Then we follow the same procedure as the one that we described at the begining of this paper, but this time in some infinite dimensional manifold: we consider the cotangent bundle to the manifold $\{(x^1, x^2, x^3) \longmapsto \varphi(x^1, x^2, x^3)\}$ and, using the Legendre transform obtained from the Lagrangian, we build a conjugate (momentum) variable $[(x^1, x^2, x^3) \longmapsto \pi(t, x^1, x^2, x^3)]$ (which in this case is just $[(x^1, x^2, x^3) \longmapsto \frac{\partial \varphi}{\partial t}(t, x^1, x^2, x^3)]$). Then one can write Hamilton equations, Poisson bracket and deduce the canonical quantization.

Note that all that is relatively formal but, after each step, it is possible to formulate a theory which makes sense mathematically. This is because of the very particular structure of the Klein–Gordon equation, being hyperbolic and linear. Similarly one can perform a quantization of Maxwell equations (with some extra work due to the gauge invariance). But everything breaks down as soon as the equation is nonlinear, even if the nonlinearity is very mild (the best that we can do is to compute physically relevant quantities by perturbation, even if there is no mathematical bases). So one can quantize only extremely particular variational problems. Beside this outstanding difficulty we are faced with a further critic which is that all this procedure does not respect relativistic invariance: indeed, we were obliged to choose a time coordinate from the begining in order to perform a Legendre transform and then to write down Hamilton's equations, and so on. Commonly, we say that this theory is not *covariant*, i.e. it does not respect the (general) relativity principle which implies the independence toward a used coordinate or reference system. Consequently, this description is not quite satisfactory, even if actually one can check that the resulting quantum field does not depend on the time coordinate which was used.

We want here to consider seriously this critic: is it possible to follow a more covariant path ? A method is well-known: it is the Feynman integral approach, based on the Lagrangian formulation, without using the Hamiltonian framework. It is much more suitable than the canonical approach for computing "correlations" for nonlinear (i.e. interacting) theories. However it is less constructive than the canonical approach, which has the advantage of providing us with a scheme to construct the Hilbert Fock space and the operators. So is there a covariant Hamiltonian approach ? In principle it should be possible and this was suggested independently by M. Born [3] and H. Weyl [37] in 1935: it would be based on using a covariant Hamiltonian formalism for variational problems with several variables quite different from the one that we described above, which used a slicing of space-time.

The first example of such a formalism was built by C. Carathéodory [4] and another version was proposed later independently by H. Weyl [37] and T. de Donder [8]. Note that in constrast with the 1-dimensional calculus of variations, there are actually infinitely many possible theories. There were described by T.H.J. Lepage [30] and H. Boerner [2]. We shall see later how to understand within a global picture this multiplicity of formalisms. Before that we will expound the de Donder–Weyl theory, since it is the simplest one.

2. The de Donder–Weyl theory

The historical background of the formalism expounded in this section is deeply rooted in the work of C. Carathéodory, H. Weyl and T. de Donder, followed by the observation by E. Cartan [5] (who called θ^{dDW} the *de Donder form*) in 1933. But it seems really to have been developped under the impulsion of W. Tulczyjew [36] in 1968 and the Polish school of mathematical physics: J. Śniatycki [34], K. Gawędski [12], J. Kijowski [26], J. Kijowski and W. Szczyrba [28], J. Kijowski and W.M. Tulczyjew [29], and through the papers of P.L. García and A. Pérez-Rendón [11] and H. Goldschmidt and S. Sternberg [14]. Various descriptions and points of view about the foundations of these theories can also be found in the books [1], [13], [32] and [33] or in the papers [15], [16], [17], [25], [31].

Let us consider the 4-dimensional space-time $\mathbb{R}^{1,3}$ as a source domain and $\mathbb{R}^k$ as a target space. A first order action functional on maps $u : \mathbb{R}^{1,3} \longrightarrow \mathbb{R}^k$ is defined by means of a Lagrangian density L: it is a function on the variables (x, y, v), where $x \in \mathbb{R}^{1,3}$, $y \in \mathbb{R}^k$ and $v \in \left(\mathbb{R}^{1,3}\right)^* \otimes \mathbb{R}^k$ is a linear map from $\mathbb{R}^{1,3}$ to $\mathbb{R}^k$ (alternatively we may consider v as a $n \times k$ real matrix). Then we define the functional

$$\mathcal{L}[u] := \int_{\mathbb{R}^{1,3}} L(x, u(x), du(x))\omega,$$

where $\omega := dx^0 \wedge dx^1 \wedge dx^2 \wedge dx^3$. The Euler–Lagrange equation for critical points is

$$\frac{\partial}{\partial x^\mu}\left(\frac{\partial L}{\partial v_\mu^i}(x, u(x), du(x))\right) = \frac{\partial L}{\partial y^i}(x, u(x), du(x)).$$

The de Donder–Weyl theory is simply based on the change of variables

$$\pi_i^\mu(x) := \frac{\partial L}{\partial v_\mu^i}(x, u(x), du(x)),$$

i.e. exchanging the variables (x, y, v) with (x, y, p) where $p \in \mathbb{R}^{1,3} \otimes \left(\mathbb{R}^k\right)^*$ is given by $p_i^\mu = \partial L / \partial v_\mu^i (x, y, v)$. This works of course if we make the assumption that $(x, y, v) \longmapsto (x, y, p)$ is a diffeomorphism, an analog of the Legendre condition. In the following we shall suppose that this Legendre hypothesis is satisfied (but very interesting situations occur when precisely this Legendre condition fails, as for instance in the case of gauge theories). Then we can consider the Hamiltonian function on $\mathbb{R}^{1,3} \times \mathbb{R}^k \times \left(\mathbb{R}^{1,3} \otimes \left(\mathbb{R}^k\right)^*\right)$

$$H(x, y, p) := p_i^\mu v_\mu^i - L(x, y, v), \quad \text{where } v \text{ is defined implicitly by} \quad p_i^\mu = \frac{\partial L}{\partial v_\mu^i}(x, y, v).$$

A simple computation shows that the Euler–Lagrange equations are equivalent to the system of generalized Hamilton equations

$$(2) \qquad \sum_\mu \frac{\partial \pi_i^\mu(x)}{\partial x^\mu} = -\frac{\partial H}{\partial y^i}(x, u(x), \pi(x)), \quad \frac{\partial u^i(x)}{\partial x^\mu} = \frac{\partial H}{\partial p_i^\mu}(x, u(x), \pi(x)).$$

This is a simple example of a more general situation. We can replace for example $\mathbb{R}^{1,3}$ by a smooth n-dimensional oriented manifold $\mathcal{X}$ and $\mathbb{R}^k$ by another k-dimensional manifold $\mathcal{Y}$. Then the Lagrangian density L is a function on variables $x \in \mathcal{X}$, $y \in \mathcal{Y}$ and $v \in T_y\mathcal{Y} \otimes T_x^*\mathcal{X}$. Thus L can be seen as a smooth function defined on the bundle over $\mathcal{X} \times \mathcal{Y}$ with fiber over (x, y) equal to $T_y\mathcal{Y} \otimes T_x^*\mathcal{X}$. We denote by $T\mathcal{Y} \otimes_{\mathcal{X} \times \mathcal{Y}} T^*\mathcal{X}$ this bundle. Using some volume n-form ω on $\mathcal{X}$ we hence define the functional $\mathcal{L}[u] := \int_{\mathcal{X}} L(x, u(x), du(x))\omega$ on the set of maps $u : \mathcal{X} \longrightarrow \mathcal{Y}$. A similar Legendre transform can be defined, leading to a Hamiltonian function defined on the *multisymplectic manifold* $T^*\mathcal{Y} \otimes_{\mathcal{X} \times \mathcal{Y}} T\mathcal{X}$.

We shall see now that this manipulation has some geometrical interpretation in a way analogous to the 1-dimensional calculus of variations. The more naive way to formulate it consists to associate to any pair of maps $x \longmapsto (u(x), \pi(x))$ its graph Γ in $T^*\mathcal{Y} \otimes_{\mathcal{X} \times \mathcal{Y}} T\mathcal{X}$, i.e. the image of $\mathcal{X} \ni x \longmapsto (x, u(x), \pi(x))$. Γ is the n-dimensional analog of a curve in a symplectic manifold. Then at each point $(x, u(x), \pi(x))$ of Γ we attach the tangent n-multivector $X \in \Lambda^n T_{(x, u(x), \pi(x))} \left(T^*\mathcal{Y} \otimes_{\mathcal{X} \times \mathcal{Y}} T\mathcal{X}\right)$ defined by

$$X := X_1 \wedge \cdots \wedge X_n, \quad \text{where} \quad X_\mu := \frac{\vec{\partial}}{\partial x^\mu} + \frac{\partial u^i(x)}{\partial x^\mu}\frac{\vec{\partial}}{\partial y^i} + \frac{\partial \pi_i^\nu(x)}{\partial x^\mu}\frac{\vec{\partial}}{\partial p_i^\nu}.$$

Now we define on the multisymplectic manifold $T^*\mathcal{Y} \otimes_{\mathcal{X} \times \mathcal{Y}} T\mathcal{X}$ the *multisymplectic $(n+1)$-form*

$$\Omega^* := \sum_\mu \sum_i dp_i^\mu \wedge dy^i \wedge \omega_\mu,$$

where $\omega_\mu := \vec{\partial}/\partial x^\mu \, \lrcorner \, \omega$, i.e. ω_μ is the unique $(n+1)$-form such that $\forall V_1, \cdots, V_{n-1} \in T_{(x,y,p)}\left(T^*\mathcal{Y} \otimes_{\mathcal{X} \times \mathcal{Y}} T\mathcal{X}\right)$, $\omega_\mu(V_1, \cdots, V_{n-1}) = \omega(\vec{\partial}/\partial x^\mu, V_1, \cdots, V_{n-1})$.

We also define the interior product of X by Ω^* to be the unique 1-form $X \lrcorner \Omega^*$ such that $\forall V \in T_{(x,y,p)}\left(T^*\mathcal{Y} \otimes_{\mathcal{X} \times \mathcal{Y}} T\mathcal{X}\right)$, $X \lrcorner \Omega^*(V) = \Omega^*(X_1, \cdots, X_n, V)$. Then we can compute that

$$X \lrcorner \Omega^* = (-1)^n \left(\frac{\partial u^i(x)}{\partial x^\mu} dp_i^\mu - \frac{\partial \pi_i^\mu(x)}{\partial x^\mu} dy^i\right).$$

Hence, by comparing this last expression with the value of $dH = \frac{\partial H}{\partial x^\mu}dx^\mu + \frac{\partial H}{\partial y^i}dy^i + \frac{\partial H}{\partial p_i^\mu}dp_i^\mu$ at $(x, u(x), \pi(x))$, we see that Γ is the graph of a solution of the Hamilton system of equations (2) if and only if

$$(3) \qquad X \lrcorner \, \Omega^* = (-1)^n dH_{(x,u(x),\pi(x))} \quad \text{mod} \quad dx^\mu.$$

Here "mod dx^μ" means that the equality holds between the coefficients of dy^i and dp_i^μ in both sides. This looks like the analog of the classical Hamilton equation (1) (except that here we do have the "mod dx^μ" restriction). This suggests us to use Ω^* as a replacement for the standard symplectic form. Note that Ω^* is the differential of the generalized *Poincaré–Cartan n-form* $\theta^* := \sum_\mu \sum_i p_i^\mu dy^i \wedge \omega_\mu$. Moreover, as in the case of classical mechanics, the generalized Poincaré–Cartan form encodes all the informations that we need in order to perform the Legendre transform.

Let us discuss now the second question — about the Poisson bracket. It has to be defined on functionals on the space of solutions to the generalized Hamilton equations. Thus we first consider the space $\mathcal{E}$ of all smooth n-dimensional submanifolds Γ of the multisymplectic manifold which are graphs of the generalised Hamilton equations, i.e. such that for any $m \in \Gamma$, there exists an n-multivector X which is tangent to Γ at m and which satisfies equation (3). We call a *Hamiltonian n-curve* any such submanifold. Among the functionals $\mathcal{E} \longrightarrow \mathbb{R}$ we shall restrict ourself to a particular class: this choice is motivated by the particular observable quantities used by physicists in quantum field theory. Indeed the observable functionals in the canonical field theory are integrals (smeared with smooth test functions) over a spacelike hypersurface of the space-time of either the values of fields components or the value of their time first derivative (in the latter case the time derivative appear because we are actually considering momenta). And it turns out that (almost) all such observable functionals (see the next section) can be described by a pair (Σ, F), where Σ is a codimension 1 submanifold of the multisymplectic manifold and F is a $(n-1)$-differential form on the multisymplectic manifold such that there exists a tangent vector field ξ_F such that

$$dF + \xi_F \lrcorner \, \Omega^* = 0.$$

F is then called an *algebraic observable* $(n-1)$-form. Then, if Σ satisfies suitable transversality conditions (see [**21**]), $\Sigma \cap \Gamma$ is a $(n-1)$-dimensional manifold and we define the functional $\int_\Sigma F$ to be

$$\begin{array}{ccc} \mathcal{E} & \longrightarrow & \mathbb{R} \\ \Gamma & \longmapsto & \int_{\Sigma \cap \Gamma} F. \end{array}$$

Then a bracket can be defined between two observable functionals $\int_\Sigma F$ and $\int_\Sigma G$ by

$$\left\{ \int_\Sigma F, \int_\Sigma G \right\} := \int_\Sigma \{F, G\}, \quad \text{where} \quad \{F, G\} := \xi_F \wedge \xi_G \lrcorner \, \Omega^*.$$

Here $\xi_F \wedge \xi_G \lrcorner \, \Omega^*$ is the unique $(n-1)$-form such that for all tangent vectors $V_1, \cdots, V_{n-1}$, $\xi_F \wedge \xi_G \lrcorner \, \Omega^*(V_1, \cdots, V_{n-1}) = \Omega^*(\xi_F, \xi_G, V_1, \cdots, V_{n-1})$. It can be checked that this Poisson bracket satisfies "good" properties: first of all it coincides with the standard Poisson bracket ot the canonical theory of fields, second it satisfies the Jacobi identity, if for instance the observable $(n-1)$-forms decrease to zero at infinity[3].

The relatively naive description presented here deserves some critics:

- Equation (3) holds only "mod dx^μ", which is not very aesthetic: this reflects a disymmetry between the space-time variables and the field component variables.
- Very important "observable" quantities are the components of the energy-momentum tensor $T_{\mu\nu}$. This tensor is linked through the Noether theorem to space-time translations symmetries in special relativity or to diffeomorphism invariance in general relativity. It is important in quantum field theory, where it helps to construct the Hamiltonian in the standard canonical picture, but also crucial in general relativity, since it models the way the distribution of energy and matter bend the space-time geometry through the Einstein equation $R_{\mu\nu} - \frac{1}{2}Rg_{\mu\nu} = T_{\mu\nu}$. In the previous setting there is no clear representation of the energy-momentum tensor.
- Most important variational problems in physics involve fiber bundles: the Maxwell–Dirac, the Yang–Mills–Dirac or the Yang–Mills–Higgs theories for the electrodynamics, the electroweak and the strong forces and also the general relativity. They does not fit into the above formalism: indeed the field components cannot be treated as coordinates on an independant manifold and one needs a more general framework.

The two first critics can be cured by adding to the set of variable (x, y, p) a further variable $e \in \mathbb{R}$, canonically conjugate to the space-time volume form ω. Then we consider on $(T^*\mathcal{Y} \otimes_{\mathcal{X}\times\mathcal{Y}} T\mathcal{X}) \times \mathbb{R}$ the multisymplectic form

$$\Omega^{dDW} := de \wedge \omega + \sum_\mu \sum_i dp_i^\mu \wedge dy^i \wedge \omega_\mu = de \wedge \omega + \Omega^*.$$

Any solution of the Hamilton equations can be represented by a smooth n-dimensional submanifold Γ in $(T^*\mathcal{Y} \otimes_{\mathcal{X}\times\mathcal{Y}} T\mathcal{X}) \times \mathbb{R}$ such that, instead of equation (3) we have

(4)
$$X \lrcorner \Omega^{dDW} = (-1)^n d\mathcal{H}_{(x,u(x),\epsilon(x),\pi(x))}, \quad \text{where} \quad \mathcal{H}(x, y, e, p) := e + H(x, y, p).$$

Indeed it can be achieved by choosing $\epsilon(x)$ such that $\mathcal{H}(x, u(x), \epsilon(x), \pi(x))$ is constant along Γ. We can hence avoid the "mod dx^μ" restriction. Moreover Ω^{dDW} is the differential of a Poincaré–Cartan form $\theta^{dDW} := e\,\omega + p_i^\mu dy^i \wedge \omega_\mu$ and the component of the stress-energy tensor can be recovered as coefficients of the observable $(n-1)$-forms $\frac{\partial}{\partial x^\mu} \lrcorner \theta^{dDW}$.

Thus it remains to understand better the third point: to find a geometrical framework for generalizing the above construction to more general variational problems. This question is even more accurate now since we added a further variable e whose geometrical sense needs to be understood. For all variational problems on fields which are sections of a bundle $\mathcal{F}$, the usual approach is to consider the affine first jet bundle associated to these sections and to build the multisymplectic manifold

---

[3]nevertheless a notable difference with the Poisson bracket of classical mechanics is that we cannot easily make sense of the product of two such forms. Attempts in this direction are proposed in [**22**]; we suggest an alternative point of view at the end of Paragraph 5.4 in this text

as a dual of this affine jet bundle. References concerning this approach are [18] or [10].

3. A general framework: multisymplectic manifolds

The above theory is an example of a multisymplectic manifold $(\mathcal{M}, \Omega)$: a differential manifold $\mathcal{M}$ equipped with a multisymplectic $(n+1)$-form Ω. An $(n+1)$-form Ω is *multisymplectic* if and only if

- it is closed: $d\Omega = 0$
- it is nondegenerate: $\forall \xi \in T_m \mathcal{M}$, $\xi \lrcorner \Omega = 0 \implies \xi = 0$.

Given a multisymplectic manifold $(\mathcal{M}, \Omega)$ we can define the notion of algebraic observable $(n-1)$-forms and use them, together with hypersurfaces Σ to define observable functionals as in the preceeding section. Poisson brackets are defined in a similar way.

There is also a notion which leads to a generalization of the standard relation $\frac{dF}{dt} = \{H, F\}$ of classical mechanics. Given the Hamiltonian function $\mathcal{H}$, for any algebraic observable $(n-1)$-form F we let

$$(5) \qquad \{\mathcal{H}, F\} := -d\mathcal{H}(\xi_F).$$

Although this definition and the notation used here are reminiscent from those of Poisson bracket, it is not clear whether we may consider this operation a Poisson bracket (because in particular, $\mathcal{H}$ is not an observable form nor any observable quantity). Thus we prefer to call this operation a Poisson *pseudobracket*. Anyway it is related to the following useful result: assume that Γ is a Hamiltonian n-curve and F is an algebraic observable $(n-1)$-form, then we have

$$(6) \qquad dF_{|\Gamma} = \{\mathcal{H}, F\}\omega_{|\Gamma},$$

where $dF_{|\Gamma}$ is the restriction of dF to Γ (see [22], [20]). The proof is straightforward: for all $q \in \Gamma$ we let $(X_1, \cdots, X_n)$ be a basis of $T_q\Gamma$ such that $\omega(X_1, \cdots, X_n) = 1$. Then $dF(X_1, \cdots, X_n) = -\xi_F \lrcorner \Omega(X_1, \cdots, X_n) = -(-1)^n X_1 \wedge \cdots \wedge X_n \lrcorner \Omega(\xi_F) = -d\mathcal{H}(\xi_F) = \{\mathcal{H}, F\}$.

However in a work in collaboration with J. Kouneiher in [21] we point out that the class of algebraic observable $(n-1)$-forms can be enlarged as follows: an $(n-1)$-form F is called an *observable* $(n-1)$-form if and only if, $\forall q \in \mathcal{M}$, if X and $\widetilde{X}$ are two decomposable n-multivectors in $\Lambda^n T_q\mathcal{M}$ such that $X \lrcorner \Omega = \widetilde{X} \lrcorner \Omega$, then $dF(X) = dF(\widetilde{X})$ (see [21] for details). One can then define the pseudobracket $\{\mathcal{H}, F\}$ to be equal to $dF(X)$, where X is any decomposable n-multivector such that $X \lrcorner \Omega = (-1)^n d\mathcal{H}$. It is easy to show that this pseudobracket agrees with the previous one (5) for algebraic observable $(n-1)$-forms and that the dynamical relation (6) can hence be generalized to (non necessarily algebraic) observable $(n-1)$-forms. This definition and the underlying point of view differ from the one previously used in the litterature (which corresponds to algebraic observable $(n-1)$-forms). We believe that this point of view is more natural, being directly related to the fundamental identity (6). When we shall follow this point of view in the next Section to the question of observable $(p-1)$-forms, for $1 \leq p < n$, it will lead also to a new definition of these observable $(p-1)$-forms.

Now what is the difference between the two definitions ? On the one hand every algebraic observable $(n-1)$-form F is an observable $(n-1)$-form since $X \lrcorner \Omega = \widetilde{X} \lrcorner \Omega$ implies $dF(X) = -\xi_F \lrcorner \Omega(X) = -\xi_F \lrcorner \Omega(\widetilde{X}) = dF(\widetilde{X})$. On the other hand the converse is false: consider for example the de Donder–Weyl theory expounded above for maps $u : \mathbb{R}^n \longrightarrow \mathbb{R}^k$, for $k > 1$. Then $y^1 dy^2 \wedge dx^3 \wedge \cdots \wedge dx^n$ is observable but not algebraic observable. We propose to call a multisymplectic manifold on which the set of algebraic observable $(n-1)$-forms coincide with the set of observable $(n-1)$-forms a *pataplectic manifold*. We shall see examples in the next section.

4. Generalizations

The formalism that we have presented in Section 2 is based on the de Donder–Weyl theory, which is a particular case among infinitely many theories which were classified in 1936 by T.H.J. Lepage [30]. A geometric and universal setting for describing simultaneously all these theories was expounded first by P. Dedecker in 1953 [7]. Here the idea is that we can view each first order variational problem as a variational problem on a class of n-dimensional submanifolds G of some manifold $\mathcal{N}$ of dimension $n + k$: G could be the graph of a map between two manifolds, the image of a section of a fiber bundle or something more general. Then the Lagrangian density can be identified with a function L defined on the Grassmannian bundle $Gr^n\mathcal{N}$, i.e. the bundle over $\mathcal{N}$ whose fiber at any point $q \in \mathcal{N}$ is the set of oriented n-dimensional vector subspaces of $T_q\mathcal{N}$. This is a kind of analog of the (projective) tangent bundle of a manifold that is used in Lagrangian Mechanics. Then the analog of the cotangent bundle is the bundle of differential n-forms on $\mathcal{N}$, $\Lambda^n T^*\mathcal{N}$. Note that, in contrast with classical mechanics (which corresponds to $n = 1$) or with the de Donder–Weyl theory, $1 + \dim Gr^n\mathcal{N} = 1 + n + k + nk$ is in general strictly less than $\dim \Lambda^n T^*\mathcal{N} = n + k + \frac{(n+k)!}{n!k!}$. So the Legendre transform is replaced by a Legendre correspondence which — generically — associates to each "multivelocity" $T \in Gr_q^n\mathcal{N}$ an affine subspace of $\Lambda^n T_q^*\mathcal{N}$ of dimension $\frac{(n+k)!}{n!k!} - nk - 1$ called *pseudofiber* by Dedecker. Now each Lepage theory (one instance being the de Donder–Weyl one, when it makes sense) corresponds to choosing a submanifold of $\Lambda^n T^*\mathcal{N}$ which intersects transversally all pseudofibers through exactly one point (if the Legendre condition holds).

Note that, to my knowledge, almost all the literature on the subject focuses on the de Donder–Weyl theory, excepted [7] and [27]. It seems however important (if in particular we are interested in gravitation theories) to understand all the development expounded in the previous paragraph in the Lepage–Dedecker framework. This has been addressed in collaboration with J. Kouneiher in our papers [20] and [21].

In order to understand the difference let us see a very simple example: variational problems on maps $u : \mathbb{R}^2 \longrightarrow \mathbb{R}^2$. Let us denote by $\omega := dx^1 \wedge dx^2$ the volume form on the domain space $\mathbb{R}^2$. A map is pictured by its graph, a 2-dimensional submanifold G of $\mathbb{R}^4 = \mathbb{R}^2 \times \mathbb{R}^2$, whose projection on the first factor $\mathbb{R}^2$ is a diffeomorphism (or equivalently s.t. $\omega_{|G} \neq 0$). Given a point $(x, y) \in G \subset \mathbb{R}^4$ the tangent plane to G at (x, y) is spanned by vectors $X_1 = \frac{\vec{\partial}}{\partial x^1} + v_1^i \frac{\vec{\partial}}{\partial y^i}$ and $X_2 = \frac{\vec{\partial}}{\partial x^2} + v_2^i \frac{\vec{\partial}}{\partial y^i}$, where $(v_1^1, v_2^1, v_1^2, v_2^2) \in \mathbb{R}^4$. So the set of all possible tangent planes to such G's is parametrized by the variables v_μ^i and we can use the local coordinates x^μ, y^i and v_μ^i on $Gr^2\mathbb{R}^4$. Now the analog of the cotangent bundle in this situation is $\Lambda^2 T^*\mathcal{N}$. Using coordinates x^μ and y^i on $\mathbb{R}^4$, a basis of the 6-dimensional space $\Lambda^2 T^*_{(x,y)}\mathbb{R}^4$ is $(dx^1 \wedge dx^2, dy^i \wedge dx^\mu, dy^1 \wedge dy^2)$. Thus any 2-form $P \in \Lambda^2 T^*_{(x,y)}\mathbb{R}^4$ can be identified with the coordinates (e, p_i^μ, r) such that $P = e\, dx^1 \wedge dx^2 + \epsilon_{\mu\nu} p_i^\mu dy^i \wedge dx^\nu + r\, dy^1 \wedge dy^2$, where $\epsilon_{12} = -\epsilon_{21} = 1$ and $\epsilon_{11} = \epsilon_{22} = 0$.

Now the Legendre correspondence is obtained by the following. Given some $P \simeq (e, p_i^\mu, r) \in \Lambda^2 T^*_{(x,y)}\mathbb{R}^4$ and some tangent space $T \simeq (v_\mu^i) \in Gr^2_{(x,y)}\mathcal{N}$, we define the pairing $\langle T, P \rangle := P(X_1, X_2)$, where (X_1, X_2) forms a basis of T such that $\omega(X_1, X_2) = 1$. Using local coordinates we have here $\langle T, P \rangle = e + p_i^\mu v_\mu^i + r(v_1^1 v_2^2 - v_2^1 v_1^2)$. We also define the function $W(x, y, T, P) = \langle T, P \rangle - L(x, y, T)$, where L is the Lagrangian density (identified here with a function on $Gr^2\mathbb{R}^4$). Then we say that T is in correspondence with P if and only if $\frac{\partial W}{\partial T}(x, y, T, P) = 0$. This relation writes in local coordinates

$$p_i^\mu + \epsilon^{\mu\nu}\epsilon_{ij}v_\nu^j\, r = \frac{\partial L}{\partial v_\mu^i}(x^\mu, y^i, v_\mu^i).$$

As announced previously, given some (x^μ, y^i, v_μ^i) the solution to this equation is in general not unique, but it is actually an affine plane (the pseudofiber) inside $\Lambda^2 T^*_{(x,y)}\mathbb{R}^4$, parallel to the vector plane spanned by

$$dx^1 \wedge dx^2 \quad \text{and} \quad \left(v_1^1 v_2^2 - v_1^2 v_2^1\right) dx^1 \wedge dx^2 - \epsilon_{ij}v_\nu^j dy^i \wedge dx^\nu + dy^1 \wedge dy^2.$$

Hence, inside $\Lambda^2 T^*_{(x,y)}\mathbb{R}^4$, pseudofibers form a 4-parameters family of non parallel affine planes. The subset $\mathcal{M}_{(x,y)}$ of $\Lambda^2 T^*_{(x,y)}\mathbb{R}^4$ filled by all these pseudofibers is always dense — meaning that the Legendre correspondence is "almost surjective" — and it can inverted on a dense subset. This contrasts strongly with situations in classical mechanics (where the Legendre transform may degenerate, i.e. its image may be reduced to a strict submanifold, in particular if the variational problem is parametrization invariant) or in fields theory, if we restrict ourself to the de Donder–Weyl theory or if we use the standard canonical approach (here the Legendre transform degenerates as soon as we have a gauge invariance, a phenomenon known as *Dirac's constraints*). A Hamiltonian function $\mathcal{H}$ can be defined on $\mathcal{M} := \cup_{(x,y)\in\mathbb{R}^4}\mathcal{M}_{(x,y)}$ by setting $\mathcal{H}(x, y, P) := W(x, y, T, P)$, where T is an implicit funtion of (x, y, P) through the relation $\frac{\partial W}{\partial T}(x, y, T, P) = 0$.

The more convincing example is the trivial variational problem: We just take $L = 0$, so that any map from $\mathbb{R}^2$ to $\mathbb{R}^2$ is a critical point of our variational problem ! Then the image $\mathcal{M}_{(x,y)}$ of the Legendre correspondence is the union of the complementary of the hyperplane $r = 0$ and of $\{(e, p_i^\mu, r) = (e, 0, 0)/e \in \mathbb{R}\}$. The Hamiltonian function is given by $\mathcal{H}(x, y, e, 0, 0) = e$ and

$$\mathcal{H}(x, y, e, p_i^\mu, r) = e - \frac{p_1^1 p_2^2 - p_2^1 p_1^2}{r},$$

if $r \neq 0$. One can then check that all Hamiltonian 2-curves are of the form

$$\Gamma = \left\{\left(x, u(x), e(x)dx^1 \wedge dx^2 + \epsilon_{\mu\nu}p_i^\mu(x)dy^i \wedge dx^\nu + r(x)dy^1 \wedge dy^2\right)/x \in \mathbb{R}^2\right\},$$

where $u : \mathbb{R}^2 \longrightarrow \mathbb{R}^2$ is an *arbitrary* smooth function, $r : \mathbb{R}^2 \longrightarrow \mathbb{R}^*$ is also an arbitrary smooth function and

$$e(x) = r(x) \left(\frac{\partial u^1}{\partial x^1}(x) \frac{\partial u^2}{\partial x^2}(x) - \frac{\partial u^1}{\partial x^2}(x) \frac{\partial u^2}{\partial x^1}(x) \right) + h,$$

for some constant $h \in \mathbb{R}$, and

$$p_i^\mu(x) = -r(x) \epsilon^{\mu\nu} \epsilon_{ij} \frac{\partial u^j}{\partial x^\nu}(x).$$

Note that in this setting the de Donder–Weyl theory corresponds to the further constraint $r = 0$, which here implies that $p_i^\mu = 0$: we thus recover the fact that the Legendre transform completely degenerates.

Other examples can be studied like for instance:

- The harmonic map Lagrangian $L(x, y, v) = \frac{1}{2}|v|^2$ where $|v|^2 :$ $= (v_1^1)^2 + (v_2^1)^2 + (v_1^2)^2 + (v_2^2)^2$. Then one finds that $\mathcal{H}(q, p)$ $= e + \frac{1}{1-r^2} \left(\frac{|p|^2}{2} + r(p_1^1 p_2^2 - p_2^1 p_1^2) \right)$.

- The Maxwell equations in two dimensions. We take $L(x, y, v)$ $= -\frac{1}{2}\left(v_2^1 - v_1^2 \right)^2$. Then $\mathcal{H}(q, p) = e + \frac{(p_2^1 + p_1^2)^2 - 4p_1^1 p_2^2}{4r} - \frac{1}{4}\frac{(p_2^1 - p_1^2)^2}{2+r}$.

Beside the Legendre correspondence, the Lepage–Dedecker theory differs from the de Donder–Weyl theory in many other aspects, sometime relatively subtle. For example, on $\mathcal{M} := \Lambda^n T^*\mathcal{N}$, with its standard multisymplectic form Ω, the set of algebraic observable $(n-1)$-forms coincides with the set of observable $(n-1)$-forms. Hence $(\mathcal{M}, \Omega)$ is an example of a pataplectic manifold. I refer to our paper [21] for a complete exposition.

Another question which is discussed in [20] and [21] is the possibility and the relevance of considering observable forms of degree $p - 1$, where $p < n$. This was proposed first by I. Kanatchikov in [22], [23]. For instance in the Hamiltonian system (2), one sees a disymmetry between π_i^μ and u^i (which disappears when $n = 1$): the equations on π_i^μ involve a divergence whereas equations on u^i prescribe its derivatives in all direction. This reflects the fact that the π_i^μ's are actually the components of an observable $(n-1)$-form, whereas u^i could be considered as an observable 0-form. Another beautiful example holds for gauge theories (see [22], [21]): where the gauge potential $A_\mu dx^\mu$ can be considered as an observable 1-form, whereas the Faraday form $\star(dA + A \wedge A)$ as an observable $(n-2)$-form. Moreover these forms are canonically conjugate (in a sense similar to the duality between position and momentum variables in classical mechanics). The first definition proposed in [22] (a $(p-1)$-form F is observable if and only if there exists an $(n-p)$-multivector ξ such that $dF + \xi \lrcorner \Omega = 0$) leads to a quite interesting notion of graded Poisson bracket, but causes some difficulties when one tries to generalize relation (6) to $(p-1)$-forms, for $1 \leq p < n$. In [21], starting from another point of view, namely by characterising the property which seems to be relevant in order to generalize relation (6), we proposed the following alternative definition. We define *collectively* the set of all observable $(p-1)$-forms, for $1 \leq p < n$: it is a vector subspace $\mathfrak{P}^*\mathcal{M}$ of the set of sections of $\oplus_{p=1}^n \Lambda^{p-1} T^*\mathcal{M}$ such that $\forall F_1, \cdots, F_k \in \mathfrak{P}^*\mathcal{M}$, if $dF_1 \wedge \cdots \wedge dF_k$ is a section of $\Lambda^n T^*\mathcal{M}$, then there exists a vector field ξ on $\mathcal{M}$ such that $dF_1 \wedge \cdots \wedge dF_k + \xi \lrcorner \, \Omega = 0$. This definition seems slightly unpleasant at first glance (there could be several systems of observable forms, i.e. several choices for $\mathfrak{P}^* \mathcal{M}$) but it provides the right hypothesis in order to prove generalizations of (6) (see [21]). Note also that it has some physical meaning (see the next paragraph). One drawback is that it is now more delicate to define a notion of Poisson bracket between such forms. However we were able to propose a partial definition which works for the most important cases (see [21]).

Note that the above definition of observable $(p-1)$-forms collectively is an example of a mechanism by which observable quantities (and in particular space-time coordinates) could merge out from intrinsic properties. In this spirit we also remarked in [21] that the dynamical relation (6) can be replaced by a more general one:

$$\{\mathcal{H}, F\} dG_{|\Gamma} = \{\mathcal{H}, G\} dF_{|\Gamma},$$

(a generalisation for observable forms of lower degrees also exists). The underlying idea, that nothing can be measured in an absolute sense and that we can only compare the results of two different measures, is very much in the spirit of general relativity.

5. Dynamical observable forms and perturbation theory

5.1. Dynamical observable $(n-1)$-forms and their motivations. When building a quantum field theory, in order to achieve a relativistic invariance and in particular to be free from any choice of time coordinate, one is led to use the *Heisenberg point of view*. There a vector in the Hilbert (Fock) space of quantum states represents the complete history over all space-time of a quantized field, and observable operators act on this Hilbert space with eigenvalues which are smeared integrals of functions of the values of the fields and their space-time derivatives on space-time. The classical counterpart of this point of view, sometime called Einstein point of view, is to consider the set $\mathcal{E}$ of solutions to the Euler–Lagrange equations of motion over all space-time to be the set of physical states and to consider the set of functionals defined on $\mathcal{E}$ to be the set of observables. If we wish to understand in a covariant way how to quantize these fields it seems to be crucial to be able to define a Poisson bracket between observable functionals and in particular between two observable functionals of the type $\int_\Sigma F : \Gamma \longmapsto \int_{\Gamma \cap \Sigma} F$ and $\int_{\widetilde{\Sigma}} G : \Gamma \longmapsto \int_{\Gamma \cap \widetilde{\Sigma}} G$, even if Σ and $\widetilde{\Sigma}$ are two different hypersurfaces. This is however not clear in general. One possiblity arises when, for instance, F is such that

$$(7) \qquad\qquad \{\mathcal{H}, F\} = -d\mathcal{H}(\xi_F) = 0.$$

We then say that F is a *dynamical* observable $(n-1)$-form. Assume furthermore that Σ and $\widetilde{\Sigma}$ are homologous hypersurfaces, so that $\widetilde{\Sigma} - \Sigma$ is the boundary of an open subset D. Then by applying first Stokes' theorem and second (6) we obtain that

$$\int_{\Gamma \cap \widetilde{\Sigma}} F - \int_{\Gamma \cap \Sigma} F = \int_{\Gamma \cap D} dF = \int_{\Gamma \cap D} \{\mathcal{H}, F\} \omega = 0.$$

Hence the two functionals $\int_\Sigma F$ and $\int_{\widetilde{\Sigma}} F$ coincide on $\mathcal{E}$. We can thus pose

$$\left\{ \int_\Sigma F, \int_{\widetilde{\Sigma}} G \right\} := \left\{ \int_{\widetilde{\Sigma}} F, \int_{\widetilde{\Sigma}} G \right\} = \int_{\widetilde{\Sigma}} \{F, G\}.$$

Thus it remains to find dynamical observable forms. Here comes a surprise and a relative deception. We quote H. Goldschmidt and S. Sternberg in [**14**]: *"For the free fields (i.e. quadratic Lagrangians) that arise in quantum field theory, the algebra[4] P is infinite dimensional and provides enough elements to yield the operators of the associated free quantum fields. However computations done jointly with S. Coleman to whom we are very grateful, seem to indicate that of $n \geq 3$, then for "interacting Lagrangians", i.e. those containing higher order terms, the algebra P is finite dimensional, and hence does not provide enough operators for quantization"*. A more detailed computation with basically the same conclusion can be found in J. Kijowski's paper [**26**]. It is interesting here that this question meet the same kind of difficulties as the quantization problem for fields: the quantization procedure works when the classical equation is linear but fails as soon as the problem become nonlinear (*interacting fields* in the language of physicists). This is perhaps an indication that the two questions are related (although there is no doubt that the quantization problem is *much more* difficult).

There are however some ways to escape from this dead end. One is to remark that the set of dynamical observable forms is roughly speaking in correspondence with the set of symmetries of the problem (Noether theorem). In particular in the presence of a gauge symmetry we can produce an infinite dimensional family of dynamical observable forms, which corresponds to the set of all current densities smeared with any test function. We have discussed this approache in [**21**]. It suggests the question whether a gauge symmetry could improve the quantization of a field theory. In the same spirit it could be interesting to explore integrable systems, which possesses infinitely many symmetries. A third possibility is when the problem is nonlinear but close to a linear one: one can then hope to build observable functionals by perturbations. We expound here a simple example which illustrates this idea.

5.2. The interacting field: obstructions to dynamical observables forms. We let $\eta_{\mu\nu}$ be a constant metric on $\mathbb{R}^n$ (which could be Euclidean or Minkowskian), with inverse $\eta^{\mu\nu}$, and we consider the following functional on the set of maps $\varphi : \mathbb{R}^n \longrightarrow \mathbb{R}$:

$$\mathcal{L}[\varphi] := \int_{\mathbb{R}^n} \left(\frac{1}{2} \eta^{\mu\nu} \frac{\partial \varphi}{\partial x^\mu} \frac{\partial \varphi}{\partial x^\nu} + \frac{m^2}{2} \varphi^2 + \frac{\lambda}{3} \varphi^3 \right) \omega,$$

where $\omega = dx^1 \wedge \cdots \wedge dx^n$ and λ is a scalar constant, that we suppose to be small. Denoting by $\Delta := -\eta^{\mu\nu} \frac{\partial^2}{\partial x^\mu \partial x^\nu}$, the Euler–Lagrange equation is

$$\Delta \varphi + m^2 \varphi + \lambda \varphi^2 = 0.$$

Since here the target space is 1-dimensional there is no difference between the de Donder–Weyl theory and the other Lepage theories. The multisymplectic manifold $\mathcal{M}$ can hence either be constructed as the dual of the first affine jet bundle or with $\Lambda^n T^*(\mathbb{R}^n \times \mathbb{R})$. We identify it with $\mathbb{R}^{2n+2}$, with the coordinates (x^μ, ϕ, e, p^μ) and with the multisymplectic (or pataplectic) form

$$\Omega := de \wedge \omega + dp^\mu \wedge d\phi \wedge \omega_\mu,$$

---

[4]P is here the quotient of the set of algebraic observable $(n-1)$-forms by the subset of exact $(n-1)$-forms

where $\omega_\mu := \frac{\vec{\partial}}{\partial x^\mu} \, \lrcorner \, \omega$. Then Hamiltonian n-curves are n-dimensional submanifolds Γ of $\mathcal{M} \simeq \mathbb{R}^{2n+2}$ such that, for any point $p \in \Gamma$ there exists a unique n-multivector X tangent to Γ at p such that $X \, \lrcorner \, \Omega = (-1)^n d\mathcal{H}$, where

$$\mathcal{H}(x, \phi, e, p) := e + \frac{1}{2}\eta_{\mu\nu}p^\mu p^\nu - \frac{m^2}{2}\phi^2 - \frac{\lambda}{3}\phi^3.$$

The search for all dynamical algebraic observable $(n-1)$-forms consists in the following: one looks at vector fields ξ on $\mathcal{M}$ such that

$$(8) \qquad\qquad\qquad d\left(\xi \, \lrcorner \, \Omega\right) = 0,$$

which will implies that there exists an $(n-1)$-form F such that $\xi = \xi_F$, i.e. $dF + \xi \, \lrcorner \, \Omega = 0$, and such that

$$(9) \qquad\qquad\qquad \{\mathcal{H}, F\} = -d\mathcal{H}(\xi) = 0.$$

This has been done in [**26**]. The results are that: if $\lambda \neq 0$ the only solutions are $\xi = X^\mu \frac{\vec{\partial}}{\partial x^\mu}$, where X^μ are constants, if $\lambda = 0$ the solutions are $\xi = X^\mu \frac{\vec{\partial}}{\partial x^\mu} + \eta^{\mu\nu}\frac{\partial \Phi}{\partial x^\nu}(x)\frac{\vec{\partial}}{\partial p^\mu} + \left(m^2\phi\Phi(x) - p^\mu\frac{\partial \Phi}{\partial x^\mu}(x)\right)\frac{\vec{\partial}}{\partial e} + \Phi(x)\frac{\vec{\partial}}{\partial \phi}$, where Φ is a solution of $\Delta\Phi + m^2\Phi = 0$.

Let us revisit partially this analysis: we assume for simplicity that $dx^\mu(\xi) = 0$ (i.e. we throw away the X^μ's which correspond to parts of the stress-energy-tensor). First one finds that such a ξ satisfies (8) if and only if

$$(10)$$
$$\xi = \left(P^\mu(x, \phi) - p^\mu\frac{\partial \Phi}{\partial \phi}(x, \phi)\right)\frac{\vec{\partial}}{\partial p^\mu} + \left(E(x, \phi) - p^\mu\frac{\partial \Phi}{\partial x^\mu}(x, \phi)\right)\frac{\vec{\partial}}{\partial e} + \Phi(x, \phi)\frac{\vec{\partial}}{\partial \phi},$$

where Φ, E and P^μ are arbitrary functions of (x, ϕ) subject to the condition

$$(11) \qquad\qquad\qquad \frac{\partial E}{\partial \phi} - \frac{\partial P^\mu}{\partial x^\mu} = 0.$$

Second the substitution of the value of ξ in (9) leads to the system of equations

$$(12) \qquad \frac{\partial \Phi}{\partial \phi}(x, \phi) = 0 \quad , \quad \frac{\partial \Phi}{\partial x^\mu}(x, \phi) - \eta_{\mu\nu}P^\nu(x, \phi) = 0,$$

(which implies that Φ and P^μ depend only on x and $P^\mu(x) = \eta^{\mu\nu}\frac{\partial \Phi}{\partial x^\nu}(x)$) and

$$(13) \qquad\qquad \left(m^2\phi + \lambda\phi^2\right)\Phi(x) - E(x, \phi) = 0.$$

The system (12), (13) has no nontrivial solution when $\lambda \neq 0$, because by (11) it would contradict the fact that $P^\mu(x) = \eta^{\mu\nu}\frac{\partial \Phi}{\partial x^\nu}(x)$.

So let us forget condition (13) and assume only (10), (11) and (12). By (11) and (12) we deduce that $\frac{\partial E}{\partial \phi}(x, \phi) = -\Delta\Phi(x)$ and so there exists a function $A : \mathbb{R}^n \longrightarrow \mathbb{R}$ such that $E(x, \phi) = A(x) - \phi\Delta\Phi(x)$. We shall assume $A = 0$ in the following (since A does not help in anything), we deduce that

$$\xi = \eta^{\mu\nu}\frac{\partial \Phi}{\partial x^\nu}(x)\frac{\vec{\partial}}{\partial p^\mu} - \left(\phi\Delta\Phi(x) + p^\mu\frac{\partial \Phi}{\partial x^\mu}(x)\right)\frac{\vec{\partial}}{\partial e} + \Phi(x)\frac{\vec{\partial}}{\partial \phi},$$

(then $\xi \, \lrcorner \, \Omega = -dF$, where $F = \left(p^\mu\Phi(x) - \eta^{\mu\nu}\phi\frac{\partial \Phi}{\partial x^\nu}(x)\right)\omega_\mu$) and

$$d\mathcal{H}(\xi) = -\phi\left(\Delta\Phi(x) + m^2\Phi(x)\right) - \lambda\phi^2\Phi(x).$$

142 FRÉDÉRIC HÉLEIN

We now suppose that $\Phi = \Phi^{(1)}$, a solution of the equation $\Delta\Phi^{(1)} + m^2\Phi^{(1)} = 0$. We denote by $\xi^{(1)}$ the corresponding vector field and $F^{(1)}$ the associated observable $(n-1)$-form: $F^{(1)} := \left(p^\mu\Phi^{(1)}(x) - \eta^{\mu\nu}\phi\frac{\partial\Phi^{(1)}}{\partial x^\nu}(x)\right)\omega_\mu$. Then, instead of (9), we have

$$(14) \qquad \{\mathcal{H}, F^{(1)}\} = -d\mathcal{H}(\xi^{(1)}) = \lambda\phi^2\Phi^{(1)}(x),$$

and thus

$$\int_{\Gamma\cap\partial D} F^{(1)} = \int_{\Gamma\cap D} dF^{(1)} = \int_{\Gamma\cap D} \lambda\phi^2\Phi^{(1)}(x)\omega.$$

5.3. Second order correction. We now add to the functional $\int_{\partial D} F^{(1)}$ another functional of the form

$$\lambda\int_{\partial D}\int_{\partial D} F^{(2)} : \quad \mathcal{E} \longrightarrow \mathbb{R}$$
$$\Gamma \longmapsto \lambda\int_{\Gamma\cap\partial D}\int_{\Gamma\cap\partial D} F^{(2)},$$

where $F^{(2)} \in \Gamma(\mathcal{M}, \Lambda^{n-1}T^*\mathcal{M}) \otimes \Gamma(\mathcal{M}, \Lambda^{n-1}T^*\mathcal{M})$ (here $\Gamma(\mathcal{M}, \Lambda^{n-1}T^*\mathcal{M})$ is the set of sections over $\mathcal{M}$ of the bundle $\Lambda^{n-1}T^*\mathcal{M}$, i.e. the set of $(n-1)$-forms over $\mathcal{M}$). We shall make the following hypotheses on $F^{(2)}$: we let $\Omega^{\otimes 2} := \Omega \otimes \Omega$ be in $\Gamma(\mathcal{M}, \Lambda^{n+1}T^*\mathcal{M}) \otimes \Gamma(\mathcal{M}, \Lambda^{n+1}T^*\mathcal{M})$ and we assume that there exists a "bivector" $\xi^{(2)}$, i.e. an element of $\Gamma(\mathcal{M}, T\mathcal{M}) \otimes \Gamma(\mathcal{M}, T\mathcal{M})$, such that

$$(15) \qquad d^{\otimes 2} F^{(2)} = (-1)^2 \xi^{(2)} \lrcorner\, \Omega \otimes \Omega.$$

It deserves some definitions about notations. Let us abbreviate by z^I the system of coordinates (x^μ, y^i, e, p^μ) on $\mathcal{M}$ and by $(z_1^I; z_2^J) = (x_1^\mu, y_1^i, e_1, p_1^\mu; x_2^\mu, y_2^i, e_2, p_2^\mu)$ coordinates on $\mathcal{M} \times \mathcal{M}$. Then any $d^{\otimes 2}$ is the unique linear operator from $\Gamma(\mathcal{M}, \Lambda^{n-1}T^*\mathcal{M})\otimes\Gamma(\mathcal{M}, \Lambda^{n-1}T^*\mathcal{M})$ to $\Gamma(\mathcal{M}, \Lambda^n T^*\mathcal{M})\otimes\Gamma(\mathcal{M}, \Lambda^n T^*\mathcal{M})$ such that if $\alpha^{(2)} \in \Gamma(\mathcal{M}, \Lambda^{n-1}T^*\mathcal{M}) \otimes \Gamma(\mathcal{M}, \Lambda^{n-1}T^*\mathcal{M})$ writes

$$\alpha^{(2)} = \alpha(z_1^I, z_2^J)dz_1^{I_1} \wedge \cdots \wedge dz_1^{I_{n-1}} \otimes dz_2^{J_1} \wedge \cdots \wedge dz_2^{J_{n-1}},$$

then

$$d^{\otimes 2}\alpha^{(2)} := \sum_{I,J} \frac{\partial^2\alpha}{\partial z_1^I \partial z_2^J}(z_1^I, z_2^J)dz_1^I \wedge dz_1^{I_1} \wedge \cdots \wedge dz_1^{I_{n-1}} \otimes dz_2^J \wedge dz_2^{J_1} \wedge \cdots \wedge dz_2^{J_{n-1}}.$$

(Remark: similar tensor product of differential forms and operators $d^{\otimes 2}$ were used in [19] for the purpose of proving isoperimetric inequalities through calibrations.) Similarly $\lrcorner$ is an operation with standard linear properties such that if $\xi^{(2)} \in \Gamma(\mathcal{M}, T\mathcal{M}) \otimes \Gamma(\mathcal{M}, T\mathcal{M})$ writes

$$\xi^{(2)} = \xi(z_1^I, z_2^J)\frac{\vec{\partial}}{\partial z_1^I} \otimes \frac{\vec{\partial}}{\partial z_2^J},$$

then

$$\xi^{(2)} \lrcorner\, \Omega^{\otimes 2} := \xi(z_1^I, z_2^J)\left(\frac{\vec{\partial}}{\partial z_1^I} \lrcorner\, \Omega\right) \otimes \left(\frac{\vec{\partial}}{\partial z_2^J} \lrcorner\, \Omega\right).$$

Then, denoting by $d_1 := d \otimes Id$ and $d_2 := Id \otimes d$, so that $d^{\otimes 2} = d_1 \circ d_2$, we have

$$
\int_{\Gamma \cap \partial D} \int_{\Gamma \cap \partial D} F^{(2)} = \int_{\Gamma \cap D} \int_{\Gamma \cap \partial D} d_1 F^{(2)} = \int_{\Gamma \cap D} \int_{\Gamma \cap D} d^{\otimes 2} F^{(2)}
$$
$$
= \int_{\Gamma \cap D} \int_{\Gamma \cap D} \xi^{(2)} \lrcorner \, \Omega \otimes \Omega
$$
$$
= \int_{\Gamma \cap D} \int_{\Gamma \cap D} \left(\xi^{(2)} \lrcorner \, \Omega \otimes \Omega \right) (X(z_1) \otimes X(z_2)) \, \omega \otimes \omega
$$

$(X(z)$ is there a n-multivector tangent to

Γ at z, s.t. $\omega_z(X(z)) = 1)$

$$
= \int_{\Gamma \cap D} \int_{\Gamma \cap D} (-1)^{2n} \left(X(z_1) \otimes X(z_2) \lrcorner \, \Omega \otimes \Omega \right) \left(\xi^{(2)} \right) \omega \otimes \omega
$$
$$
= \int_{\Gamma \cap D} \int_{\Gamma \cap D} d\mathcal{H}_{z_1} \otimes d\mathcal{H}_{z_2} \left(\xi^{(2)} \right) \omega \otimes \omega,
$$

where we have used the fact that $\left(X(z_1) \otimes X(z_2) \lrcorner \, \Omega \otimes \Omega \right)_{|\Gamma \times \Gamma} = (-1)^{2n} \left(d\mathcal{H}_{z_1} \otimes d\mathcal{H}_{z_2} \right)_{|\Gamma \times \Gamma}$. The idea is to look for an $F^{(2)}$ such that

$$
(16) \quad \{\mathcal{H}^{\otimes 2}, F^{(2)}\} := d\mathcal{H}_{z_1} \otimes d\mathcal{H}_{z_2} \left(\xi^{(2)} \right) = -\Phi^{(1)}(x_1)\delta(x_1 - x_2)\phi_1\phi_2 + \mathcal{O}(\lambda),
$$

where δ is the Dirac distribution on $\mathbb{R}^n$. Then

$$
\lambda \int_{\Gamma \cap \partial D} \int_{\Gamma \cap \partial D} F^{(2)} = -\lambda \int_{\Gamma \cap D} \Phi^{(1)}(x)\phi^2 \omega + \mathcal{O}(\lambda^2),
$$

so that $\int_{\Gamma \cap \partial D} F^{(1)} + \lambda \int_{\Gamma \cap \partial D} \int_{\Gamma \cap \partial D} F^{(2)} = \mathcal{O}(\lambda^2)$.

We choose $F^{(2)}$ of the form

$$
F^{(2)} = \left(p_1^\mu - \eta^{\mu\lambda}\phi_1 \frac{\partial}{\partial x_1^\lambda} \right) \left(p_2^\nu - \eta^{\nu\sigma}\phi_1 \frac{\partial}{\partial x_2^\sigma} \right) \Phi^{(2)}(x_1, x_2)\omega_\mu \otimes \omega_\nu,
$$

where $\Phi^{(2)} : \mathbb{R}^n \times \mathbb{R}^n \longrightarrow \mathbb{R}$ is a function to be precised later. One can check that such an $F^{(2)}$ satisfies (15) with

$$
\xi^{(2)} = \left(\left(p_1^\mu \frac{\partial}{\partial x_1^\mu} + \phi_1 \Delta_1 \right) \frac{\vec{\partial}}{\partial e_1} - \eta^{\mu\lambda} \frac{\partial}{\partial x_1^\lambda} \frac{\vec{\partial}}{\partial p_1^\mu} - \frac{\vec{\partial}}{\partial \phi_1} \right)
$$
$$
\otimes \left(\left(p_2^\nu \frac{\partial}{\partial x_1^\nu} + \phi_2 \Delta_2 \right) \frac{\vec{\partial}}{\partial e_2} - \eta^{\nu\sigma} \frac{\partial}{\partial x_2^\sigma} \frac{\vec{\partial}}{\partial p_2^\nu} - \frac{\vec{\partial}}{\partial \phi_2} \right) \Phi^{(2)}(x_1, x_2).
$$

Here we denote by $\Delta_1 := \eta^{\mu\nu} \frac{\partial^2}{\partial x_1^\mu \partial x_1^\nu}$ and $\Delta_2 := \eta^{\mu\nu} \frac{\partial^2}{\partial x_2^\mu \partial x_2^\nu}$ and we have introduced a symbolic notation in order to shorten the expression of $\xi^{(2)}$ (which is quite long to write): one should understand that this expression should be developp using the rule $\left(K_1 \frac{\vec{\partial}}{\partial z_1^I} \otimes K_2 \frac{\vec{\partial}}{\partial z_2^J} \right) A(x_1, x_2) = (K_1 K_2 A(x_1, x_2)) \frac{\vec{\partial}}{\partial z_1^I} \otimes \frac{\vec{\partial}}{\partial z_2^J}$, for all linear differential operator K_1 (resp. K_2) acting on the variables x_1^μ (resp. x_2^μ) (for instance K_1 can be $p_1^\mu \frac{\partial}{\partial x_1^\mu} + \phi_1 \Delta_1$, $\frac{\partial}{\partial x_1^\lambda}$ or 1).

Then one compute that

$$\{\mathcal{H}^{\otimes 2}, F^{(2)}\} = d\mathcal{H}_{z_1} \otimes d\mathcal{H}_{z_2}\left(\xi^{(2)}\right) = \phi_1\phi_2\left(\Delta_1 + m^2\right)\left(\Delta_2 + m^2\right)\Phi^{(2)}(x_1, x_2)$$
$$+ \lambda\left(\phi_1^2\phi_2\left(\Delta_2 + m^2\right) + \phi_1\phi_2^2\left(\Delta_1 + m^2\right)\right)$$
$$\Phi^{(2)}(x_1, x_2) + \lambda^2\phi_1^2\phi_2^2\Phi^{(2)}(x_1, x_2).$$

Thus in order to achieve (16) it suffices to choose $\Phi^{(2)}$ such that

$$\left(\Delta_1 + m^2\right)\left(\Delta_2 + m^2\right)\Phi^{(2)}(x_1, x_2) = -\Phi^{(1)}(x_1)\delta(x_1 - x_2).$$

A solution to this equation is *formally*

$$\Phi^{(2)}(x_1, x_2) = -\int_{\mathbb{R}^n}\Phi^{(1)}(t)G(t, x_1)G(t, x_2)dt,$$

where G is the Green function on $\mathbb{R}^n$ of the operator $\Delta + m^2$.

5.4. Perturbation series. We have produced two observables, $\int_{\partial D} F^{(1)}$ and $\int_{\partial D} F^{(1)} + \lambda\left(\int_{\partial D}\right)^2 F^{(2)}$, which are approximately vanishing, up to order λ (resp. λ^2). This looks clearly as the beginning of an infinite expansion series defining the functional

$$(17) \qquad \Gamma \longmapsto \sum_{k=1}^{\infty}\lambda^{k-1}\left(\int_{\Gamma\cap\partial D}\right)^k F^{(k)}.$$

Here we let $F^{(k)} := \left[\prod_{j=1}^k\left(p_j^\mu - \eta^{\mu\lambda}\phi_j\frac{\partial}{\partial x_j^\lambda}\right)\right]\Phi^{(k)}(x_1, \cdots, x_k) \in \Gamma(\mathcal{M}, \Lambda^{n-1}T^*\mathcal{M})^{\otimes k}$, where $\Phi^{(k)}$ is a function on $(\mathbb{R}^n)^k$. We need to choose functions $\Phi^{(k)}$ in such a way that

$$\sum_{k=1}^{\infty}\lambda^{k-1}\left(\int_{\Gamma\cap\partial D}\right)^k \{\mathcal{H}^{\otimes k}, F^{(k)}\}\omega^{\otimes k} = 0,$$

where $\{\mathcal{H}^{\otimes k}, F^{(k)}\} := (-1)^k d^{\otimes k}\mathcal{H}^{\otimes k}(\xi^{(k)}) = \left[\prod_{j=1}^k\left(\phi_j(\Delta_j + m^2) + \lambda\phi_j^2\right)\right]\Phi^{(k)}(x_1, \cdots, x_k)$. Of course the computation of the further terms should more and more complicated but is possible in principe and should be described by graphs analog to Feynman's diagramm. Note that these graphs should all be trees (i.e. without loops), since they describe classical observable functionals.

Then (17) provides us with a vanishing functional, if λ is sufficiently small. This was not exactly our original motivation, which was to find non vanishing dynamical functionals. These can be obtained as follows. Assume for instance that $\eta^{\mu\nu}$ is Minkowskian, i.e. Δ is hyperbolic, and $\partial D = \widetilde{\Sigma} - \Sigma$, where Σ is a fixed space-like hypersurface (say $\{x^0 = t_0\}$ for some fixed t_0) and $\widetilde{\Sigma}$ is a parallel space-like hypersurface (say $\{x^0 = t\}$, where $t \neq t_0$). Then we prescribe all functions $\Phi^{(k)}$, for $k \geq 2$, in such a way that they vanish and their first time derivative vanish along Σ. This implies that

$$\sum_{k=1}^{\infty}\lambda^{k-1}\left(\int_{\Gamma\cap\partial D}\right)^k F^{(k)} = \sum_{k=1}^{\infty}\lambda^{k-1}\left(\int_{\Gamma\cap\widetilde{\Sigma}}\right)^k F^{(k)} - \int_{\Gamma\cap\Sigma} F^{(1)}.$$

Hence we get the coincidence of the two functionals $\sum_{k=1}^{\infty}\lambda^{k-1}\left(\int_{\widetilde{\Sigma}}\right)^k F^{(k)}$ and $\int_\Sigma F^{(1)}$ on $\mathcal{E}$. Then it remains of course to define the bracket

$\left\{ \sum_{k=1}^{\infty} \lambda^{k-1} \left(\int_{\widetilde{\Sigma}} \right)^k F^{(k)}, \int_{\widetilde{\Sigma}} G \right\}$, in order to set

$$\left\{ \int_{\Sigma} F^{(1)}, \int_{\widetilde{\Sigma}} G \right\} := \left\{ \sum_{k=1}^{\infty} \lambda^{k-1} \left(\int_{\widetilde{\Sigma}} \right)^k F^{(k)}, \int_{\widetilde{\Sigma}} G \right\}.$$

In principle there should be no difficulty in defining the above Poisson bracket, either by coming back to the Poisson bracket obtained by the standard canonical theory of physicists or by using for instance the theory expounded in [9]. Note that in order to make connection with quantum field theory, which provides us quantum scattering amplitudes, it is more appropriate to choose the slice Σ at infinity, i.e. such that $t_0 = -\infty$ or $t_0 = \infty$.

Lastly we also remark that replacing the set of functionals $\int_{\Sigma} F$ by the set of functionals of the type (17) has another advantage: it is then possible to define a product law between such functionals, by the rule

$$\left(\sum_{k=1}^{\infty} \left(\int_{\Sigma} \right)^k F^{(k)} \right) \left(\sum_{k=1}^{\infty} \left(\int_{\Sigma} \right)^k G^{(k)} \right) = \sum_{k=1}^{\infty} \sum_{l=0}^{k} \left(\int_{\Sigma} \right)^k F^{(l)} \otimes G^{(k-l)}.$$

6. Conclusion

We have tried here to introduce the Reader to multisymplectic formalisms, mainly through the de Donder–Weyl theory, we have explained quickly a more general framework based on Lepage–Dedecker developed in [20] and [21] (that is, we believe, supported by a more relativistic point of view) and we have explained how a perturbative theory analog to the theory of Feynman and Schwinger could be be built for classical solutions. This theory need of course to be developed and it should be interesting to understand whether it could lead to the perturbative quantum field theory by a direct quantization.

We have not discussed many other important questions like the construction of a non perturbative quantum field theory in this framework (interesting results have been obtained by I. Kanatchikov, see [24]), or how such theories could help in understanding hyperbolicity of variational nonlinear hyperbolic partial differential equations as done in D. Christodoulou's book [6]. We should also mention the existence of other covariant theories as for example F. Takens' one [35] (see also [38] or [9]).

Acknowledgements — I thank Joseph Kouneiher for useful discussions and comments on this text.

References

[1] E. Binz, J. Śniatycki and H. Fisher, *Geometry of Classical Fields*, (North-Holland, Amsterdam, 1989)

[2] H. Boerner, *Carathéodory's Eingang zur Variationsrechnung*, Jber. dt. Math.-Vereinig. 56 (1953), 31–58; *Darstellung von Gruppen*, Springer, Berlin–Göttingen–Heidelberg (1955).

[3] M. Born, *On the quantum theory of the electromagnetic field*, Proc. Roy. Soc. London A143 (1934), 410–437.

[4] C. Carathéodory, *Über die Extremalen und geodätischen Felder in der Variationsrechnung der mehrfachen Integrale*, Acta Sci. Math. (Szeged) 4 (1929) 193-216.

[5] E. Cartan, *Les espaces métriques fondés sur la notion d'aire*, 1933.

[6] D. Christodoulou, *The action principle and partial differential equations*, Annals of Mathematics Studies, 146, Princeton University Press 2000.

[7] P. Dedecker, *Calcul des variations, formes différentielles et champs géodésiques*, in *Géométrie différentielle*, Colloq. Intern. du CNRS LII, Strasbourg 1953, Publ. du CNRS, Paris, 1953, p. 17-34; *On the generalization of symplectic geometry to multiple integrals in the calculus of variations*, in *Differential Geometrical Methods in Mathematical Physics*, eds. K. Bleuler and A. Reetz, Lect. Notes Maths. vol. 570, Springer-Verlag, Berlin, 1977, p. 395-456.

[8] T. de Donder, *Théorie invariante du calcul des variations*, Nuov. éd. (Gauthiers–Villars, Paris 1935).

[9] P. Deligne, D. Freed, *Classical field theory*, in *Quantum fields and strings: a course for mathematicians, Volume 1*, P. Deligne, P. Etingof, D.S. Freed, L.C. Jeffrey, D. Kazhdan, J.W. Morgan, D.R. Morrison and E. Witten, editors, American Mathematical Society, 1999.

[10] A. Echeverria-Enriquez, M. Muñoz-Lecanda, N. Roman-Roy, *Multivector field formulation of Hamiltonian field theories*, J.Phys.A, v.32 (1999) 8461, math-ph/9907007

[11] P.L. García, A. Pérez-Rendón, *Symplectic approach to the theory of quantized fields, II*, Archive Rat. Mech. Anal. 43 (1971), 101–124.

[12] K. Gawędski, *On the generalization of the canonical formalism in the classical field theory*, Rep. Math. Phys. 3 (1972) 307-326.

[13] M. Giaquinta, S.Hildebrandt, *Calculus of variations*, Vol. 1 and 2, Springer, Berlin 1995 and 1996.

[14] H. Goldschmidt, S. Sternberg, *The Hamilton–Cartan formalism in the calculus of variations*, Ann. Inst. Fourier 23, p. 203–267 (1973).

[15] M.J. Gotay, *An exterior differential systems approach to the Cartan form*, in *Symplectic Geometry and Mathematical Physics*, eds. P. Donato, C. Duval, e.a. (Birkhäuser, Boston, 1991) p. 160-188

[16] M.J. Gotay, *A multisymplectic framework for classical field theory and the calculus of variations I. Covariant Hamiltonian formalism*, in *Mechanics. Analysis and Geometry: 200 Years after Lagrange*, ed. M. Francaviglia (North Holland, Amsterdam, 1991) p. 203-235

[17] M.J. Gotay, *A multisymplectic framework for classical field theory and the calculus of variations II. Space + time decomposition*, Diff. Geom. and its Appl. 1 (1991) 375-390

[18] M.J. Gotay, J. Isenberg, J. Marsden (with the collaboration of R. Montgomery, J.Śniatycki, P.B. Yasskin), *Momentum Maps and Classical Relativistic Fields. Part I: covariant field theory*, arXiv:physics/9801019.

[19] F. Hélein, *Inégalité isopérimétrique et calibration*, Ann. Inst. Fourier 44, fasc. 4 (1994), 1211–1218; *Isoperimetric inequalities and calibrations*, in *Progress in partial differential equations: the Metz surveys 4*, M. Chipot and I. Shafrif ed., Pitman Research Notes in Mathematics Series 345, Longman 1996.

[20] F. Hélein, J. Kouneiher, *Finite dimensional Hamiltonian formalism for gauge and quantum field theory*, J. Math. Physics, vol. 43, No. 5 (2002).

[21] F. Hélein, J. Kouneiher, *Lepage–Dedecker general multisymplectic formalisms*, preprint 2002, CMLA, ENS de Cachan.

[22] I. V. Kanatchikov *Canonical structure of classical field theory in the polymomentum phase space*, hep-th/9709229

[23] I. V. Kanatchikov, *On the canonical structure of the De Donder-Weyl covariant Hamiltonian formulation of field theory I. Graded Poisson brakets and the equation of motion*, hep-th/9312162

[24] I. V. Kanatchikov, *Covariant geometric prequantization of fields*, (arXiv:gr-qc/0012038)Proc. 19th Marcel Grossmann Meeting, Rome (Italy), July 2000, World Scientific, Singapore 2000; *Precanonical quantization and the Schrödinger wave functional*, Phys. Letters A 283 (2001), 25–36.

[25] H. Kastrup, *Canonical theories of Lagrangian dynamical systems in physics*, Phys. Rep. 101 (1983) 1-167

[26] J. Kijowski, *A finite dimensional canonical formalism in the classical field theory*, Comm. Math. Phys. 30 (1973), 99-128.

[27] J. Kijowski, *Multiphase spaces and gauge in the calculus of variations*, Bull. de l'Acad. Polon. des Sci., Série sci. Math., Astr. et Phys. XXII (1974), 1219-1225.

[28] J. Kijowski, W. Szczyrba, *A canonical structure for classical field theories*, Comm. Math. Phys. 46(1976), 183-206.

[29] J. Kijowski, W.M. Tulczyjew, *A symplectic framework for field theories*, Springer-Verlag, Berlin, 1979.

[30] T.H.J. Lepage, *Sur les champs géodésiques du calcul des variations*, Bull. Acad. Roy. Belg. Cl. Sci. V. Sér. 22 (1936), 716–729; 1036–1046.

[31] D. H. Martin, *Canonical variables and geodesic fields for the calculus of variations of multiple integrals in parametric form*, Math. Z. 104 (1968), 16-27.

[32] H. Rund, *The Hamilton-Jacobi Theory in the Calculus of Variations*, D. van Nostrand Co. Ltd., Toronto, etc. 1966 (Revised and augmented reprint, Krieger Publ., New York, 1973).

[33] G. Sardanashvily, *Generalized Hamiltonian Formalism for Field Theory*, ed. World Scientific, Singapore, 1995.

[34] J. Śnyatycki, *On the geometric structure of classical field theory in Lagrangian formulation*, Proc. Camb. Phil. Soc. 68 (1970), 475–483.

[35] F. Takens, *A global version of the inverse problem of the calculus of variations*, J. Diff. Geom. 14 (1979), 543–569.

[36] W.M. Tulczyjew, Warsaw seminar in *Geometry of phase space*, unpublished, 1968.

[37] H. Weyl, *Geodesic fields in the calculus of variations*, Ann. Math. (2) 36 (1935) 607-629.

[38] G.J. Zuckerman, *Action principles and global geometry*, Mathematical aspects of string theory, ed. S.T. Yau, World Scientific Publishing, 1987, 259–284.

http://www.cmla.ens-cachan.fr/Utilisateurs/helein/

CMLA, ENS DE CACHAN AND INSTITUT UNIVERSITAIRE DE FRANCE, 64 AVENUE DU PRÉSIDENT WILSON, 94235 CACHAN CEDEX, FRANCE

E-mail address: `helein@cmla.ens-cachan.fr`

Contemporary Mathematics
Volume **350**, 2004

Revisit the Topology of Sobolev Maps

Fang Hua Lin

This paper is dedicated to F. Browder and H. Brezis

ABSTRACT. This paper summarizes some main results obtained in joint works [HL 1,2,3] concerning topological and analytical properties of Sobolev maps between manifolds. In particular, the questions of density of smooth maps in the spaces of Sobolev maps in both strong and weak sequential topologies are discussed.

This write up covers the author's lecture at the International Conference on Nonlinear P.D.E. held in October 2001 at Rutgers University in honor of H. Brezis and F. Browder. The author would also like to take this opportunity to describe some very recent joint work with F. B. Hang [HL 1, 2, 3].

Background

Given two smooth, compact connected Riemann manifolds M, N, and let $n = \dim M$, $1 \leq p < \infty$, and N be isometrically embedded in $\mathbb{R}^\ell$. For discussions throughout this article, any smooth (even Lipschitz) embeddings of N into $\mathbb{R}^\ell$ will serve the same purpose. We let $W^{1,p}(M, N) = \{ u \in W^{1,p}(M, \mathbb{R}^\ell) : u(x) \in N$ for a.e. $x \in M \}$

It is classically well-known that for any $f \in C(M, N)$, one can define not only the homotopy class, $[f]$, associated with f but also $f_* : \pi_k(M) \to \pi_k(N)$ a conjugate class of homomorphisms (for every $k \geq 1$). If f is smooth, then one can further introduce homomorphisms $f_\# : H_k(M) \to H_k(N)$ and homomorphisms $f^\# : H^k(N, R) \to H^k(M, R)$. A natural question would be what can one say for a map $f \in W^{1,p}(M, N)$? It was shown by Schoen-Yau [SY], that if $f \in W^{1,p}(M, N)$, $n = \dim M$, then $f_* : \pi_{n-1}(M) \to \pi_{n-1}(N)$ is a well-defined conjugate class of homomorphisms. A related result of Burstall [Bu], says that: any map in $W^{1,2}(M, N)$ defines a conjugate class of homomorphisms of $\pi_1(M) \to \pi_1(N)$. In an unpublished note in 1984, Schoen-Uhlenbeck proved that any map $f \in W^{1,p}(M, N)$, one has a well-defined $f^\# : H^k(N, \mathbb{R}) \to H^k(M, R)$, for $o \leq k \leq [p-1]$. It would be interesting to examine whether $f_\# : H_k(M) \to H_k(N)$ makes sense. An obvious

1991 *Mathematics Subject Classification.* Primary 58D15; Secondary 46E35.
Key words and phrases. Sobolev Maps, k-homotopy class, weak and strong density.
The research is partially supported by DMS 0201443 .

requirement would be $[p] \geq k$ and $n - k \leq [p]$. Hence p has to be not less than $n/2$. The author, however, is not aware of any result concerning $f_{\#}$ on integral homology classes when $f \in W^{1,p}(M, N)$.

A systematic study was made by B. White for topological informations carried by a Sobolev map [Wh 1, 2]. He showed, in particular, each map $u \in W^{1,p}(M, N)$ has a well-defined $k = [p] - 1$ homotopy type and it is preserved under the sequentially weak convergence of maps in $W^{1,p}(M, N)$. He also proved:

(i) Let $H_w^{1,p}(M, N) = \left\{ u \in W^{1,p}(M, N) : \text{there is a sequence } u_i \in C^\infty(M, N) \right.$
$\left. \text{such that } u_i \text{ converses weakly to } u \text{ in } W^{1,p}(M, N) \right\}$, and let

$$k = \begin{cases} [p] & \text{if } p \text{ is not an integer} \\ [p-1] & \text{if } p \text{ is an integer} \end{cases}, \text{ then each } u \in H_w^{1,p}(M, N) \text{ has a}$$

well-defined k homotopy type which is preserved under sequentially weak convergence.

(ii) Let $H_s^{1,p}(M, N) = $ strong closure of $C^\infty(M, N)$ in $W^{1,p}(M, N)$, then every $u \in H_s^{1,p}(M, N)$ has a well-defined $[p]$ homotopy type that is preserved under the strong convergence of maps in $W^{1,p}(M, N)$.

The spaces $H_s^{1,p}(M, N)$ and $H_w^{1,p}(M, N)$ are rather important from the analysis point of view. It is easy to see $H_s^{1,p}(M, N) = W^{1,p}(M, N)$ for $p > n$ via Sobolev embedding theorem. It can be shown (see [SU]) that $H_s^{1,n}(M, N) = W^{1,n}(M, N)$ $n = \dim M$ (a generalization to VMO—maps was carried out by Brezis-Nirenberg in [BN]). On the other hand, it is also easy to see $H_s^{1,2}(B^3, \mathbb{S}^2) \neq W^{1,2}(B^3, \mathbb{S}^2)$. The local topological obstruction in this case is that $\pi_2(\mathbb{S}^2) \neq 0$. Bethuel-Zheng [BZ] generalized this situation by showing that $H_s^{1,p}(M, N) \neq W^{1,p}(M, N)$ whenever $\pi_{[p]}(N) \neq 0$ and $1 \leq p < n$. An elegant result proved by Bethuel [B] is the following characterization:

$$u \in H_s^{1,2}(B^3, \mathbb{S}^2) \iff d\left(u^{\#} w_{\mathbb{S}^2}\right) = 0.$$

Here $w_{\mathbb{S}^2}$ is the area form of $\mathbb{S}^2$. A related fact is the following statement due to Bethuel-Brezis-Coron [BBC], $H_w^{1,2}(B^3, \mathbb{S}^2) = W^{1,2}(B^3, \mathbb{S}^2)$¿

Classically, it is also interesting to study the topology of $C(M, N)$. The topology of the space $W^{1,p}(M, N)$ would also be important. It will provide some insights in variational problems for $W^{1,p}(M, N)$ maps. Recently, Brezis-Li [BL] initiated the study whether $W^{1,p}(M, N)$ is connected or not. Among many results they proved are the following:

(a) Assume $1 \leq p < n$, and N is $[p-1]$-connected, then $W^{1,p}(M, N)$ is path connected, i.e., for any $u, v \in W^{1,p}(M, N)$, there is a continuous path $\gamma : [0,1] \to W^{1,p}(M, N)$ such that $\gamma(0) = u$, $\gamma(1) = v$. We shall denote such u, v by $u \underset{p}{\sim} v$.

(b) If $1 \leq p < n$, then $W^{1,p}(\mathbb{S}^n, N)$ is path connected.

(c) If $2 \leq p < n$, then the map

$$i_p : \mathrm{Lip}\left(M, \mathbb{S}^1\right) \sim_{M, \,\mathrm{lip}} \to W^{1,p}\left(M, \mathbb{S}^1\right)/\sim_p$$

is a bijection. (Left-side denotes the Lipschitz homotopies.)

(d) For any $u \in W^{1,p}(M, \mathbb{S}^1)$, there is a $v \in C^\infty(M, \mathbb{S}^1)$ such that $u \underset{p}{\sim} v$.

They also conjectured that:

(A) For any $u \in W^{1,p}(M, N)$, there is a $v \in C^\infty(M, N)$ such that $u \underset{p}{\sim} v$.

(B) For any $k \in N$, $k \leq q < p < k+1$, $i_{p,q}$ is a bijection. Here, for $1 \leq q < p < \infty$, $i_{p,q} : W^{1,p}(M,N)/ \sim_p \to W^{1,q}(M,N)/ \sim_q$ defined in the natural way.

Main Results

Our first result characterizes when two maps in $W^{1,p}(M,N)$ can be connected by a continuous path in $W^{1,p}(M,N)$.

THEOREM 1 (HL2). *If $1 \leq p < n$, $u,v \in W^{1,p}(M,N)$ then $u \underset{p}{\sim} v$ if and only if u and v are $[p-1]$ homotopic.*

We note that $u \underset{p}{\sim} v$ is an analytical statement, while u and v are $[p-1]$-homotopic is a topological condition. Since the $[p-1]$ homotopy classes are preserved under the weak convergence in $W^{1,p}(M,N)$, we conclude that each connect component of $W^{1,p}(M,N)$ are both open and close in the weak topology of $W^{1,p}(M,N)$ (see also [HL2] and [Wh1]). Therefore, if one studies variational problems with energy functionals growth like $|\nabla u|^p$ for maps from M into N, then one deduces that each connected component of $W^{1,p}(M,N)$ contains an energy minimizer in this component. This obviously leads to multiplicity results for solutions of such variational problems. One can also use some arguments in [L] as well as [HL2] to show partial regularity theorems for all such minimizers. They all have singular sets of Hausdorff dimension $\leq n - [p] - 1$ where p is not an integer. If p is an integer, then it is only known the singular sets are at most $n - p$ dimensional (see [L] for some better results). However, those maps minimize energy in a component contain some maps in $C^\infty(M,N)$ should have "better" singularities than those minimizers in components disjointing from $C^\infty(M,N)$ in the sense that the latter maps are not in $H_w^{1,p}(M,N)$. A further study on these regularity issues would be desirable.

We now draw some consequences from Theorem 1. First of all, the conjecture of Brezis-Li, part B, is true, that is $i_{p,q}$ is a bijection from $W^{1,q}(M,N)/ \sim_q \to W^{1,p}(M,N)/ \sim_p$ whenever $1 \leq k \leq p < q < k+1 \leq n$. Secondly, we note that $W^{1,p}(M,N)$ is connected whenever $\pi_i(M) = 0$ for $0 \leq i \leq k$ and $\pi_i(N) = 0$ for $k+1 \leq i \leq [p]-1$. Here $0 \leq k \leq [p-1]$. We note that statement (a) above follows from the last conclusion with $k = 0$, (b) follows when $k = [p-1]$. Third, we have

$$\text{Lip}\,(M,N)/ \sim_{M,\,\text{lip}} \to W^{1,p}(M,N)/ \sim_p .$$

is a bijection whenever $1 \leq p < n$, $\pi_i(N) = 0$ for $[p] \leq i \leq n$. This implies (c) above.

We now turn to the question (Conjecture (A) of Brezis-Li) whether a given map $u \in W^{1,p}(M,N)$ can be connected to a $v \in C^\infty(M,N)$ by a continuous path in $W^{1,p}(M,N)$, we have

PROPOSITION 1. *Let $1 \leq p < n$, $u \in W^{1,p}(M,N)$, and let $h : K \to M$ be an arbitrary Lipschitz triangulation. Then u can be connected to a smooth map if and only if $u_{\#p}(h)$ (i.e., the $[p-1]$-homotopy class introduced by u on a $[p-1]$-dimensional skeletons of the triangulation h) is extendible to M with respect to N.*

To explain more precisely the meaning of the above Proposition, we introduce the notion of k-extension property which was well studied in the obstruction theory of classical homotopy theory. Given a topological space X with some CW-complex structures, and let Y be a topological space. For $k \in \mathbb{N}$, if for any CW complex structure $(X^j)_{j \in \mathbb{N}}$ any $f \in C(X^{k+1}, Y)$ the restriction $f|_{X^k}$ of f on X^k has a continuous extension to X then we say X satisfies the k-extension property with respect to Y. As a consequence of Proposition 1 is the following:

(i) Every map in $W^{1,p}(M, N)$ can be connected to a smooth map if and only if M satisfies $[p-1]$-extension property with respect to N.

(ii) If either $[p] = 1$ or $[p] \geq 2$ and $\pi_i(N) = 0$ for $[p] \leq i \leq n-1$, then every map in $W^{1,p}(M, N)$ may be connected to a smooth map.

Note that (ii) implies the statement (d) due to Brezis-Li. On the other hand, by cohomology theory, we have $\mathbb{CP}^3$ does not satisfy 2-extension property $w.r.t.$ $\mathbb{CP}^2$ and $\mathbb{CP}^2$ does not satisfy 2-extension property $w.r.t.$ $\mathbb{CP}1$. Similarly, $\mathbb{RP}^3$ does not satisfy 1-extension property $w.r.t.$ $\mathbb{RP}^2$ and $\mathbb{RP}^4$ does not satisfy 1-extension property $w.r.t.$ $\mathbb{RP}^3$.

We thus obtain counter-examples to Brezis-Li conjecture (A) above. In particular, there are maps in spaces $W^{1,3}\left(\mathbb{CP}^2, \mathbb{CP}^1\right)$, $W^{1,3}\left(\mathbb{CP}^3, \mathbb{CP}^2\right)$, $W^{1,2}\left(\mathbb{RP}^3 \, \mathbb{RP}^2\right)$, $W^{1,2}\left(\mathbb{RP}^4, \mathbb{RP}^3\right)$ cannot be connected to smooth maps in these spaces, respectively.

We next turn to the question of whether smooth maps are strongly dense in $W^{1,p}(M, N)$. The issue was thought settled in the work [B2]. Unfortunately the proofs of [B2] contain serious flaws. In [HL2] we proved the following:

THEOREM 2. *Let $1 \leq p < n$, then $H_s^{1,p}(M, N) = W^{1,p}(M, N)$ if and only if $\pi_{[p]}(N) = 0$ and M satisfies the $[p-1]$ extension proerty with respect to N.*

The proof of Theorem 2 is rather involved. It also needs a delicate deformation construction. The earlier deformations constructed via Federer-Fleming [FF] and others [Wh1], [Ha] do not seem to work. We note that $\pi_3(\mathbb{CP}^2) = 0$, however, $H_w^{1,3}(\mathbb{CP}^3, \mathbb{CP}^2) \neq W^{1,3}(\mathbb{CP}^3, \mathbb{CP}^2)$ (see discussion below). This says, in particular, that $H_s^{1,3}(\mathbb{CP}^3, \mathbb{CP}^2) \neq W^{1,3}(\mathbb{CP}^3, \mathbb{CP}^2)$. Similarly, one has $\pi_2(\mathbb{RP}^3) = 0$, but $H_w^{1,2}(\mathbb{RP}^4, \mathbb{RP}^3) \neq W^{1,2}(\mathbb{RP}^4, \mathbb{RP}^3)$ (see discussion below). Hence the main result of [B2] cannot be valid.

On the other hand, assume $1 \leq p, n$, $k \in \mathbb{N}$ such that $0 \leq k \leq [p-1]$, and that $\pi_i(M) = 0$ for $1 \leq i \leq k$ while $\pi_i(N) = 0$ for $k + 1 \leq i \leq [p]$, then $H_s^{1,p}(M, N) = W^{1,p}(M, N)$. This is an easy consequence of Theorem 2. This last statement generalizes the earlier result of P. Hajlasz, [Ha]. We also note that, if $\pi_i(N) = 0$ for $1 \leq [p] \leq i \leq n-1$, then $H_s^{1,p}(M, N) = W^{1,p}(M, N)$. This is because the $[p-1]$-extension property trivially valid in this case.

Finally, we consider the delicate question regarding the sequential weak density of smooth maps in $W^{1,p}(M, N)$. We have the following partial answer:

THEOREM 3. *Let $1 \leq p < n$, $u \in W^{1,p}(M, N)$. Then a necessary condition for $u \in H_w^{1,p}(M, N)$ is the following. For any Lipschitz triangulation $h : K \to M$, the $[p-1]$-homotopy class $u_{\#, p}(h)$ defined by u is extendible to M with respect to N. In particular, such u can be connected to a smooth map by a continuous path in $W^{1,p}(M, N)$.*

We note, in particular, that $H_w^{1,2}(\mathbb{RP}^4, \mathbb{RP}^3) \neq W^{1,2}(\mathbb{RP}^4, \mathbb{RP}^3)$ and that $W^{1,3}(\mathbb{CP}^3, \mathbb{CP}^2) \neq H_w^{1,3}(\mathbb{CP}^3, \mathbb{CP}^2)$. In general, we see that for $1 \leq p < n$ a necessary condition for $W^{1,p}(M, N) = H_w^{1,p}(M, N)$ is that M satisfies $[p-1]$ extension property with respect to N. Natually one would like to conjecture that the above necessary topological condition is also sufficient. However, it is still unknown whether the conjecture is true or not. There are two special cases that are worth pointing out. The first is due to Hajlasz [Ha] (see also [PR] for another proof and some generalizations) which says if N is $[p-1]$ connected, then $H_w^{1,p}(M,N) \equiv W^{1,p}(M,N)$. The second is discovered recently by Pakzad and Riviere which says if $p = 2$ and $\pi_1(M) = 0$, then $H_w^{1,2}(M,N) = W^{1,2}(M,N)$.

Further Remarks and Conclusions

One may also consider the following question that whether maps in $W^{1,p}(M,N)$ can be connected by continuous paths in $W^{1,p}(M,N)$ to maps in $W^{1,q}(M,N)$. Here $1 \leq p < q < \infty$. If $q \geq n$, the question reduces to the one we had studied, i.e., the one for $C^\infty(M,N)$ instead of $W^{1,q}(M,N)$. If $q < n$, then one can adopt the similar arguemtns as in [HL2], (see [HL3]) to show that:

Assume $1 \leq p < q < n$, $u \in W^{1,p}(M,N)$, $h : K \to M$ is a Lipschitz rectlinear cell decomposition, then u can be connected by a continuous path in $W^{1,p}(M,N)$ to a map in $W^{1,q}(M,N)$ if and only if, $u_{\#,p}(h)$ is extendible to $[K^{(q)}]$ with respect to N. Here $[K^{(q)}]$ is the $[q]$-dimensional cell complex of K. As an immediate consequence is that every map in $W^{1,p}(M,N)$ can be connected continuously in $W^{1,p}(M,N)$ to a map in $W^{1,q}(M,N)$ if and only if, M satisfies $([p-1], q)$ extension property with respect to N. Here, suppose X and Y are topological spaces, and that X posses some CW complex structure $(X^j)_{j \in \mathbb{N}}$. If $f \in C\left(X^{k+1}, Y\right)$, $f|_{X^k}$ has a continous extension to X^ℓ, then we say X satisfies (k, ℓ) extension property with respect to Y. (Note $\ell \geq k$).

It is also possible to answer when the space $W^{1,q}(M,N)$ is strongly dense in $W^{1,p}(M,N)$ for $1 \leq p < q < \infty$. We have the following (see [HL3]).

THEOREM 4. *Assume* $1 \leq p < q < \infty$, $p < n$. *If* $[p] = [q]$, *then one always has the strong closure of* $W^{1,q}(M,N)$ *in* $W^{1,p}(M,N)$ *is the whole space* $W^{1,p}(M,N)$. *If* $[p] < [q]$, *then the strong closure of* $W^{1,q}(M,N)$ *in* $W^{1,p}(M,N)$ *coincides with* $W^{1,p}(M,N)$ *if and only if* $\pi_{[p]}(N) = 0$ *and* M *satisfies* $([p-1], [q])$ *extension property with respect to* N.

Similarly, one can establish a necessary topological condition for $W^{1,q}(M,N)$ to be weakly sequentially dense in $W^{1,p}(M,N)$. Whether or not such a topological condition is also sufficient remains as an open problem. We do have the following interesting partial answer.

THEOREM 5 (See [**HL3**]). *Assume* $2 \leq p < n$, $p \in \mathbb{N}$, $h : K \to M$ *is a Lipschitz rectilinear cell decompostion. Let* $M^i = h(|K^i|)$ *for* $i \geq 0$, L^{n-p-1} *be one of the dual* $n - p - 1$ *dimensional skeletons, and* $u \in W^{1,p}(M,N)$ *be such that* $u \in C\left(M/L^{n-p-1}, N\right)$. *Then* $u \in H_w^{1,p}(M,N)$ *if and only if* $u|_{M^{p-1}}$ *has a continuous extension to* M. *Moreover, if for some homotopy class* $\alpha \in [M,N]$, *we have* $u|_{M^{p-1}} \in \alpha|_{M^{p-1}}$, *Then we may find a sequence* $u_i \in C^\infty(M,N)$, *such that* $[u_i] = \alpha$, $u_i \rightharpoonup u$ *in* $W^{1,p}(M,N)$ *and* $du_i \to du$ *a.e. on* M.

We note that, by a theorem in [HL2] (see also [B2]) that maps $u \in W^{1,p}(M,N)$ satisfying the condition in Theorem 5 form a set which is dense in $W^{1,p}(M,N)$ in the strong topology of $W^{1,p}(M,N)$. Thus, our Theorem 5 says that for a dense subset D(in the strong topology) of $W^{1,p}(M,N)$ the necessary topological condition for a map u in D to be also in $H_w^{1,p}(M,N)$ turns out to be also sufficient. It becomes a rather difficult analytical question to control uniformly the $W^{1,p}$-norm of u_i in the statement of Theorem 5 above, otherwise, we would show this topological condition is also sufficient. In [Ha] it was shown that, if N is $[p]$-connected, then $H_s^{1,p}(M,N) =$

$W^{1,p}(M,N)$, and that if N is $[p-1]$-connected, then $H_w^{1,p}(M,N) = W^{1,p}(M,N)$ (see also [PR]). Here we have the following (see [HL3]).

PROPOSITION 2. *Assume* $1 \leq p < \dim N$, *and* N *is connected. Then there is a sequence of* $u_i \in C^\infty(N,N)$ *such that* $u_i \to id_N$ *in* $W^{1,p}(N,N)$ *and* u_i *is homotopic to a constant if and only if* N *is* $[p]$-*connected. Suppose* $2 \leq p \leq n$ *is an integer, then there is a sequence* $u_i \in C^\infty(N,N)$ *such that* $u_i \rightharpoonup id_N$ *weakly in* $W^{1,p}(N,N)$ *and that* u_i *is homotopic to a constant if and only if* N *is* $[p-1]$-*connected.*

As a consequence of this proposition, we have the weak sequential density theorem of [Ha]: Assume $2 \leq p < n$, p is an integer and N is $(p-1)$-connected then $H_w^{1,p}(M,N) = W^{1,p}(M,N)$. In fact, for any $u \in W^{1,p}(M,N)$, there is a sequence $u_i \in C^\infty(M,N)$ such that $u_i \rightharpoonup u$ in $W^{1,p}(M,N)$, $\|du_i\|_{L^p(M)} \leq C(p,M,N)\|du\|_{L^p}$ and $du_i \rightharpoonup du$ a.e. on M.

Let us now introduce a quantity (which is similar to the notion of Lebesgue area) associated with weak sequential convergence.

For $u \in H_w^{1,p}(M,N)$, we define

$$I(u) = \inf \left\{ \lim_k \frac{\|du_k\|_{L^p(M)}}{\|du\|_{L^p(M)}} \quad : \quad u_k \in C^\infty(M,N), \quad u_k \rightharpoonup u \right\}.$$

Note that $u \in H_w^{1,p}(M,N)$ implies $I(u) < \infty$.

Let $\mu_k = |du_k|^p(x)\,dx \rightharpoonup \mu = |du|^p(x) + \nu$, for $u_k \in C^\infty(M,N)$, $u_k \rightharpoonup u$ in $W^{1,p}(M,N)$, then $\nu \geq 0$ is a Randon Measure. In [L], ν is callled the defect measure, and [L] generalized and improved works [BBC] and [GMS]. We are interested in the general structure of ν. For this purpose, we first let $u \in W^{1,p}(M,N)$ satisfies the condition of Theorem 5 above (i.e., $u \in R^{p\infty}(M,N)$ in the terminology of [HL2]). Theorem 5 says that such $u \in H_w^{1,p}(M,N)$ if and only if $u|_{M^{p-1}=h([K^{p-1}])}$ has a continuous extension to M. Assume this topological condition is always satisfied, then we let $u_k \in C^\infty(M,N)$ be such that $u_k \rightharpoonup u$, and

$$\int_M |du_k|^p\,dx \to \int_M |du|^p\,dx + \nu(M) = I^p(p)\int_M |du|^p.$$

Generalizing arguments in [L] one can derive

THEOREM 6. *Let* $u \in H_w^{1,p}(M,N)$ *satisfy the condition of Theorem 5. Then* $\nu = \Theta(x)\mathcal{H}^{n-p}\lfloor\Sigma$, *where* $0 < \epsilon_0(p,M,N) \leq \Theta(x) \leq C(u,M,N)$, Σ *is closed* $n-p$ *rectifiable subset of* M. *Moreover, for* $\mathcal{H}^{n-p}$ *a.e.,*

$$x, \Theta(x) = \inf\left\{\int_{\mathbb{S}^p} |\nabla v|^p : v \in W^{1,p}(\mathbb{S}^p,N) : [v] = \alpha_x\right\} \text{ for some nontrivial } \alpha_x \in \pi_p(N).$$

From the work of [DK], we see that

$$\Theta(x) = \sum_{j=1}^{\ell} \int_{\mathbb{S}^p} |d\,\phi_j|^p.$$

Here each $\phi_j : \mathbb{S}^p \to N$ *is a* $C^{1,\alpha}$, p-*harmonic map which is energy minimizing in its homotopy class of maps from* $\mathbb{S}^p$ *into* N. *Moreover* $[\phi_j]$'*s is a decomposition of* α_x.

Note, as in [L], $\Theta(x)$ takes discrete values, and thus $\nu = v(\Sigma,\Theta)$ is an integral rectifiable varifold.

In the special case $W^{1,p}(\mathbb{B}^4,\mathbb{S}^2)$, a recent work of Hardt-Riviere [HR] introduced a new notion "Scans" in place of "minimal connections" (see [BBC]). In this case, maps from $\mathbb{B}^4$ into $\mathbb{S}^2$ with isolated singularities are strongly dense in $W^{1,3}(\mathbb{B}^4, \mathbb{S}^2)$. Let u be such a map, i.e., $u \in C^\infty(\overline{B}^4 \setminus \{x_1, \ldots, x_k\}, \mathbb{S}^4)$ for some $x_1, \ldots, x_k \in \mathbb{B}^4$. Let γ be a small positive number such that the Hopf degrees $u|_{\partial B_\gamma(x_j)}, j = 1, \ldots, k$, are all well-defined (independent on small γ's). There is a sequence $u_i \in C^\infty(\overline{B}^4, \mathbb{S}^2)$ such that $u_i \rightharpoonup u$ weakly in $W^{1,3}(B^4, \mathbb{S}^2)$ and

$$\int_{\mathbb{B}^4} |du_i|^4 \, dx \to \left(\int_{\mathbb{B}^4} |du|^4 \, dx \right) I^3(u).$$

From [HR] one knows $G_{u_i} \rightharpoonup G_u + I \times \mathbb{S}^3$ as scans. On the other hand, from [L] one may deduce that $|du_i|^3 \, dx \to |du|^3 \, dx + \nu$, $\nu = \Theta \, H^1 \lfloor \Sigma$. Here $v = v(\Sigma, \Theta)$ is one-dimensional integral rectifiable varifold. If we define $\pi = \langle \vec{T}, \Theta, \Sigma \rangle$, here $\vec{T}$ is the orientation which is chosen according to whether the Hopf degree $\alpha_x \in \pi_3(\mathbb{S}^2)$ is positive or negative. More precisely, we require $\vec{T}(x) \wedge \vec{e}, \; \vec{e}_1(x) \wedge \vec{e}_2(x) \wedge \vec{e}_3(x) =$ sign of the degree of α_x. Here $\vec{e}_1(x) \wedge \vec{e}_2(x) \wedge \vec{e}_3(x)$ is a given orientation on the 3-dimensional plan orthogonal to $\vec{T}(x)$ so that the map restricted to it gives the local Hlpf-degree of α_x. π thus defined is an integral rectifiable current. Moreover, $\Theta(x) \approx |\deg \alpha_x|^{3/4}$ by the earlier work of Riviere [R].

References

[B]   F. Bethuel, A characterization of maps in $H^1(B^3, \mathbb{S}^2)$ which can be approximated by smooth maps, *Ann. Inst. H. Poincare, Anal. Non Linaire* **7** (1990), 269–286.

[B2]   —————— , the approximation problem for Sobolev map s between two manifolds, *Acta Math.* **167** (1991), 153–206.

[BBC]   F. Bethuel, H. Brezis, and J. M. Coron, Relaxed energies for harmonic maps, *Variational Methods*, Birkhäuser, Boston, 1990, pp. 37–52.

[BL]   H. Brezis and Y. Y. Li, Topology and Sobolev spaces, *J. Functional Analysis*, **183** (2001), 321–364.

[BN]   H. Brezis and L. Nirenberg, Degree theory and BMO, I. *Sel. Math., New Series*, **1** (1995), 197–263.

[Bu]   F. E. Burstall, Harmonic maps of finite energy from noncompact manifolds, *J. London Math. Soc.* **30** (1984), 361–370.

[HL]   F. B. Hang and F. H. Lin, Topology of Sobolev Mappings, *Math. Res. Lett.*, **8** (2001), 321–330.

[HL2]   —————— , Topology of Sobolev Mappings, II, to appear in *Acta Math.*

[HL3]   —————— , Topology of Sobolev Mappings, III. Preprint (2002).

[Ha]   P. Hajlasz, Approximation of Sobolev Mappings, *Nonlinear Analysis, Theory, Methods & Applications*, **22** (1994), 1579–1591.

[GMS]   M. Giaquinta, G. Modica and J. Souček, *Cartesian currents in the Calculus of Variations*, I, Cartesian Currents, Springer-Verlag , Berlin, 1998.

[L]   F. H. Lin, Mapping Problems, fundamental groups and defect measures, *Acta Math. Sin. (Eng. ser.)*, **15** (1999), 25–52.

[PR]   M. R. Pakzad and T. Riviere, Weak density of smooth maps for the Dirichlet energy between manifolds. Preprint (2001).

[R]   T. Riviere, Minimizing fibrations and p-harmonic maps in homotopy classes from $\mathbb{S}^3$ into $\mathbb{S}^2$, *Comm. in Anal. & Geom.* **6** (1998), 427–483.

[SU]   R. Schoen and K. Uhlenbeck, Approximation theorems for Sobolev mappings. Preprint (1984).

[SY]   R. Schoen and S. T. Yau, The existence of incompressible minimal surfaces and topology of three dimension al manifolds, with non-negative scalar curvature, *Ann. of Math.* **1 10** (1979), 127–142.

[Wh1]   B. White, Infima of energy functions in homotopy classes of mappings, *J. Diff. Geom.* **23** (1986), 127–142.

[Wh2] _____________ , Homotopy classes in Sobolev spaces and the existence of energy minimizing maps, *Acta Math.* **160** (1988), 1–17.

NYU-COURANT, 251 MERCER STREET, NEW YORK, NY 10012
E-mail address: linf@math1.nyu.edu

Contemporary Mathematics
Volume **350**, 2004

Review on Blow Up and Asymptotic Dynamics for Critical and Subcritical gKdV Equations

Yvan Martel and Frank Merle

In honor of F. Browder and H. Brezis

ABSTRACT. We review recent results concerning qualitative properties (blow up and asymptotic behavior) of solutions of the critical and subcritical generalized KdV equations.

We first focus on the blow up phenomenon for the critical generalized KdV equation in the energy space for solutions close in L^2 to the soliton. This was an open problem for many years. A series of results give a precise description and understanding of the blow up phenomenon in this framework: blow up result in finite or infinite time of solutions with negative energy, description of an universal blow up profile, existence of solutions blowing up in finite time with information on the blow up rate. We provide a sketch of the proofs of these results; in particular, we recall a key rigidity theorem of the KdV flow around the solitons.

Second, we present some results concerning the subcritical KdV equations (in this case there is no blow up solutions). On the one hand, a result of asymptotic completeness of the family of solitons is obtained as a direct application of the techniques developed for the critical case. On the other hand, for $N \geq 2$, by energetic arguments and monotonicity properties of local L^2 mass, we prove stability and asymptotic stability of the sum of N solitons in the energy space.

1. Introduction

We consider the generalized Korteweg–de Vries equations, for $p \geq 2$ integer:

$$(1) \qquad \begin{cases} u_t + (u_{xx} + u^p)_x = 0, & (t, x) \in \mathbf{R}^+ \times \mathbf{R}, \\ u(0, x) = u_0(x), & x \in \mathbf{R}. \end{cases}$$

This model was originally introduced by Korteweg and de Vries [10] for $p = 2$ in the study of water waves. Other applications to Physics have been found for $p = 2$ and 3, see for example Lamb [11]. Cases $p = 2$ and $p = 3$, which correspond respectively to the KdV equation and modified KdV equation, have been studied

---

1991 *Mathematics Subject Classification.* Primary 35Q53; Secondary 35B05, 35Q51.
Key words and phrases. Blow up, KdV, asymptotics.

extensively for being completely integrable (see for example Lax [12] and Miura [24]). In this paper, we first focus on the so-called **critical case** $p = 5$:

$$(2) \qquad \begin{cases} u_t + (u_{xx} + u^5)_x = 0, & (t, x) \in \mathbf{R}^+ \times \mathbf{R}, \\ u(0, x) = u_0(x), & x \in \mathbf{R}, \end{cases}$$

for $u_0 \in H^1(\mathbf{R})$. The techniques are also applicable to the subcritical case $1 < p < 5$.

Let us point out that few results on blow up are known in the context of partial differential equations with Hamiltonian structure. For the semilinear wave equation, or more generally for hyperbolic systems, the finite speed of propagation allows one to build blowing up solutions by reducing the problem to an ordinary differential equation. For the nonlinear Schrödinger equation, the formation of singularity is related to the existence of a conformal invariance of the equation in the critical case.

Let us recall some standard facts on the gKdV equations. From the Hamiltonian structure, there are two conservation laws (for any $p > 1$)

$$(3) \qquad \int u^2(t) = \int u_0^2 \quad \text{(mass conservation)},$$

(4)
$$\frac{1}{2} \int u_x^2(t) - \frac{1}{p+1} \int u^{p+1}(t) = \frac{1}{2} \int u_{0x}^2 - \frac{1}{p+1} \int u_0^{p+1} \quad \text{(energy conservation)}.$$

Recall that for any $p > 1$ integer, (1) is well-posed in the energy space $H^1(\mathbf{R})$. Indeed, in [9], Kenig, Ponce and Vega prove the following existence and uniqueness result in $H^1(\mathbf{R})$: for $u_0 \in H^1(\mathbf{R})$, there exists $T > 0$ and a unique maximal solution $u \in C([0, T), H^1(\mathbf{R}))$ of (1) on $[0, T)$. Moreover, if $T < +\infty$ then $|u(t)|_{H^1} \to +\infty$, as $t \uparrow T$. In addition, for all $t \in [0, T)$, (3) and (4) are satisfied. Note that for equation (2), the local Cauchy problem is also well posed in $L^2(\mathbf{R})$ (see [9]). We refer to Kato [8] and Ginibre and Tsutsumi [7] for previous results on the well-posedness of the Cauchy problem for (1) and to Bourgain [3] for the periodic case.

For $1 < p < 5$ (the subcritical case), as a consequence of the Gagliardo–Nirenberg inequality, all solutions in H^1 are global (i.e. $T = +\infty$) and bounded in time. For $p \geq 5$, the situation is different and so $p = 5$ appears as a critical power for the problem of blow up.

Recall that (1) is translation invariant

$$(5) \qquad \text{if } u(t, x) \text{ is solution of (1) then } \forall x_0 \in \mathbf{R}, \ u(t, x + x_0) \text{ is solution of (1)},$$

and scaling invariant:
(6)
if $u(t, x)$ is solution of (1) then $\forall \lambda > 0$, $u_\lambda(t, x) = \lambda^{\frac{2}{p-1}} u(\lambda^3 t, \lambda x)$ is solution of (1).

Note that $|u_\lambda|_{L^2} = \lambda^{\frac{5-p}{2(p-1)}} |u|_{L^2}$ and thus $p = 5$ is also critical in an L^2 sense.

For any $p > 1$, equation (1) has special explicit solutions which are traveling waves solutions. They are of the form $u(t, x) = Q_c(x - ct)$, where

$$(7) \qquad Q_c(x) = c^{\frac{1}{p-1}} Q(\sqrt{c}\,x), \qquad Q(x) = \left(\frac{\frac{p+1}{2}}{\operatorname{ch}^2\left(\frac{p-1}{2}x\right)} \right)^{\frac{1}{p-1}}.$$

These solutions are called **solitons**.

For $p = 5$, the variational characterization of Q implies the following Gagliardo–Nirenberg inequality with best constant (see Weinstein [26]):

$$(8) \qquad \forall v \in H^1(\mathbf{R}), \quad \frac{1}{6}\int v^6 \leq \frac{1}{2}\left(\frac{\int v^2}{\int Q^2}\right)^2 \int v_x^2.$$

In particular, if $|u_0|_{L^2} < |Q|_{L^2}$ then by (3), (4) and (8), the solution $u(t)$ of (2) is global and uniformly bounded in H^1. On the contrary, for $|u_0|_{L^2} > |Q|_{L^2}$ there is no obstruction to blow up from energy type arguments and we will see that blow up is possible in H^1.

The study of the flow around the solitons is crucial to understand equation (1). Stability and asymptotic stability results on the solitons are of particular interest since they give information on the global behavior of a class of solutions of (1).

The nonlinear Schrödinger equations (NLS):

$$(9) \qquad iu_t = -\Delta u - |u|^{p-1}u, \quad \text{where } u : \mathbf{R} \times \mathbf{R}^N \to \mathbf{C},$$

have a similar structure (scaling and translation invariances, mass and energy conservation). However, for these equations, the formation of singularities is related to the existence of an additional invariance of the equation in the critical case, called the conformal invariance. Indeed, let

$$(10) \qquad iu_t = -\Delta u - |u|^{\frac{4}{N}}u, \quad (\text{for } N = 1: iu_t = -u_{xx} - |u|^4 u).$$

If $u(t,x)$ is a solution of equation (10) then

$$v(t,x) = \frac{1}{|t|^{\frac{N}{2}}}e^{\frac{i|x|^2}{4t}}\,\overline{u}\left(\frac{1}{t}, \frac{x}{t}\right)$$

is also a solution of (10) . Observe that this invariance relates regular solutions defined for all time to singular solutions. (Note that in the supercritical case, i.e. $p > 1 + \frac{4}{N}$, the Virial identity, which provides blow up solutions with negative energy, is itself a consequence of the conformal invariance in the critical case.) However, most questions on the qualitative description of blow up solutions for the Schrödinger equations are open. For more detail see for example [21].

For the generalized KdV equations, there is no such conformal invariance and the question of blow up in the energy space was completely open (apart from some numerical studies, see e.g. Bona et al. [2] and Dix and McKinney [4]).

In a series of papers concerning the critical generalized KdV equation, we introduce a new approach to study blow up phenomenon for equation (2) (see [15], [14], [22], [17], [18] and [19]). This approach was also successful in the subcritical case in [16] and [20].

The starting point is to consider solutions that are in an H^1 neighborhood of Q for all time, up to the invariances of the equation (scaling and translation). Recall that for subcritical p's, it is necessarily the case if $u(t)$ is initially close to Q since it is well known that the solitons are stable in H^1, up to translation, (see Benjamin [1] for a first result in this direction).

In the critical case, for $u_0 \in H^1(\mathbf{R})$ such that

$$(11) \qquad E(u_0) = \frac{1}{2}\int (u_0)_x^2 - \frac{1}{6}\int u_0^6 < 0, \quad \int Q^2 < \int u_0^2 < \int Q^2 + \alpha_0,$$

where α_0 is small enough, the variational characterization of Q and energetic constraints impose that $u(t)$ stays close for all time to Q in H^1 up to scaling and translation. In particular, we are able to define a continuous decomposition of the solution of the type:

$$(12) \qquad u(t,x) = \frac{1}{\sqrt{\lambda(t)}}(Q + \varepsilon)\left(t, \frac{x - x(t)}{\lambda(t)}\right),$$

with $|\varepsilon(t)|_{H^1} \leq \delta(\alpha_0)$, where $\delta(\alpha_0) \to 0$ as $\alpha_0 \to 0$.

Our aim in this framework is to give blow up results and description of the blow up behavior for solutions of (2). (Recall that a first hint in the direction of blow up solutions for (2) was given in [15], where the authors prove the instability of the solitons.) Let $u(t)$ be a solution admitting the decomposition (12), then information on $\varepsilon(t)$, $\lambda(t)$ and $x(t)$ is equivalent to information on the solution $u(t)$; for example, blow up means $\lambda(t) \to 0$. Let us note that there is no obstruction to the existence of blow up solution of the type (12) since the two conservation laws in H^1 do not see the size of the soliton (measured by $\lambda(t)$). Indeed, the L^2 norm and the energy of Q_c do not depend on c ($E(Q) = 0$).

The strategy is to apply two types of arguments:

- First, geometrical arguments constraint the long time behavior of the solutions. The idea is to construct and then classify asymptotic objects from the solution $u(t)$. Sharp properties of solutions of (2) are used to give precise information on the behavior of the asymptotic object(in particular heavy L^2 constraints). Then, classification theorems, given in different regimes, constraint the asymptotic object to be a special solution related to Q. Going back to $u(t)$, we obtain information on the asymptotic behavior of $u(t)$. In this way, one is able to prove the blow up in finite or infinite time.

- Second, dynamical arguments give more precise information on the blow up. This kind of argument is necessary to prove blow up in finite time and give refined information on the blow up rate.

Sections 2–5 are devoted to the blow up problem in the critical case. The last section is devoted to the subcritical case.

2. Regular regime; asymptotic stability result

Recall that in the problem of blow up for NLS, the long time behavior of regular solutions is linked to the singular solutions through the conformal transformation. In the case of the KdV equation, there is no such transformation, nevertheless, one can ask whether regular and singular regimes are still linked. It turns out to be the case and the study of regular regime is the first fundamental step for the proof of existence of singular solutions.

We consider global bounded solutions of (2) in a neighborhood of the soliton. We assume in particular that for some $c_1, c_2 > 0$, we have

$$(13) \qquad \forall t \geq 0, \quad c_1 \leq |u_x(t)| \leq c_2 \quad \text{or equivalently} \quad \frac{1}{\lambda_1} \leq \frac{1}{\lambda(t)} \leq \frac{1}{\lambda_2}.$$

Under these assumptions, we claim the following asymptotic result.

THEOREM 1 (Asymptotic stability of the family of solitons [14]). *Let $c_1, c_2 > 0$. There exists $\alpha_0 > 0$ such that for any $u_0 \in H^1$, if $|u_0 - Q|_{H^1} \leq \alpha_0$, and if the* *solution $u(t)$ of (2) is defined for all $t \geq 0$ and satisfies (13), then there exists $\lambda(t)$, $x(t)$ such that*

$$(14) \qquad u(t) - \frac{1}{\sqrt{\lambda(t)}} Q\Big(\frac{x - x(t)}{\lambda(t)}\Big) \rightharpoonup 0 \quad \text{in } H^1 \text{ as } t \to +\infty.$$

Remark. In the previous result, we assume $\frac{1}{\lambda_1} \leq \frac{1}{\lambda(t)} \leq \frac{1}{\lambda_2}$, however, it is not clear whether $\lambda(t)$ converges to some limit value as $t \to +\infty$. This is due to the degeneracy of the critical case : all solitons have the same L^2 mass. In the subcritical case, we do prove convergence of $\lambda(t)$ as $t \to +\infty$, see Theorem 8.

Sketch of the proof. Following the strategy outlined in the introduction, we introduce an asymptotic object from $u(t)$, and then by rigidity property, we prove that this object is exactly $Q(x - t)$.

Consider a sequence $t_n \to +\infty$, $\widetilde{u}_0 \in H^1(\mathbf{R})$ and $\widetilde{\lambda}_0 > 0$ such that $\sqrt{\lambda(t_n)}u(t_n, \lambda(t_n). + x(t_n)) \rightharpoonup \widetilde{u}_0$ in H^1 and $\lambda(t_n) \to \widetilde{\lambda}_0$ as $n \to +\infty$. Let $\widetilde{u}(t)$ be the corresponding solution of (2) defined for all $t \in \mathbf{R}$. We call such a solution $\widetilde{u}$ an asymptotic object. The fact that it represents an asymptotic behavior locally in space of the original solution $u(t)$ as $t \to +\infty$ yields strong rigidity properties on it.

1) New set of estimates for asymptotic regime.

Coming from the fact that the solution $\widetilde{u}(t)$ is recurrent in the behavior of $u(t)$, we can prove on $\widetilde{u}(t)$ new estimates of elliptic type. Note that these estimates are completely nonlinear (they do not exist for the linear problem) and are not of oscillatory integral type, as it is usually the case for this equation.

These estimates are obtained in two steps. First, we give an uniform L^2 property.

LEMMA 1. *The solution $\widetilde{u}(t)$ is L^2 compact in the following sense: there exists $\widetilde{x}(t)$ such that*

$$(15) \qquad \forall \epsilon_0, \exists A_0 > 0, \forall t \in \mathbf{R}, \quad \int_{|x|>A_0} \widetilde{u}^2(t, x + \widetilde{x}(t))dx \leq \epsilon_0.$$

Sketch of the proof. The proof is mainly based on the almost monotonicity in time of some L^2 quantity. Indeed, let

$$\mathcal{I}_{x_0}(t) = \int u^2(t, x)\psi(x - \sigma t + x_0)dx,$$

where ψ is some smooth approximation of the Heaviside function, behaving as e^x at $-\infty$ and as $1 - e^{-x}$ at $+\infty$.

Then, we prove that for $\sigma > 0$ small enough, if $\varepsilon(t, y) = \sqrt{\lambda(t)}u(t, \lambda(t)y + x(t)) - Q(y)$ (see (12)) is small uniformly in time (in L^∞ norm for example), then

$$\forall x_0 > 0, \forall t \geq 0, \quad \mathcal{I}_{x_0}(t) - \mathcal{I}_{x_0}(0) \leq Ce^{-x_0/K},$$

for some $K > 0$ related to σ. This property says that the L^2 mass essentially travels only from the right of the soliton to the left of the soliton. This turns out to be an essential property of such solution of (2).

First, applying this property directly on $\widetilde{u}$ and using conservation of L^2 norm, we prove the L^2 compactness of $\widetilde{u}(t)$ on the right in space ($x > 0$).

Second, by this L^2 property applied on $u(t)$, we see that a loss of mass to the left of the soliton is irreversible in the sense that it will not come back to the soliton for any time. Therefore, up to the transfer of a certain amount of mass from the center and the right of the soliton to left of the soliton, there is no other transfer of mass for $u(t)$. The solution $\widetilde{u}(t)$ being recurrent in the asymptotic behavior of $u(t)$, this proves that $\widetilde{u}(t)$ has to be L^2 compact (a default of compactness of $\widetilde{u}(t)$ would imply the transfer of an infinite quantity of L^2 mass from the right to the left of the soliton on $u(t)$, which is impossible).

This proves Lemma 1.

Next, we prove the following decay property.

LEMMA 2 (Exponential decay property of compact solution). *For $C, \theta > 0$,*

$$(16) \qquad \forall x \in \mathbf{R}, \forall t \in \mathbf{R}, \quad |\widetilde{u}(t, x + \widetilde{x}(t))| \leq C e^{-\theta |x|}.$$

Sketch of the proof. First, as a consequence of the properties of the Airy group and the exponential decay properties of $Q(x)$, one can show that the property of L^2 compactness (15) implies surprisingly a point wise exponential decay on $\widetilde{\varepsilon}(t, x)$ and thus on $\widetilde{u}(t, x)$ in space. The proof is as follows: we decompose $\widetilde{\varepsilon}$ in two parts. The first part, $\widetilde{\varepsilon}_1$ is solution of the critical KdV equation, with small L^2 mass. Therefore, it goes to zero on compact sets as $t \to +\infty$. Since $\widetilde{\varepsilon}(t)$ is defined for all time, we can start the decomposition at some t_0 going to $-\infty$ so that $\widetilde{\varepsilon}_1$ disappears. The second part $\widetilde{\varepsilon}_2$ has zero initial data, and satisfies an Airy equation with interaction terms. All interaction terms contain some multiplication by Q or Q_y and then, have exponential decay properties in space uniform in time. Thus by the property of the Airy group ε has exponential decay properties in space.

Note that the exponential decay property of the Airy group is only for $x > 0$. However, using the reversibility of the KdV equation (it is invariant under the transformation $x \to -x$, $t \to -t$), and the fact that the solution is defined for all time, we prove exponential decay also for $x < 0$.

2) Characterization of the soliton by dispersion.

Now, we finish the proof of Theorem 1. The general idea, which is used several times in this work, is to invoke a rigidity property of the KdV flow around the soliton which says that such a solution $\widetilde{u}(t)$ is necessarily Q up to scaling and translation. Indeed, we have the following Liouville type theorem mixing both oscillatory in time properties and elliptic type estimates.

THEOREM 2 (Liouville property close to Q [**14**]). *Let $c_1, c_2 > 0$. There exists $\alpha_0 > 0$ such that for any $u_0 \in H^1$, if $|u_0 - Q|_{H^1} \leq \alpha_0$, and if the solution $u(t)$ of (2) is defined for all $t \in \mathbf{R}$, and satisfies (13) and (15), then there exists σ_1, x_1 such that*

$$u(t, x) = \sigma_1^{1/4} Q(\sqrt{\sigma_1}((x - x_1) - \sigma_1 t))).$$

Therefore, we have necessarily that $\widetilde{u}_0 = Q$ ($\widetilde{u}_0$ is already rescaled). This concludes the proof of Theorem 1.

Remark. Theorem 2 says that any solution global in time, close to the soliton and L^2 compact is necessarily exactly the soliton up to scaling and translation. To our knowledge, it is the first result of this type for a dispersive equation.

The interest of the rigidity property stated in Theorem 2 is not only to provide an useful tool to prove Theorem 1. Indeed, it will appear clearly in the next sections that Theorem 2, as well as other forms of rigidity properties of the KdV flow are fundamental also to understand the blow up phenomenon.

Sketch of the proof of Theorem 2. Under the assumptions of Theorem 2, one proves easily using energy argument that $u(t)$ admits a decomposition as in (12), where ε is uniformly small in H^1. Moreover, by modulation theory (i.e. a right choice of $\lambda(t)$, $x(t)$ through the implicit function theorem), one chooses orthogonality conditions $\int yQ_y\varepsilon(t) = \int y(\frac{Q}{2} + yQ_y)\varepsilon(t) = 0$. In addition, with the change of variable $ds = \frac{dt}{\lambda^3}$, $\varepsilon(s)$ satisfies the following equation:

$$\varepsilon_s = (L\varepsilon)_y + \frac{\lambda_s}{\lambda}\left(\frac{Q}{2} + yQ_y\right) + \left(\frac{x_s}{\lambda} - 1\right)Q_y + \frac{\lambda_s}{\lambda}\left(\frac{\varepsilon}{2} + y\varepsilon_y\right) + \left(\frac{x_s}{\lambda} - 1\right)\varepsilon_y - (R(\varepsilon))_y,$$

where $L\varepsilon = -\varepsilon_{xx} + \varepsilon - 5Q^4\varepsilon$ and $R(\varepsilon) = 10Q^3\varepsilon^2 + 10Q^2\varepsilon^3 + 5Q\varepsilon^4 + \varepsilon^5$.

As before, by the assumption of L^2 compactness, $u(t)$ has a uniform in time point wise exponential decay in space. Next, the proof of Theorem 2 proceeds in two steps.

(i) We argue by contradiction, assuming that no value of α_0 is giving the desired result, we construct a sequence of solutions (u_n), satisfying $|u_n(0) - Q|_{H^1} \to 0$ as $n \to +\infty$ and the assumptions of Theorem 2. Let $\varepsilon_n \not\equiv 0$ be associated to u_n by (12).

By using Virial type arguments, we prove a regularity type property:

$$\forall n, \quad \sup_{t\in\mathbf{R}} |\varepsilon_n(t)|_{H^1} \leq C \sup_{t\in\mathbf{R}} |\varepsilon_n(t)|_{L^2}.$$

This allows us to prove that a renormalized version of (ε_n) converges to a solution w of the following linear Airy equation

$$w_s - (Lw)_y = \alpha(s)\left(\frac{Q}{2} + yQ_y\right) + \beta(s)Q_y,$$

where $\alpha(s)$ and $\beta(s)$ are given functions of s. Moreover, w has point wise exponential decay

$$\forall y \in \mathbf{R}, \forall s \in \mathbf{R}, \quad |w(s,y)| \leq Ce^{-\theta|y|},$$

and satisfies the same orthogonality conditions as ε_n.

(ii) To conclude the proof, we prove a Liouville property on this linear equation. Indeed, we prove that we have necessarily $w \equiv 0$, which leads to a contradiction.

Note that the operator appearing in the equation of w is the linearized operator around Q and since $Q^4(x) = \frac{3}{\mathrm{ch}^2(2x)}$, it is a classical operator. We can describe its spectrum and then give algebraic properties of the equation. Using nonlinear tools (such as Virial type identities), and suitable orthogonality conditions on ε_n and thus w, we prove the linear Liouville property. We just mention the key quantities in this proof:

(a) A quantity related to L^1 dispersion

$$J(s) = \int w(s,y)\left(\int_0^y \frac{Q}{2} + yQ_y\right)dy.$$

Note that $\int_0^{+\infty} \frac{Q}{2} + yQ_y > 0$, so that the quantity J is not L^2 controlled.

(b) A quantity related to L^2 dispersion and Virial type relation

$$I(s) = \int yw^2(s,y)dy.$$

(c) A quantity related to energy relation

$$(Lw(s), w(s)).$$

(d) A invariant scalar product

$$\int w(s)Q.$$

The key contradiction argument is based on sharp dispersive property related to $I(s)$, which contains, at the linear level, some monotony properties of the equation (the orthogonality conditions satisfied by w are chosen in relation to this property).

3. Blow up in finite or infinite time; blow up profile

3.1. Blow up in finite or infinite time. In this section, we recall the first blow up result in the energy space for the critical KdV equation.

THEOREM 3 (Blow-up result for critical KdV equation [**22**]). *There exists $\alpha_0 > 0$ such that the following property is true. Let $u_0 \in H^1(\mathbf{R})$, and $u(t)$ be the solution of (2). Assume that*

$$E(u_0) < 0 \quad and \quad \int u_0{}^2 < \int Q^2 + \alpha_0,$$

then the solution $u(t)$ blows up in H^1 in finite or infinite time, i.e. there exists $0 < T \le +\infty$, such that $|u(t)|_{H^1} \to +\infty$ as $t \to \infty$.

Remark. Since from Pohozaev identity, we have $E(Q) = 0$ and by direct calculation $\nabla E(Q) = -Q$, we have produced a large class of blow-up solutions close to soliton. Note that the smallness condition is reasonable from the physical point of view since the generic behavior of a blow-up solution is conjectured to be a local perturbation the function Q (up to scaling and translation) plus some residual mass far in space from it.

Sketch of the proof of Theorem 3. The proof of this result is not obtained by a direct obstruction for existence of a global solution of equation (2), as it is the case in some previous results of this type in PDE. Rather, it is in some sense one of the first result in the direction of understanding a blow-up mechanism in the Hamiltonian context (except in the case of the critical nonlinear Schrödinger equation, where the conformal invariance yields explicit blow-up solutions, see the introduction). We aim to understand the blow up mechanism of the critical KdV equation by relating the nonlinear dynamic and the mechanism of dispersion. The proof shows in fact, that the blow-up is a combination of two different effects:
  - a process of ejection of mass at infinity in some suitable coordinates system,
  - the conservation of energy.

Indeed, from dynamical properties, the solution converges locally to the soliton (up to scaling and translation). Therefore, when $E(u_0) < 0$, we have necessarily $\lambda(t) \to 0$ so that the rescaled energy $\lambda^2(t)E(u_0)$ goes to zero ($E(Q) = 0$).

First, by the assumptions on u_0 and from the characterization of Q, we note that $u(t)$ stays close to Q in H^1 up to scaling and translation. Thus, we have the decomposition (12), where ε is uniformly small in H^1. Moreover, as in the previous section, we have some choice of orthogonality conditions on ε.

We claim now that under the assumptions of Theorem 3, the solution $u(t)$ has to blow-up in finite time or infinite time in H^1. That is, for T finite or infinite, we have

$$|u(t)|_{H^1} \to +\infty \quad \text{or equivalently} \quad \lambda(t) \to 0 \quad as \ \ t \to T.$$

We argue by contradiction. We assume that $u(t)$ is defined for all $t > 0$ and for a sequence $t_n \to +\infty$, we have that for a constant $C(u_0)$, $|u(t_n)|_{H^1} \leq C(u_0)$. It corresponds in the variable s (recall $ds = \frac{dt}{\lambda^3}$) to a sequence $s_n \to +\infty$ such that for a constant $\lambda(u_0) > 0$,

$$\lambda(s_n) \geq \lambda(u_0).$$

1) Introduction of a limit object and first properties.

The first idea is to use in various ways that the Airy equation (the linear part of the generalized KdV equation) pushes the mass on the left hand side, and that the nonlinear soliton travels to the right, which means that in some sense, linear and nonlinear effects are decoupled. From this fact, a key quantity is the local L^2 norm of $\varepsilon(s)$, which is a monotone function of s. Then the assumption on $\lambda(s_n)$, and energy identities imply that the L^2 local norm of $\varepsilon(s)$ has a limit as s goes to infinity which is not zero. Consider now a limit object as time goes to infinity, as in the proof of Theorem 2. From the recurrence in time of this object, we are able to prove that it satisfies a surprising exponential decay property. Indeed, in this problem related to oscillatory integral we obtain a decay related to elliptic type problem.

Let us give a more precise statement. Consider $\widetilde{\varepsilon}_0 \in H^1(\mathbf{R})$ and $\widetilde{\lambda}_0 > 0$ such that

$$(17) \qquad \varepsilon(s_n) \rightharpoonup \widetilde{\varepsilon}_0 \quad \text{in } H^1, \quad \text{and} \quad \lambda(s_n) \to \widetilde{\lambda}_0,$$

the associated functions $\widetilde{\varepsilon}(s)$, $\widetilde{u}(t)$ and $(-\widetilde{T}_1, \widetilde{T}_2)$ its maximal time existence interval (note that at this stage, we do not know whether $\widetilde{T}_1$ or $\widetilde{T}_2$ are infinite). By rescaling, we assume $\widetilde{\lambda}_0 = 1$. We observe that

$$E(\widetilde{u}(t)) < 0.$$

Very delicate estimates involving front type estimates and monotony of mass imply

$$(18) \qquad \forall x \in \mathbf{R}, \quad |\widetilde{\varepsilon}(t, x)| \leq c(\delta_0)^{\frac{1}{4}} e^{-c_2 \widetilde{\lambda}_0(t)|x|}.$$

The proof uses the same L^2 functional $\mathcal{I}$ as in the proof of Theorem 1 but with sharper estimates on the nonlinear term. Note that so far, there is no control of the H^1 norm of the solution $\widetilde{u}(t)$, nevertheless, by a nonlinear technique, an exponential estimate can be derived in local L^2 norm. Estimate (18) is then obtained by interpolation. This is the key part of the proof, saying that independently of the possible oscillations in time of the function $\varepsilon(s)$, the L^2 compactness of $\widetilde{u}(t)$ implies some exponential estimates on $\widetilde{u}(t)$ on both sides.

2) Control of the size of the asymptotic solution.

Now, we see that the exponential estimates will give a rigidity in the time oscillation of $\widetilde{u}(t)$. Indeed, by the decay in space, we clearly have

$$\widetilde{u}(t) \in L^1.$$

In L^1, equation (2) has a third invariant which is the following

$$\forall t \in (-\widetilde{T}_1, \widetilde{T}_2), \quad \int \widetilde{u}(t, x)dx = \int \widetilde{u}(0, x)dx.$$

Replacing formally $u(t, x)$ by $\frac{1}{\sqrt{\lambda(t)}} Q(\lambda(t)(x - x(t)))$, the third invariant sees $\frac{1}{\lambda(t)}$ in a certain regime, which gives a strong control of $\lambda(t)$. Using the exponential estimates to control the terms in ε, we obtain in particular that by a priori estimates, $\widetilde{u}(t)$ is defined for all time and

$$\forall t \in \mathbf{R}, \quad \widetilde{\lambda}_1 \leq \widetilde{\lambda}(t) \leq \widetilde{\lambda}_2,$$

for some $\widetilde{\lambda}_2 > \widetilde{\lambda}_1 > 0$. Therefore, $\widetilde{u}(t)$ is in a regular regime.

3) Conclusion using the result in regular regime.

Since the solution $\widetilde{u}(t)$ is L^2 compact, it follows from the Liouville Theorem 2 that

$$\forall s \in \mathbf{R}, \quad \widetilde{\varepsilon}(s) = 0$$

which is a contradiction with the energy condition on the function $\widetilde{u}$ since $E(Q) = 0$. This concludes the proof.

3.2. Blow up profile. We present the result of universality of blow up profile for blow up solutions close to Q. Note that this result, which is in itself physically relevant, also has applications. First, it will be applied directly to give a first lower estimate on the blow up rate (see below). Second, we will see in the next section that the blow up profile result, as well as the first blow up result, are fundamental to the construction of solutions blowing up in finite time.

THEOREM 4 (Stability of Q as a blow up profile [**17**]). *There exists $\alpha_0 > 0$ such that if $u_0 \in H^1(\mathbf{R})$ satisfies*

$$\int u_0^2 < \int Q^2 + \alpha_0$$

and if the solution $u(t)$ of (2) blows up in finite or infinite time $T > 0$, then for all $0 \leq t < T$, there exists $\lambda(t) > 0$ and $x(t) \in \mathbf{R}$ such that either

$$\lambda^{1/2}(t)u(t, \lambda(t)x + x(t)) \rightharpoonup Q \quad as\ t \uparrow T\ in\ H^1(\mathbf{R}),$$

or

$$-\lambda^{1/2}(t)u(t, \lambda(t)x + x(t)) \rightharpoonup Q \quad as\ t \uparrow T\ in\ H^1(\mathbf{R}).$$

Remark. Note that the alternative in Theorem 4 comes from the fact that if $u(t, x)$ is solution of (2) then $-u(t, x)$ is also solution of (2). There is no such result of determination of blow up profile in Hamiltonian systems or more generally in evolution partial differential equations except for diffusion equations where the existence of Liapounov functions plays a fundamental role. Indeed, in the case of the nonlinear heat equations $u_t = \Delta u + u^p$ in $\mathbf{R}^N$, with some restriction on p, the blowup rate and profile (and their stability) have been determined, see for example Giga and Kohn [**6**] and Fermanian–Kammerer, Merle and Zaag [**5**].

We give an application of this result to obtain a refined lower bound on the blow up rate. In particular we exclude some candidates (deduced from scaling argument) of the blow up rate. Indeed, by scaling argument and the resolution of the Cauchy problem, if $u(t)$ is a solution blowing up at some finite time $T > 0$, then for some $C > 0$,

$$\forall t_0 \in [0, T), \quad |u_x(t_0)|_{L^2} \geq \frac{C}{(T - t_0)^{1/3}}.$$

Indeed, consider

$$v_{t_0}(t, x) = |u_x(t_0)|_{L^2}^{-1/2} u(t_0 + |u_x(t_0)|_{L^2}^{-3}t, |u_x(t_0)|_{L^2}^{-1}x);$$

v_{t_0} is a solution of (2) by scaling invariance. We have $|v_{t_0 x}|_{L^2} + |v_{t_0}|_{L^2} \leq C$, and so by the resolution of the Cauchy problem locally in time by fixed point argument (see [**9**]), there exists $\tau > 0$, independent of t_0, such that $v_{t_0}(t)$ is defined on $[0, \tau]$. Therefore, $t_0 + |u_x(t_0)|_{L^2}^{-3} \tau < T$, which is the desired result. Theorem 4 implies that this lower bound represents the exact blow up rate for no solution with small L^2 mass. Indeed, we have the following theorem, which is obtained directly from Theorem 4.

THEOREM 5 (Lower bound on the blow up rate [**17**]). *There exists $\alpha_0 > 0$ such that if $u_0 \in H^1(\mathbf{R})$ satisfies $\int u_0^2 < \int Q^2 + \alpha_0$ and if the solution $u(t)$ of (2) blows up in finite time $T > 0$, then*

$$\lim_{t \uparrow T}(T - t)^{1/3}|u_x(t)|_{L^2} = +\infty.$$

Remark. Theorem 2 excludes the self similar blow up rate for solutions with L^2 mass close to the minimal mass allowing blow up, and we expect this result to extend to all initial data in H^1.

Note that there are solutions with self-similar behavior that are not in L^2. This type of behavior does not appear in the energy space.

Sketch of the proof of Theorem 4. First, as before, let

$$(19) \qquad \varepsilon(t, y) = \lambda^{1/2}(t)u(t, \lambda(t)y + x(t)) - Q(y)$$

satisfy orthogonality conditions and smallness. Let $s = \int_0^t \frac{dt'}{\lambda^3(t')}$ Theorem 4 is equivalent to prove $\varepsilon(s) \rightharpoonup 0$ as $s \to +\infty$. The proof is by contradiction. Assume that we have a solution for α_0 small such that

$$\varepsilon(s) \nrightarrow 0 \quad \text{in } H^1(\mathbf{R}), \text{ as } s \to +\infty.$$

As in the previous section, the idea is to define a recurrent object as $t \to T$. From the property of recurrence of this object, and the almost monotone in time functional $\mathcal{I}$, this object has more properties (decay properties as $y \to -\infty$). The proof of nonexistence of such object then concludes the proof of the Theorem.

More precisely, define $s_n \to +\infty$ such that $\varepsilon(s_n) \rightharpoonup \widetilde{\varepsilon}(0) \not\equiv 0$, and $\widetilde{u}$ solution of (2) with $\widetilde{u}(0) = Q + \widetilde{\varepsilon}(0)$. This limit solution $\widetilde{u}(t)$ is associated to $\widetilde{\varepsilon}$, $\widetilde{\lambda}$, $\widetilde{x}$, and is such that $\widetilde{\lambda}(0) = 1$. Define $0 < \tau \le +\infty$ such that $\forall s \in [0, \tau)$, $\frac{1}{1.1} \le \widetilde{\lambda}(t) \le 1$. We consider the two possible cases:

Case 1: $\tau = +\infty$. In this case, the solution is in a regular regime. Proving some decay estimate, we are reduced to an L^2 compact solution and we find a contradiction with the Liouville Theorem.

Case 2: $\tau < +\infty$ and $\widetilde{\lambda}(\tau) = \frac{1}{1.1}$. Here, we need to prove another rigidity result in a focusing regime (removing the possibility of self similar solution in this framework).

(i) We first prove exponential decay on the left for $\widetilde{\varepsilon}$, in the sense that

$$\forall s \in [0, \tau), \ \forall y < 0, \quad |\widetilde{\varepsilon}(s, y)| \le C(\alpha_0)e^{-\frac{|y|}{12}},$$

where $C(\alpha_0) \to 0$ as $\alpha_0 \to 0$. This property is proved by using the same L^2 functional and the same refined technique as in the proof of Theorem 3. It consists in understanding the variation of the amounts of L^2 mass localized respectively at the right hand side, the left hand side and around the soliton.

(ii) We then conclude by using the following proposition.

PROPOSITION 1 (Nonexistence of focusing solution with exponential decay). *There exists $\alpha_I > 0$ such that there exists no solution $u(t)$ of (2) satisfying*

(i) $\int u_0^2 \leq \int Q^2 + \alpha_I$,

(ii) $E(u_0) \leq 0$,

(iii) There exist $s_1 < s_2$ such that

$$(20)$$

$$\forall s > s_1, \ \lambda(s) \leq \lambda(s_1), \quad \lambda(s_2) = \frac{\lambda(s_1)}{1.1}, \quad \forall s \in [s_1, s_2], \ \frac{\lambda(s_1)}{1.1} \leq \lambda(s) \leq \lambda(s_1),$$

(iv) $\varepsilon(s)$ is such that $\forall y < 0, \ \forall s \in [s_1, s_2], \ |\varepsilon(s,y)| \leq C_1 \alpha_0^{1/4} e^{-C_2|y|}$.

Sketch of the proof of Proposition 1. We just recall the main argument of the proof. We work on the time interval $[s_1, s_2]$.

The problem in this proof is to understand the size of different quantities such as

$$\int_{s_1}^{s_2} \int \varepsilon^2, \quad \int_{s_1}^{s_2} \int \varepsilon_y^2, \quad \int_{s_1}^{s_2} \int \varepsilon^2 e^{-\frac{|y|}{2}}$$

(and also at some point $\int_{s_1}^{s_2} \int \varepsilon Q$.) One difficulty is that we have no control of the size of the interval $[s_1, s_2]$ (the "doubling time") in terms of α_0, and that the terms we integrate in time are oscillatory integrals.

We claim that for α_0 small, the following two identities hold:

- There exists $C_I > 0$ (independent of α_0) such that

$$(21) \qquad C_I \int_{s_1}^{s_2} \int \varepsilon^2 e^{-\frac{|y|}{2}} \geq 1 + \int_{s_1}^{s_2} \int \varepsilon_y^2 + |E_0| \int_{s_1}^{s_2} \lambda^2.$$

This is a consequence of a dispersion relation in L^1 using the following quantity

$$J(s) = \int \varepsilon(s,y) \int_y^{+\infty} \left(\frac{Q}{2} + zQ_z \right) - \frac{1}{4} \left(\int Q \right)^2$$

and the exponential decay assumption.

- There exist $A_0 > 2$ and $C_{II} > 0$ (independent of α_0) such that

$$(22) \qquad \int_{s_1}^{s_2} \int \varepsilon^2 e^{-\frac{|y|}{A_0}} \leq C_{II} \alpha_0^{1/2} \left[1 + \int_{s_1}^{s_2} \int \varepsilon_y^2 + |E_0| \int_{s_1}^{s_2} \lambda^2 \right].$$

This is a consequence of a local Virial type relation (which requires suitable orthogonality conditions on ε) describing L^2 dispersion, involving a regularized version of the quantity

$$I(s) = \int y\varepsilon^2(s).$$

Since $A_0 > 2$, the contradiction is obvious for $\alpha_0 < \left(\frac{1}{C_I C_{II}} \right)^2$. Note that the same quantities I and J were used in the proof of the Liouville theorem.

4. Blow up in finite time and blow up dynamics

In this section, we describe the result of blow up in finite time in the energy space.

THEOREM 6 (Blow up in finite time and dynamics of blow up solutions [18]).
There exists $\alpha_0 > 0$ such that the following is true. Let $u_0 \in H^1(\mathbf{R})$ be such that $\int u_0^2 \leq \int Q^2 + \alpha_0$ and $E(u_0) = E_0 < 0$. Let $u(t)$ be the corresponding solution of (2). Assume in addition that for some $\theta > 0$,

$$(23) \qquad \forall x_0 > 0, \quad \int_{x \geq x_0} u_0^2(x)dx \leq \frac{\theta}{x_0^6}.$$

(i) Then $u(t)$ blows up in finite time, i.e. there exists $0 < T < +\infty$, such that

$$\lim_{t \uparrow T} |u_x(t)|_{L^2} = +\infty.$$

(ii) Moreover, let $t_n \to T$ be the sequence defined as follows

$$(24) \qquad |u_x(t_n)|_{L^2} = 2^n |Q_x|_{L^2} \quad and \quad \forall t \in (t_n, T), \; |u_x(t)|_{L^2} > 2^n |Q_x|_{L^2}.$$

Then, there exists $n(u_0)$ such that

$$(25) \qquad \forall n \geq n(u_0), \quad |u_x(t_n)|_{L^2} \leq \frac{C_0}{|E_0|(T - t_n)},$$

where $C_0 = 4(\int Q)^2 |Q_x|_{L^2}$.

Comments and remarks on Theorem 6

1. Polynomial decay (23) Note first that α_0 given in Theorem 6 does not depend on θ. Note also that, for example, condition (23) is implied by

$$\int_{x>0} u_0^2(x)\, x^6 dx < +\infty.$$

This is related to the classical assumption $\int |x|^2 |u_0(x)|^2 dx < +\infty$ in the context of the nonlinear Schrödinger equations. Recall that in this case, $E(u_0) < 0$ and $\int |x|^2 |u_0|^2 < +\infty$ imply blow up in finite time by the Virial identity. Remark that the power 6 is not optimal from the proof of Theorem 6.

2. Note that the only requirement on the initial data is the fact that the blow up dynamics is close to Q, up to scaling and translation parameters, which is implied by some energy conditions on the initial data. In particular, the proof of Theorem 6 is not based on linearization close to a formal asymptotic profile and we do not require that the initial data is close in a certain sense to a formal blow up solution.

3. We shall note that there is rather few control on how long it takes for the solution to reach its asymptotic dynamics, where $|u_x(t_n)|_{L^2}$ is controlled. Indeed, from the proof, it takes a time $t(u_0)$ which depends mainly on θ and $E(u_0)$. Note that the same problem exists for the critical nonlinear Schrödinger equation (at least in the non radial case). This remark may explain why numerical computations do not suggest any blow up rate.

4. Under the assumptions of Theorem 6, one can also prove by the same arguments the following integral estimate:

$$\forall t \in (t_0, T), \quad \int_{t_0}^{t} |u_x(t')|_{L^2}^2 dt' \leq \frac{C_1}{|E_0|^2(T - t)},$$

for some $0 < t_0 < T$.

Sketch of the proof of Theorem 6. We point out that the proof of this result relies on the preceding two results Theorems 3 and 4 and is of different nature.

It is a direct (but delicate) analysis of the variation of $\lambda(s)$ (measuring the size of the solution in H^1).

Consider an initial data with negative energy which is close to Q in L^2 norm and has an additional decay property for $x > 0$, such as (23). Define the sequence (t_n) as in the statement of Theorem 6.

$$|u_x(t_n)|_{L^2} = 2^n |Q_x|_{L^2} \quad \text{and} \quad \forall t \in (t_n, T), \ |u_x(t)|_{L^2} > 2^n |Q_x|_{L^2}.$$

The existence of such a sequence is given by the blow up result Theorem 3.

First, we reduce the proof of Theorem 6 to the proof of the following estimate of $t_{n+1} - t_n$ for n large ($n > n_0$ where n_0 is such that for all $t > t_{n_0}$, the local L^2 norm of the solution is small, the existence of n_0 is a consequence of the blow up profile result Theorem 4):

$$(26) \qquad t_{n+1} - t_n \leq \frac{(\int Q)^2}{|E_0|} \tilde{\lambda}(t_n).$$

By elementary computations, from (26) it is easy to see that the sequence (t_n) is bounded, and thus the blow up time T is finite, and then estimate (25) follows again from (26). Note that the proof is not based on a linearization close to a formal solution and that there is no initialization of the estimates.

To prove (26), we argue by contradiction, assuming that for some n arbitrary large, $t_{n+1} - t_n > \frac{(\int Q)^2}{|E_0|} \tilde{\lambda}(t_n)$. Under the constraint of closeness to Q in L^2, we have as usual the decomposition of the solution

$$\lambda^{1/2}(t) u(t, \lambda(t) y + x(t)) = Q(y) + \varepsilon(t, y),$$

where the function $\varepsilon(t, y)$ is small in H^1 and satisfies for all time t two suitable orthogonality conditions.

The assumption of polynomial decay on the initial data in L^2 at the right, through the use of the stability in time of such property on $u(t)$ solution of (2) (use the functional $\mathcal{I}(t)$), allows us to obtain decay at the right on ε. This decay for $y > 0$ on ε allows us to control $\int_{y>0} |\varepsilon|$, by a *nonlinear* estimate. Note that the space in which this estimate holds is $L^2(\mathbf{R}) \cap L^1(\mathbf{R}^+)$.

From the proof, four quantities are important: $\int \varepsilon Q$, $\int \varepsilon^2 e^{-\frac{|y|}{100}}$, $\frac{\lambda_s}{\lambda}$, $\int \varepsilon_y^2$. The proof is based on the study of conservation laws (and in fact, local conservation laws close to Q), which will give relations between these four quantities. For each conservation law, a suitable orthogonality condition is adapted.

(i) By the additional decay property for $y > 0$, we are able to choose the orthogonality conditions on ε so that we have a relation of the following type

$$-\frac{\lambda_s}{\lambda} = \int \varepsilon Q + O\left(\int \varepsilon^2 e^{-\frac{|y|}{100}}\right),$$

where $\int \varepsilon Q$ has a slow variation in time. (Indeed, at the linear level, this quantity is invariant, and seems to contain no oscillatory integrals)

(ii) Next, for other orthogonality conditions, related to another decomposition

$$\tilde{\lambda}^{1/2}(t) u(t, \tilde{\lambda}(t) y + \tilde{x}(t)) = Q(y) + \tilde{\varepsilon}(t, y),$$

quadratic terms $\int_{t_n}^{t_{n+1}} \int \varepsilon^2 e^{-\frac{|y|}{100}}$ are controlled in some sense by $\int_{t_n}^{t_{n+1}} \int \tilde{\varepsilon} Q$. This result is obtained by a local Virial identity on $\tilde{\varepsilon}$. This points out the fact that time oscillations, involving oscillatory integrals are controlled by quantities that are not oscillatory such as $\int \widetilde{\varepsilon} Q$.

(iii) Under this regime, we are able to compare the functions ε and $\widetilde{\varepsilon}$ coming from the two decompositions, by estimates involving L^2 exponential decay at the right of the soliton, and an additional surprising degeneracy in the relation between ε and $\widetilde{\varepsilon}$ in the relation giving $\frac{\lambda_s}{\lambda}$ (first order cancellation are given by the criticality, the more remarkable fact is coming from second order cancellations)

Thus, in some sense on the time interval (t_n, t_{n+1}), we obtain

$$-\frac{\lambda_s}{\lambda} = \int \widetilde{\varepsilon} Q + O\left(\int \widetilde{\varepsilon}^2 e^{-\frac{|y|}{100}}\right).$$

The small term at the right can again be controlled when $\int \widetilde{\varepsilon}^2 e^{-\frac{|y|}{100}}$ is small (implied for a sufficiently large time by the result of asymptotic profile [22]).

(iv) Now using the energy estimate for $\widetilde{\varepsilon}$ and the fact that $\int \widetilde{\varepsilon}_y^2 \geq 0$, we obtain

$$-\frac{\lambda_s}{\lambda} \geq C|E_0|\lambda^2,$$

in some integral in time sense between t_n, t_{n+1}. This allows us to obtain a contradiction, and thus to prove (26).

5. Nonexistence of minimal mass blow up solution

From these results, a natural question is the existence of a blow up solution of (2) in the case $\int u_0^2 = \int Q^2$. We have the following Theorem.

THEOREM 7 (Nonexistence of minimal mass blow up solutions [19]).
Let $u_0 \in H^1(\mathbf{R})$ be such that

$$\int u_0^2 = \int Q^2.$$

Assume that for some $C > 0$ and $\theta > 3$, we have

$$(27) \qquad \forall x_0 > 0, \qquad \int_{x > x_0} u_0^2(x)dx \leq \frac{C}{x_0^\theta}.$$

Then, the corresponding solution $u(t)$ of (2) does not blow up in $H^1(\mathbf{R})$ neither in finite nor in infinite time, and in particular the solution $u(t)$ is global in time in $H^1(\mathbf{R})$.

Remark. Note that condition (27) is implied by $\int_{x>0} u_0^2(x)\, x^\theta dx < +\infty$.

A similar problem has been studied for NLS (nonlinear Schrödinger) type equations. The critical NLS equation (10) has the same energetic invariants in $H^1(\mathbf{R}^N)$. Thus we also have for $\int u_0^2 < \int Q^2$ that the solution is global and bounded in H^1. For $\int u_0^2 = \int Q^2$, the conformal invariance yields an explicit blow up solution

$$\frac{1}{|t|^{N/2}} Q\left(\frac{x}{t}\right) e^{\frac{i|x|^2}{4t} - \frac{i}{t}}.$$

In Merle [23], it is proved that this explicit solution is in fact the unique blow up solution with minimal L^2 mass, up to the invariances of the equation.

Note that for the critical Zakharov equation, in space dimension 2,

$$(28) \qquad iu_t = -\Delta u + nu, \quad \text{where } u : \mathbf{R} \times \mathbf{R}^2 \to \mathbf{C}, \, n : \mathbf{R} \times \mathbf{R}^2 \to \mathbf{R},$$

$$(29) \qquad \frac{1}{c_0^2} n_{tt} = \Delta n + \Delta |u|^2,$$

the conformal invariance does not exist anymore, and there is no minimal mass blow up solution, see Glangetas and Merle [**7**].

The result of nonexistence of minimal mass blow up solutions for the critical generalized KdV equation points out the nonexistence of additional invariance (such as the conformal invariance for the critical NLS equation) and illustrates the relation between blow up and dispersion.

In view of some results in §3 (nonexistence of self similar blow up solution), it seems that dispersion of part of the mass is needed to allow blow up. The result of the present paper gives another illustration of this fact. Indeed, for a minimal mass solution, all the L^2 mass of a blow up solution would have to concentrate at the blow up time, and thus the solution does not have dispersion.

Sketch of the proof of Theorem 7.   The argument is by contradiction. Consider a solution $u(t)$ of (2) with $\int u_0^2 = \int Q^2$ and blowing up in finite or infinite time T, i.e. $|u_x(t)|_{L^2} \to +\infty$ as $t \to T$. Assume in addition that for some $C > 0$ and $\theta > 3$, (27) holds.

We establish some preliminary properties of $u(t)$. Using variational arguments, we prove that as t approaches T, $u(t)$ converges to Q in H^1 up to scaling and translation (the two invariances of the KdV equation). This allows us to choose a decomposition of $u(t)$ of the form

$$(30) \qquad u(t,x) = \frac{1}{\widetilde{\lambda}^{1/2}(t)}(Q + \widetilde{\varepsilon})\left(t, \frac{x - \widetilde{x}(t)}{\widetilde{\lambda}(t)}\right),$$

where $\widetilde{\lambda}(t) \to 0$ and $\widetilde{\varepsilon}(t) \to 0$ in $H^1(\mathbf{R})$ as $t \to T$ (this important fact is straightforward in the case of the minimal mass). This implies in particular that

$$u^2(t, x + \widetilde{x}(t)) \rightharpoonup \left(\int Q^2\right)\delta_{x=0} \quad \text{as } t \to T.$$

From this, we have a property of exponential decay in space of $u(t)$ on the left of the soliton ($x < \widetilde{x}(t)$) uniformly in time, following similar argument as in [**16**]. In addition, we observe that the decay condition on the right in L^2, (27) is preserved through time.

The proof is based on two main estimates.

(i) First, we are able to match these decay properties and global space–time information on the solution. Indeed, we obtain control on an invariant quantity and a Virial type quantity in L^1 ($\int u(t,x)dx$ and $\int xu(t,x)dx$). It follows that blow up occurs in finite time ($T < +\infty$) and we obtain a surprising control from above of the blow up rate.

CLAIM 1 (Upper control of the blow up rate).

$$(31) \qquad T < +\infty \quad \text{and} \quad \int_0^T |u_x(t)|_{L^2}^{3/2}dt < +\infty.$$

These estimates are of the same nature as the ones used in the case of the critical NLS equation, where control of the Virial identity in L^2 (see [**23**]) gives a precise information on the blow up rate.

However, for what follows, the situation is completely different from the critical NLS equation, where the conformal invariance allows one to conclude. For the critical KdV equation, in the absence of such invariance, we need rigidity on the asymptotic dynamics of $\widetilde{\varepsilon}$.

(ii) Indeed, energy, Virial type estimate and a degeneracy property on the equation of $\widetilde{\varepsilon}$, imply the following lower bound on the blow up rate as $t \to T$.

CLAIM 2 (Lower bound on the blow up rate). *There exists $C > 0$ and $t_0 \in [0, T)$ such that*

$$(32) \qquad \forall t \in [t_0, T), \quad |u_x(t)|_{L^2} \geq \frac{C}{E(u_0)(T - t)}.$$

This estimate is in the spirit of the proof of finite time blow up and upper bound, Theorem 6 (in the proof of Theorem 6, under the condition $E(u_0) < 0$, we prove the reverse inequality; of course here we have $E(u_0) > 0$). This is a consequence of a local in space analysis.

A contradiction then follows from Claims 1 and 2 and thus Theorem 7 is proved.

6. Subcritical case; Asymptotic stability of the sum of N solitons

It is well-known that for $p = 2$ and $p = 3$, (1) is completely integrable. Indeed, for suitable u_0 (u_0 and its derivatives with exponential decay at infinity) there exist an infinite number of conservation laws, see e.g. Lax [**12**] and Miura [**24**]. Moreover, many results on these equations rely on the inverse scattering method, which transform the problem in a sequence of linear problems (but requires strong decay assumption on the solution). *In our results, we do not use integrability.*

Here, we give result of stability and asymptotic stability of N-solitons in the energy space for $p = 2, 3$ and 4.

6.1. Asymptotic stability of the family of solitons. In this subsection, we state a result of asymptotic stability of the family of solitons for the subcritical KdV equation. Its proof is very similar to the proof of Theorem 1 in the critical case, in particular by the use of a fundamental rigidity property around Q.

THEOREM 8 (Asymptotic stability for $p = 2, 3, 4$ [**16**]). *Let $c_0 > 0$. There exists $\alpha_0 > 0$ such that the following is true. Let $u_0 \in H^1(\mathbf{R})$ be such that $|u_0 - Q_{c_0}|_{H^1} < \alpha_0$ and $u(t)$ be the solution of (1) on $\mathbf{R}^+ \times \mathbf{R}$. Then there exist $c_{+\infty} > 0$ and a function $x(t)$ such that*

$$u(t, . + x(t)) \rightharpoonup Q_{c_{+\infty}} \quad in \ H^1(\mathbf{R}) \ as \ t \to +\infty.$$

Remark. A result of strong convergence in $L^2(\mathbf{R})$ is false. For example, for $p = 2$, using integrability, we can construct a solution such that for a small $\epsilon > 0$,

$$|u(t, . + x(t)) - Q_1 - Q_\epsilon(. - x_\epsilon(t))|_{L^\infty} \underset{t \to +\infty}{\to} 0, \quad \text{where } x_\epsilon(t) \to -\infty.$$

Moreover, for $p = 2, 3, 4$, convergence in L^2 strong as $t \to +\infty$ in Theorem 8 implies that the solution is exactly a soliton by using the characterization of the soliton given in Theorem 9, see below (indeed, in this case, the solution is L^2 compact).

Note that, from the proof of Theorem 8, the L^2 dispersion occurs only at the left of the soliton.

Remark. For other results of this type for KdV equation, we refer to Pego and Weinstein [**25**]. Their approach is based on linear theory around Q which allows only initial data with fast decay at $+\infty$ (exponential or polynomial decay). Therefore, the results are obtained in spaces different from the energy space.

Remark. From the proof of Theorem 1, we have $x_t(t) \to c_{+\infty}$ as $t \to +\infty$.

Recall that in the critical case, the result of asymptotic stability is stated as follows: $\exists c(t), x(t)$ such that

$$u(t, . + x(t)) - Q_{c(t)} \rightharpoonup 0 \quad \text{in } H^1(\mathbf{R}) \text{ as } t \to +\infty,$$

without convergence of the parameter $c(t)$. (see Theorem 1). This can be explained as follows. In both the critical and subcritical cases, the L^2_{loc} norm of the solution is monotone in time in some sense. In the subcritical case, the limit as $t \to +\infty$ of the L^2_{loc} norm of $u(t)$ selects the asymptotic soliton since for a given L^2 norm, there is only one soliton up to translation; this is crucial to obtain the limit of $c(t)$. In the critical case, the monotonicity of the L^2_{loc} norm does not prevent oscillations of $c(t)$ as $t \to +\infty$ since all solitons have same L^2 norm; this is an illustration of degeneracy.

As in the critical case, the proof of Theorem 8 is based on the following Liouville Theorem.

THEOREM 9 (Liouville property close to Q_{c_0} for $p = 2, 3, 4$ [**16**]). *Let $p = 2, 3$ or 4, and let $c_0 > 0$. There exists $\alpha_0 > 0$ such that the following is true. Let $u_0 \in H^1(\mathbf{R})$ be such that $|u_0 - Q_{c_0}|_{H^1} < \alpha_0$ and let $u(t)$ be the solution of (1) for all time $t \in \mathbf{R}$. If there exists $x(t)$ such that*
(33)

$$\forall \delta_0 > 0, \exists A_0 > 0, \forall t \in \mathbf{R}, \quad \int_{|x| > A_0} u^2(t, x(t) + x)dx \le \delta_0, \quad (L^2 \text{ compactness})$$

then there exists $c_1 > 0$, $x_1 \in \mathbf{R}$ such that $\forall t \in \mathbf{R}$, $\forall x \in \mathbf{R}$, $u(t, x) = Q_{c_1}(x - x_1 - c_1 t)$.

Remark. One could expect that the Liouville property is still true without the smallness of $|u_0 - Q_{c_0}|_{H^1}$. However, at least for $p = 3$, a counter example (see references in Lamb [**11**]) proves that the smallness condition is necessary. This question for other values of p is open. Indeed, for $p = 3$, the following solutions of (1) (called breather solutions)

$$u_{N,\omega}(t, x) = -\frac{2\sqrt{6}\,\omega}{\mathrm{ch}(\omega x + \gamma t)} \left(\frac{\cos(Nx + \delta t) - (\omega/N)\sin(Nx + \delta t)\tanh(\omega x + \gamma t)}{1 + (\omega/N)^2\sin^2(Nx + \delta t)/\mathrm{ch}^2(\omega x + \gamma t)} \right),$$

where $\delta = N(N^2 - 3\omega^2)$ and $\gamma = \omega(3N^2 - \omega^2)$, are example of global in time solutions, without dispersion, which are not of the type $Q_c(x - ct)$.

6.2. Asymptotic stability of the sum of N solitons. For the KdV equation ($p = 2$), there is a much wider class of special explicit solutions for (1), called N–solitons. They correspond to the superposition of N traveling waves with different speeds that interact and then remain unchanged after interaction. The N–solitons behave asymptotically in large time as the sum of N traveling waves, and as for the single solitons, there is no dispersion. We refer to [**24**] for explicit expressions and further properties of these solutions. Maddocks and Sachs [**13**] prove the stability in $H^N(\mathbf{R})$ of N–solitons for any initial data u_0 close in $H^N(\mathbf{R})$ to an N–soliton, the solution $u(t)$ of the KdV equation remains uniformly close in $H^N(\mathbf{R})$ for all time to an N soliton profile with same speeds. Their proof involves N conserved quantities for the KdV equation, and this is the reason why they need to impose closeness in high regularity spaces. Note that this result is known only with $p = 2$ and with this regularity assumption of the initial data. Asymptotic stability is unknown in this context.

For $p \neq 2$, even the existence of solutions behaving asymptotically as the sum of N solitons was not known.

For $p = 2, 3, 4$, using techniques developed for the critical and subcritical cases in [14] and [16] as well as a direct variational argument in H^1, we claim the stability and asymptotic stability of the sum

$$\sum_{j=1}^{N} Q_{c_j^0}(x - x_j), \quad \text{where} \quad 0 < c_1^0 < \ldots < c_N^0, \quad x_1 < \ldots < x_N,$$

in $H^1(\mathbf{R})$, for $t \geq 0$.

THEOREM 10 (Asymptotic stability of the sum of N solitons [20]). *Let $p = 2$, 3 or 4. Let $0 < c_1^0 < \ldots < c_N^0$. There exist $\gamma_0, A_0, L_0, \alpha_0 > 0$ such that the following is true. Let $u_0 \in H^1(\mathbf{R})$ and assume that there exist $L > L_0$, $\alpha < \alpha_0$, and $x_1^0 < \ldots < x_N^0$, such that*

$$(34) \quad \left| u_0 - \sum_{j=1}^{N} Q_{c_j^0}(\cdot - x_j^0) \right|_{H^1} \leq \alpha, \quad \text{and } x_j^0 > x_{j-1}^0 + L, \text{ for all } j = 2, \ldots, N.$$

Let $u(t)$ be the solution of (1). Then, there exist $x_1(t), \ldots, x_N(t)$ such that
(i) Stability of the sum of N decoupled solitons.

$$(35) \quad \forall t \geq 0, \quad \left| u(t) - \sum_{j=1}^{N} Q_{c_j^0}(x - x_j(t)) \right|_{H^1} \leq A_0 \left(\alpha + e^{-\gamma_0 L} \right).$$

(ii) Asymptotic stability of the sum of N solitons. Moreover, there exist $c_1^{+\infty}, \ldots, c_N^{+\infty}$, with $|c_j^{+\infty} - c_j^0| \leq A_0 \left(\alpha + e^{-\gamma_0 L} \right)$, such that

$$(36) \quad \left| u(t) - \sum_{j=1}^{N} Q_{c_j^{+\infty}}(x - x_j(t)) \right|_{L^2(x > c_1^0 t/10)} \to 0, \quad \dot{x}_j(t) \to c_j^{+\infty} \quad \text{as } t \to +\infty.$$

Remark. In Theorem 10 (ii), we cannot have convergence to zero in $L^2(x > 0)$. Indeed, assumption (34) on the initial data allows the existence in $u(t)$ of an additional soliton of size less that α (thus traveling at arbitrarily small speed). For $p = 2$, an explicit example can be constructed using the N–soliton solutions.

Recall that for $p = 2$ any N-soliton solution has the form $v(t, x) = U^{(N)}(x; c_j, x_j - c_j t)$, where $\{U^{(N)}(x; c_j, y_j); c_j > 0, y_j \in \mathbf{R}\}$ is the family of explicit N-soliton profiles (see e.g. [13], §3.1). As a direct corollary of Theorem 1, for $p = 2$, we prove stability and asymptotic stability of this family.

COROLLARY 1 (Asymptotic stability in H^1 of N-solitons for $p = 2$). *Let $p = 2$. Let $0 < c_1^0 < \ldots < c_N^0$ and $x_1^0, \ldots, x_N^0 \in \mathbf{R}$. For all $\delta_1 > 0$, there exists $\alpha_1 > 0$ such that the following is true. Let $u(t)$ be a solution of (1). If $|u(0) - U^{(N)}(\,.\,; c_j^0, -x_j^0)|_{H^1} \leq \alpha_1$, then there exist $x_j(t)$ such that*

$$(37) \quad \forall t > 0, \quad |u(t) - U^{(N)}(\,.\,; c_j^0, -x_j(t))|_{H^1} \leq \delta_1.$$

Moreover, there exist $c_j^{+\infty} > 0$ such that

$$(38) \quad \left| u(t) - U^{(N)}(\,.\,; c_j^{+\infty}, -x_j(t)) \right|_{L^2(x > c_1^0 t/10)} \to 0, \quad \dot{x}_j(t) \to c_j^{+\infty} \quad \text{as } t \to +\infty.$$

Note that the above corollary improves the result in [13] in two ways. First, stability is proved in H^1 instead of H^N. Second, we also prove asymptotic stability as $t \to +\infty$.

We sketch the proof of Theorem 10. Using modulation theory, $u(t) = \sum_{j=1}^N Q_{c_j(t)}(x - x_j(t)) + \varepsilon(t, x)$, where $\varepsilon(t)$ is small in H^1, and $x_i(t)$, $c_i(t)$ are geometrical parameters. The stability result is equivalent to control both the variation of $c_j(t)$ and the size of $\varepsilon(t)$ in H^1.

Our main arguments are based on L^2 properties of the solution. Arguing as usual with the functional $\mathcal{I}$, the L^2 norm of the solution at the right of each soliton is almost decreasing in time. This property together with energy argument allows us to prove that the variation of $c_j(t)$ is quadratic in $|\varepsilon(t)|_{H^1}$, which is a key of the problem.

Let us explain the argument formally by taking $\varepsilon = 0$ and so $u(t) = \sum Q_{c_j(t)}(x - x_j(t))$. The energy conservation becomes

$$\sum c_j^{\beta + \frac{1}{2}}(t) = \sum c_j^{\beta + \frac{1}{2}}(0),$$

where $\beta = \frac{2}{p-1}$. The monotonicity of the L^2 norm at the right of each soliton gives us

$$\Delta_j(t) = \sum_{k=j}^N c_k^{\beta - \frac{1}{2}}(t) - c_k^{\beta - \frac{1}{2}}(0) \leq 0.$$

We claim that $c_j(t) = c_j(0)$ by a convexity argument. Indeed,

$$\begin{aligned}
0 &= \sum c_j^{\beta + \frac{1}{2}}(t) - c_j^{\beta + \frac{1}{2}}(0) \sim \frac{2\beta + 1}{2\beta - 1} \sum c_j(0)(c_j^{\beta - \frac{1}{2}}(t) - c_j^{\beta - \frac{1}{2}}(0)) \\
&= \frac{2\beta + 1}{2\beta - 1} \sum (c_j(0) - c_{j+1}(0))\Delta_j(t) \geq \sigma_0 \sum |\Delta_j(t)| \geq \sigma_1 \sum |c_j(t) - c_j(0)|.
\end{aligned}$$

Thus $c_j(t)$ is a constant at the first order. In fact, we prove that the variation in time of $c_j(t)$ is of order 2 in $\varepsilon(t)$.

Then, we control the variation of $\varepsilon(t)$ in H^1 by a refined version of this argument, using suitable orthogonality conditions on ε.

The asymptotic stability result then follows directly from a rigidity property of the flow of equation (1) around the solitons (Theorem 9) and monotonicity properties of the mass.

References

[1] T.B. Benjamin, The stability of solitary waves, Proc. Roy. Soc. London A **328**, (1972) 153—183.

[2] J.L. Bona, V.A. Dougalis, O.A. Karakashian and W.R. McKinney, Conservative, high order numerical schemes, Phil. Trans. Roy. Soc. London Ser. A. **351**, (1995) 107—164.

[3] J. Bourgain, Harmonic analysis and nonlinear partial differential equations, Proceedings of the International Congress of Mathematicians, **1,2**, (Zurich, 1994) 31—44, Birkhäuser, Basel, 1995.

[4] D.B. Dix and W.R. McKinney, Numerical computations of self-similar blow up solutions of the generalized Korteweg-de Vries equation, Diff. Int. Eq., **11** (1998), 679—723.

[5] C. Fermanian–Kammerer, F. Merle and H. Zaag, Stability of the blow-up profile of non-linear heat equations from the dynamical system point of view, Math. Ann. **317** (2000), 347-387.

[6] Y. Giga and R. Kohn, Nondegeneracy of blowup for semilinear heat equations, Comm. Pure Appl. Math. **42** (1989), 845—884.

[7] J. Ginibre and Y. Tsutsumi, Uniqueness of solutions for the generalized Korteweg–de Vries equation, SIAM J. Math. Anal. **20**, (1989) 1388—1425.

[8] T. Kato, On the Cauchy problem for the (generalized) Korteweg–de Vries equation, Advances in Mathematical Supplementary Studies, Studies in Applied Math. **8**, (1983) 93—128.

[9] C.E. Kenig, G. Ponce and L. Vega, Well-posedness and scattering results for the generalized Korteweg–de Vries equation via the contraction principle, Comm. Pure Appl. Math. **46**, (1993) 527—620.

[10] D.J. Korteweg and G. de Vries, On the change of form of long waves advancing in a rectangular canal, and on a new type of long stationary waves, Philos. Mag. **539**, (1895) 422—443.

[11] G.L. Lamb Jr., *Element of soliton theory* (John Wiley & Sons, New York 1980).

[12] P. D. Lax, Integrals of nonlinear equations of evolution and solitary waves, Comm. Pure Appl. Math. **21**, (1968) 467—490.

[13] J.H. Maddocks and R.L. Sachs, On the stability of KdV multi-solitons, Comm. Pure Appl. Math. **46**, (1993) 867—901.

[14] Y. Martel and F. Merle, A Liouville Theorem for the critical generalized Korteweg–de Vries equation, J. Math. Pures Appl. **79**, (2000) 339—425.

[15] Y. Martel and F. Merle, Instability of solitons for the critical generalized Korteweg–de Vries equation, Geom. Funct. Anal., **11**, (2001) 74—123.

[16] Y. Martel and F. Merle, Asymptotic stability of solitons for subcritical generalized KdV equations, Arch. Ration. Mech. Anal. **157**, (2001) 219—254.

[17] Y. Martel and F. Merle, Stability of the blow up profile and lower bounds on the blow up rate for the critical generalized KdV equation, Ann. of Math. **155**, (2002) 235—280.

[18] Y. Martel and F. Merle, Blow up in finite time and dynamics of blow up solutions for the L^2-critical generalized KdV equation, J. Amer. Math. Soc. **15**, (2002) 617—663.

[19] Y. Martel and F. Merle, Nonexistence of blow up solution with minimal L^2 mass for the critical gKdV equation, Duke Math. J. **115** (2002), no. 2, 385–408.

[20] Y. Martel, F. Merle and Tai–Peng Tsai, Stability and asymptotic stability in the energy space of the sum of N solitons for subcritical gKdV equations, Comm. Math. Phys. **231** (2002), no. 2, 347–373.

[21] F. Merle, Blow-up phenomena for critical nonlinear Schrödinger and Zakharov equations, Proceeding of the International Congress of Mathematicians, (Berlin, 1998), Doc. Math. J. DMV.

[22] F. Merle, Existence of blow-up solutions in the energy space for the critical generalized KdV equation, J. Amer. Math. Soc. **14**, (2001) 555—578.

[23] F. Merle, Determination of blow up solution with minimal mass for nonlinear Schrödinger equations with critical power, Duke Math. J., **69** (1993), 427—454.

[24] R.M. Miura, The Korteweg–de Vries equation: a survey of results, SIAM Review **18**, (1976) 412—459.

[25] R.L. Pego and M.I. Weinstein, Asymptotic stability of solitary waves, Comm. Math. Phys. **164**, (1994) 305—349.

[26] M.I. Weinstein, Lyapunov stability of ground states of nonlinear dispersive evolution equations, Comm. Pure. Appl. Math. **39**, (1986) 51—68.

UNIVERSITÉ DE CERGY–PONTOISE
E-mail address: `martel@math.polytechnique.fr`

UNIVERSITÉ DE CERGY–PONTOISE, INSTITUT UNIVERSITAIRE DE FRANCE
E-mail address: `Frank.MERLE@math.u-cergy.fr`

Contemporary Mathematics
Volume **350**, 2004

Recent Advances in the Global Theory of Constant Mean Curvature Surfaces

Rafe Mazzeo

This paper is dedicated to H. Brezis and F. Browder

ABSTRACT. The theory of complete surfaces of (nonzero) constant mean curvature in $\mathbb{R}^3$ has progressed markedly in the last decade. This paper surveys a number of these developments in the setting of Alexandrov embedded surfaces; the focus is on gluing constructions and moduli space theory, and the analytic techniques on which these results depend. The last section contains some new results about smoothing the moduli space and about CMC surfaces in asymptotically Euclidean manifolds.

1. Introduction

The study of global minimal and constant mean curvature (CMC) surfaces in $\mathbb{R}^3$ has always occupied a central position in classical differential geometry, but the former class of surfaces has undoubtedly received more attention than the latter. While partly historical accident, this is also due to the classical Weierstraß representation formula for minimal surfaces, through which many interesting examples of minimal surfaces were known. In contrast, the most famous results about CMC surfaces prior to 1980 were the rigidity theorems of Hopf and Alexandrov, which state that the only compact CMC surfaces in $\mathbb{R}^3$ which are immersed and genus 0, respectively embedded of arbitrary genus, are spheres; in some sense this is a negative result. There has been an explosive development in both these fields during the past twenty years, centering on many new examples and methods of construction, and also including various classification and structure theorems. The recent volume [4] contains up-to-the-minute accounts of many of these developments, but mostly focussing on minimal surfaces. My goal here is to survey some of the advances in the global study of CMC surfaces from my own biased perspective.

Let Σ be a surface in $\mathbb{R}^3$. Its mean curvature H at a point p is the sum of the two principal curvatures (the customary factor of $1/2$ is omitted), and Σ has constant

1991 *Mathematics Subject Classification.* Primary 53A10, 53C21.
Key words and phrases. Constant Mean Curvature, Gluing Constructions, Moduli Space Theory.
Supported by NSF Grant # DMS 0204730.

mean curvature if this quantity is constant. In this case we can normalize so that either $H \equiv 0$ or $H \equiv 1$. We shall restrict attention exclusively to the second case. In addition, we shall only be concerned with complete, properly immersed surfaces of finite topology. There is a rich theory for the Plateau problem for minimal and CMC surfaces, cf. [35], [2], but we shall not discuss this at all; in particular, we shall understand all surfaces below to be complete, unless otherwise stated. On the other hand, unlike the situation for minimal surfaces, cf. [6] for example, the subject of CMC surfaces with infinite topology is mostly unexplored.

The only complete CMC surfaces known classically are the sphere, the cylinder, and the one-parameter family of rotationally invariant Delaunay surfaces, which we describe in the next section. (We should also mention a method, due originally to Bonnet but most clearly enunciated by Lawson, for associating a 'cousin' CMC surface in $\mathbb{R}^3$ to a minimal surface in $\mathbb{S}^3$.) The first modern breakthrough was Wente's discovery, detailed in his 1983 paper [36], of immersed CMC tori. Not long afterwards, Kapouleas used transcendental PDE methods to construct many new CMC surfaces, including compact ones with arbitrary genus [13], [14], and noncompact ones with finitely many ends [12]. Große-Brauckman [9] subsequently constructed certain of these noncompact surfaces with large discrete symmetry groups using more classical methods based on Schwarz reflection.

The theory has developed in two fairly distinct directions. From the techniques used for Wente's construction ultimately has emerged the DPW (Dorfmeister-Pedit-Wu) method, which draws on the theory of integrable systems, and serves as a replacement for the Weierstraß representation; on the other hand, Kapouleas' construction has engendered many new PDE approaches to the problem. These theories have distinct flavors, and in some senses illuminate quite different classes of surfaces. For example, the former method seems to be more successful in describing general immersed CMC surfaces, and is closely related to many of the computer experiments and simulations of CMC surfaces, cf. [15], while the latter has been particularly good in describing CMC surfaces which are 'nearly embedded'. We specialize once again, and for the last time, to the second PDE set of methods.

It turns out that requiring CMC surfaces to be embedded is too restrictive, and in some sense even geometrically unnatural. Thus, we shall relax this hypothesis and require instead that our CMC surfaces be *Alexandrov embedded*. To define this, first recall that a complete CMC surface Σ with finite topology is diffeomorphic to a punctured Riemann surface $\overline{\Sigma} \setminus \{p_1, \ldots, p_k\}$; we shall always assume that Σ and $\overline{\Sigma}$ are oriented. Writing $\overline{\Sigma} = \partial Y^3$, then we say that Σ is Alexandrov embedded if the immersion $\Sigma \hookrightarrow \mathbb{R}^3$ extends to an immersion $Y \hookrightarrow \mathbb{R}^3$. This condition is naturally suited to the technique of Alexandrov reflection, cf. [18].

The basic questions we shall address are whether there are any Alexandrov embedded CMC surfaces, and if so, how might we go about constructing them and describing the totality of them. The next section discusses the main elementary examples and the basic structure theory of complete Alexandrov embedded CMC surfaces, and lays out the framework for the subsequent develoment. In §3 we describe a variety of analytic constructions, centered on the basic technique of Cauchy data matching. This is followed, in §4, by the ramifications of these constructions for the moduli space theory. Finally, in §5, we generalize this theory to CMC surfaces in arbitrary asymptotically Euclidean 3-manifolds and suggest some possible directions of future research.

This survey has some overlap with one given in the recent paper by Kusner [19], which I recommend strongly to the interested reader; his emphasis is rather different than mine and I hope that these two papers complement one another in a useful way. I should also mention that the forthcoming thesis of M. Jleli at Université Paris XII contains the analogues of essentially all of the results here for CMC hypersurfaces in $\mathbb{R}^n$.

My papers in this area have been written in various collaborations with Rob Kusner, Frank Pacard, Dan Pollack and Jesse Ratzkin. I owe special thanks to Frank Pacard for all he has taught me over the years! Kusner, Pacard and Pollack each read and gave me many helpful comments on a preliminary draft of this paper, but I happily take credit for the remaining errors and solecisms. As described in a previous survey [27], most of the results in this theory have parallels for solutions of the singular Yamabe problem with isolated singularities. In particular, the general moduli space results in §4 originated in my work with Pollack and Uhlenbeck [26] on deformations of solutions of the Yamabe equation with isolated singularities on the sphere. I am also grateful to Nick Korevaar and Rick Schoen for many illuminating discussions. I thank the Mittag-Leffler Institute and Universita di Roma "La Sapienza" for their hospitality while this paper was written. Finally, this paper is based on a talk given at the conference in honour of H. Brezis and F. Browder at Rutgers University in October 2001, and I would like to acknowledge that my first exposure to CMC surfaces was in a course taught by Professor Brezis at MIT in the early 1980's.

2. Basic properties

In this section we introduce most of the basic definitions. Here and in the remainder of this paper we denote by $\mathcal{M}_{g,k}$ the set of all proper Alexandrov embedded constant mean curvature 1 surfaces in $\mathbb{R}^3$ of genus g with k ends, and when we say that a surface Σ is CMC, we mean that it lies in one of these spaces. We begin by discussing the most elementary examples of complete noncompact CMC surfaces, the Delaunay surfaces. This is followed by a description of the basic structure theory of (Alexandrov embedded) CMC surfaces. We go on to discuss the Jacobi operator and its mapping properties, and the fundamental notion of nondegeneracy of a CMC surface.

2.1. Delaunay surfaces. The CMC surface with the largest symmetry group is, of course, the sphere, but the most familiar one after that is the cylinder. The sphere is the only surface (CMC or not) with full rotational symmetry, and by the classical results of Hopf and Alexandrov mentioned earlier, it is the only compact Alexandrov embedded CMC surface; on the other hand, there is an interesting family of axially symmetric CMC surfaces discovered by Delaunay [7], in which the cylinder lies as an extreme element. To describe these, consider the cylindrical graph $(t, \theta) \mapsto (\rho(t) \cos\theta, \rho(t) \sin\theta, t)$. The CMC equation for this graph is an ODE,

$$(1) \qquad \rho_{tt} - \frac{1}{\rho}(1 + \rho_t^2) + (1 + \rho_t^2)^{3/2} = 0.$$

All positive solutions of this equation are periodic; this follows from the facts that the function $H(\rho, \rho_t) = \rho^2 - 2\rho(1 + \rho_t^2)^{-1/2}$ is an integral of motion of this ODE and has compact level curves in $\{\rho > 0, H < 0\}$. (Kusner's balancing formula, described below, gives a nice geometric interpretation of the fact that H is constant on solutions.) For any $0 < \epsilon < 1$ let $\rho_\epsilon(t)$ be the solution which attains the minimum value ϵ, and normalize by translating the independent variable so that $\rho_\epsilon(0) = \epsilon$. Then $\epsilon \leq \rho_\epsilon(t) \leq 2 - \epsilon$ for all t, and since ρ_ϵ remains positive, the surfaces D_ϵ which are the cylindrical graphs of these functions are embedded; they are called Delaunay unduloids.

There are two limiting cases: in the first, $\epsilon = 1$ and D_1 is the cylinder of radius 1; the other occurs when $\epsilon \searrow 0$, and then D_ϵ converges to an infinite array of mutually tangent spheres of radius 2 with centers along the z-axis. The family D_ϵ interpolates between these two extremes. The number ϵ measures the size of the 'neck' region.

The Delaunay family continues past the singular limit at $\epsilon = 0$ to a family of immersed (not Alexandrov embedded!) CMC surfaces known as the Delaunay nodoids. These will only play a minor role in this survey.

There is an alternate, conformal, parametrization of the D_ϵ which is more convenient for most calculations. This isothermal parametrization involves two functions σ and κ: σ is the unique smooth nonconstant solution to the initial value problem

$$\left(\frac{d\sigma}{ds}\right)^2 + \tau^2 \cosh^2 \sigma = 1, \qquad \partial_s\sigma(0) = 0, \quad \sigma(0) < 0, \quad \text{when } \tau \in (0,1]$$

while $\kappa(s)$ is determined by

$$\frac{d\kappa}{ds} = \tau^2 e^\sigma \cosh \sigma, \qquad \kappa(0) = 0.$$

Setting $t = \kappa(s)$ and $\rho(\kappa(s)) = \tau e^{\sigma(s)}$, and determining τ by $\epsilon = 1 - \sqrt{1 - \tau^2}$, then D_ϵ is parametrized by

$$(2) \qquad X_\epsilon : \mathbb{R} \times S^1 \ni (s, \theta) \mapsto \left(\tau\, e^{\sigma_\tau(s)} \cos \theta, \tau\, e^{\sigma_\tau(s)} \sin \theta, \kappa_\tau(s)\right).$$

The metric coefficients in this coordinate system are $g_{ss} = g_{\theta\theta} = \tau^2 e^{2\sigma}$, and $g_{s\theta} = g_{\theta s} = 0$.

While this definition may seem ad hoc, it is motivated by a systematic line of reasoning in the theory of integrable systems; [21] contains a more complete discussion of this parametrization.

2.2. Structure theory. There are several important facts about the global structure of Alexandrov-embedded CMC surfaces of finite topology, and we review these now.

The story begins with an important theorem of Meeks [28], which states that any proper Alexandrov embedded end of a CMC surface is cylindrically bounded, i.e. contained in some tube of radius $R > 0$ around a ray. This result reflects the 'smallness at infinity' of the ambient Euclidean space and fails definitively, for example, in hyperbolic space. (Simple counterexamples are the equidistant surfaces around a totally geodesic copy of $\mathbb{H}^2 \subset \mathbb{H}^3$.) Meeks also proves that there are no one-ended CMC surfaces, i.e. $\mathcal{M}_{g,1} = \emptyset$.

Building on this, Korevaar, Kusner and Solomon [18] obtained a much sharper picture of the behaviour of CMC ends. More precisely, for any such end E there is a half Delaunay unduloid $D_\epsilon^+ = D_\epsilon \cap \{z \geq 0\}$ and a rigid motion A of $\mathbb{R}^3$ such that E converges exponentially to $A(D_\epsilon^+)$. More precisely, if ν is the outward unit normal to $A(D_\epsilon^+)$, then any surface $\mathcal{C}^1$ close to this Delaunay end can be written as a normal graph, i.e. as the image of

$$A(D_\epsilon^+) \ni p \longrightarrow p + \phi(p)\nu(p).$$

Then E is the normal graph of a function $\phi \in \mathcal{C}^\infty(A(D_\epsilon^+))$ which decays exponentially at infinity. The proof ultimately devolves to the exponential decay of Jacobi fields (see below) on a Delaunay surface and an adaptation of a beautiful 'approximation improvement' lemma due to L. Simon. Using Alexandrov reflection they prove that $\mathcal{M}_{g,2} = \emptyset$ if $g \geq 1$, while $\mathcal{M}_{0,2}$ contains only the Delaunay surfaces.

Amongst the many corollaries of this result is the fact that we can associate an asymptotic necksize parameter ϵ to any end of $\Sigma \in \mathcal{M}_{g,k}$. There are other ways to measure this necksize. For example, there is a 'balancing formula' for CMC surfaces, discovered by Kusner, which states the following. Let Γ be a simple closed curve on Σ and C an immersed surface in $\mathbb{R}^3$ with $\partial C = \Gamma$; suppose that n and ν are the unit normals to Γ in $T\Sigma$ and C, respectively. (Thus n is orthogonal to the unit normal to Σ in $\mathbb{R}^3$.) Then, with an appropriate choice of orientations of these normals, the quantity

$$\frac{1}{\pi} \left(\int_\Gamma n\, d\sigma - \int_C \nu\, dA \right)$$

depends only on the homology class of Γ in Σ. Thus, for example, if Γ lies in an end E of Σ and represents a generator of $H_1(E)$, then it is homologous to the curve Γ_s which is the normal graph in E of the circle $\theta \to X_\epsilon(s,\theta)$ in the model Delaunay end, for any $s \geq s_0$. Using the exponential decay of E to this Delaunay end, this quantity can be computed explicitly, and in fact just equals $\epsilon(2-\epsilon)\vec{v}$, where $\vec{v}$ is the unit vector in the direction of the Delaunay axis. This is called a balancing formula because if $\Gamma_1, \ldots, \Gamma_k$ encircle the k ends of $\Sigma \in \mathcal{M}_{g,k}$, and are oriented so that their sum is null homologous, then the sum of the corresponding force vectors must vanish.

In a subsequent paper [17], these authors and Meeks establish an analog of this asymptotics result for proper Alexandrov embedded CMC surfaces in the hyperbolic space $\mathbb{H}^3$, in the case that the mean curvature $H > 2$. We note that CMC (hypersurface) theory in $\mathbb{H}^n$ is quite different in the three cases, $H > n-1$, $H < n-1$ and $H = n-1$, the latter having the most similarities with minimal surface theory (via the cousin correspondence mentioned earlier), and the first having the most similarities with CMC theory in $\mathbb{R}^n$. [17] also gives Delauany asymptotics for ends of CMC hypersurfaces in $\mathbb{H}^n$, but only under the *a priori* assumption of cylindrical boundedness. There is a growing literature on CMC surfaces in hyperbolic space, cf. [3], [31], [8], and the work of H. Rosenberg and his collaborators. It is quite likely that the results of [18] and [17] remain true for CMC surfaces in manifolds which are only asymptotically Euclidean or hyperbolic, and indeed, the proofs from these papers may require only moderate alterations. We come back to this point in §5.

These asymptotics results definitely require Alexandrov-embeddness, or something near to it. For example, the Delaunay family D_ϵ continues beyond $\epsilon = 0$ to a family of immersed (not Alexandrov embedded!) rotationally invariant CMC surfaces called nodoids, and Pacard and I show [22] that these become increasingly unstable as $\epsilon \to -\infty$: there are infinitely many values $\epsilon_j \to -\infty$ where new CMC surfaces which are not rotationally symmetric bifurcate away from this nodoid family. This bifurcation bears some relationship to the classical Rayleigh instability of the cylinder. It only occurs when $\epsilon \leq \epsilon_0 < 0$, and it is possible that there may be some global structure theory, akin to that described above, for properly immersed CMC surfaces for which all ends have force vector corresponding to some Delaunay nodoid D_ϵ with $\epsilon_0 < \epsilon < 0$. We leave this as an interesting unexplored direction.

One can interpret this asymptotics theorem by thinking of $\Sigma \in \mathcal{M}_{g,k}$ as decomposing into a large compact piece, about which we know very little *a priori*, with k asymptotically Delaunay ends emerging from it. All of the gluing constructions described in §3 involve "filling in the black box in the middle" in different ways. Kapouleas' original construction builds CMC surfaces around (suitably balanced) simplicial graphs where k of the edges are rays tending to infinity. The finite edges and the rays are replaced by finite and semi-infinite segments of Delaunay surfaces, respectively, and the vertices become spheres. (Balancing here means that there are force vectors associated with each edge and these must cancel at each vertex. An additional condition regarding the flexibility of this arrangement is also needed.) As we discuss later, there are various other geometric possibilities for the vertices, but Korevaar and Kusner [16] show that this picture is qualitatively correct in general. For any $\Sigma \in \mathcal{M}_{g,k}$, there is a simplicial graph, with explicit bounds on the numbers of edges and vertices in terms of g and k, and a (fairly large) tubular neighbourhood around it which contains Σ.

2.3. The Jacobi operator: its mapping properties, nullspace and non-degeneracy.

If Σ is a complete Alexandrov embedded CMC surface with finite topology, then we can consider all surfaces which are $\mathcal{C}^2$ close to Σ as normal graphs, i.e. for any $\phi \in \mathcal{C}^2(\Sigma)$, sufficiently small, we set

$$(3) \qquad \Sigma_\phi = \{x + \phi(x)\nu(x) : x \in \Sigma\}$$

where $\nu(x)$ is the unit normal to Σ at x. The equation which determines that Σ_ϕ also has CMC equal to 1 is a quasilinear elliptic equation for ϕ. Its linearization at $\phi = 0$ is called the Jacobi operator L (or L_Σ). It is well known that

$$L_\Sigma = \Delta_\Sigma + |A_\Sigma|^2,$$

where A_Σ is the second fundamental form. Solutions of $L\phi = 0$ are called Jacobi fields. If u_t is a 1-parameter family of functions such that all of the surfaces Σ_{u_t} are all CMC (with the same mean curvature), then $\phi = du_t/dt|_{t=0}$ is a Jacobi field on Σ. In general, Jacobi fields need only correspond to deformations of Σ which are CMC to second order.

One very important consequence of the structure theory described in the last subsection is that because CMC surfaces have very well-controlled geometry at infinity, the behaviour of their Jacobi operators is governed by the behaviour of Jacobi operators on (half) Delaunay surfaces. Indeed, as we discuss in a moment, the asymptotics of any (tempered) Jacobi field on Σ are determined by the asymptotics of tempered Jacobi fields on the model Delaunay ends. More generally, mapping properties for L_Σ may be deduced from those for the Jacobi operators L_ϵ on D_ϵ. This motivates a closer study of these model operators L_ϵ.

In cylindrical coordinates the expression for L_ϵ is quite complicated, but using the isothermal parametrization, the Jacobi operator L_ϵ on D_ϵ takes the much simpler form

$$L_\epsilon = \frac{1}{\tau^2 e^{2\sigma}} \left(\partial_s^2 + \partial_\theta^2 + \tau^2 \cosh(2\sigma) \right).$$

Since $\tau^2 e^{2\sigma}$ is bounded away from zero and infinity, it is sufficient to analyze the operator

$$\partial_s^2 + \partial_\theta^2 + \tau^2 \cosh(2\sigma(s)),$$

and this is what we do in practice. This analysis relies on the rotational invariance of this operator in θ and its periodicity in s. Thus we can simultaneously reduce by Fourier series in the $\mathbb{S}^1$ factor and use an adapted form of Flocquet theory to handle the resulting ODEs with periodic coefficients. With these techniques we are able to determine that L_ϵ is Fredholm on certain natural exponentially weighted function spaces, and to characterize all global tempered solutions. We return to this last point shortly. The analysis of L_Σ itself is accomplished by patching the parametrices for each of these model problems, corresponding to each of the ends E_j of Σ, together with an interior parametrix. These arguments are presented in detail (for a slightly different geometric problem) in [**26**], cf. also [**24**].

We shall discuss Jacobi fields on Σ in more detail in §4. For the present we note that one may generate a distinguished set of Jacobi fields on any CMC surface using rigid motions of the ambient space. The corresponding 'geometric Jacobi fields' thus correspond to infinitesimal translations and rotations of Σ in $\mathbb{R}^3$. For a Delaunay surface D_ϵ, the three independent translations induce three Jacobi fields, and the two infinitesimal rotations of the axis of symmetry induce two more. D_ϵ has a final Jacobi field, induced by change of the necksize parameter ϵ. (The case $\epsilon = 1$ is handled slightly differently.) These all have the form $e^{i\ell\theta}\psi(s)$, where $\ell = 0, \pm 1$ (or rather, are real-valued combinations of these); for the Jacobi fields coming from translations, $\psi(s)$ is periodic, while for each of the others it grows linearly as $s \to \pm\infty$. These six Jacobi fields are the only global solutions of $L_\epsilon \phi = 0$ on D_ϵ which do not grow exponentially on one end or the other. In fact, there are solutions of the form $e^{i\ell\theta}\psi(s)$ for any $\ell \in \mathbb{Z}$, $|\ell| \geq 2$, but for each of these, $\psi(s+S_\epsilon) = e^{\gamma_\ell}\psi(s)$, where S_ϵ is the period of σ and where $\gamma_\ell \in \mathbb{R}$, $\gamma_\ell \neq 0$. In particular, these other solutions have precise exponential rates of growth or decay. This set of exponents constitutes a discrete set Λ_ϵ, called the indicial set of L_ϵ.

As for the mapping properties of L_Σ, one's first temptation might be to let this operator act on $L^2(\Sigma)$. However, this mapping does not have closed range. In fact, the parametrix method sketched above shows that the spectrum of L_Σ on L^2 is comprised of bands of continuous spectrum and locally finite discrete spectrum. Unfortunately, 0 is *always* on the end of one of these bands of continuous spectrum. In the language of scattering theory, we say that it is a threshold value. To see this, recall from above that Σ always admits nontrivial global bounded and linearly growing Jacobi fields on Σ, corresponding to translations and rotations in $\mathbb{R}^3$. This is already enough to imply that 0 is in the continuous spectrum by Weyl's criterion; a closer examination of the Flocquet theory, together with the existence of linearly growing (as opposed to bounded, periodic) Jacobi fields, shows that 0 is a threshold value.

In any case, one must look elsewhere to realize L_Σ as an operator with closed range. As indicated earlier, this may be done using spaces with exponential weights. We can use either Sobolev or Hölder spaces, and each has its advantages, but we shall use the latter. Decompose Σ into the union of a compact piece, K, and a finite number of ends, $E_1, \ldots, E_k$. Fix isothermal coordinates (s_j, θ_j) on each end.

DEFINITION 1. *For $\ell \in \mathbb{N}$, $0 < \alpha < 1$ and $\mu \in \mathbb{R}$, we define $\mathcal{C}^{\ell,\alpha}_\mu(\Sigma)$ to be the set of all functions $u \in \mathcal{C}^{\ell,\alpha}(K)$ and on each end E_j satisfy*

$$\sup_{s_j \geq 0} e^{-\mu s_j} \|w\|_{\mathcal{C}^{\ell,\alpha}([s_j, s_j+1]\times S^1)} < \infty.$$

It is immediate that

$$(4) \qquad L : \mathcal{C}^{\ell+2,\alpha}_\mu(\Sigma) \longrightarrow \mathcal{C}^{\ell,\alpha}_\mu(\Sigma)$$

is a bounded mapping for any $\mu \in \mathbb{R}$. However, if ϵ_j is the asymptotic necksize of the end E_j, then it is not hard to see that whenever $\mu \in \Lambda_{\epsilon_j}$, (4) does not have closed range. However, this is the only bad case:

PROPOSITION 1. *Let $\Lambda = \Lambda_\Sigma = \cup_{j=1}^k \Lambda_{\epsilon_j}$. Then if $\mu \notin \Lambda$, the mapping (4) is Fredholm. When $\mu \notin \Lambda$ is sufficiently large, this mapping is surjective, while if μ is sufficiently negative, this mapping is injective.*

We conclude this subsection by stating the fundamental

DEFINITION 2. *The CMC surface Σ is said to be **nondegenerate** if it admits no nontrivial global Jacobi fields which decay along all of the ends E_j, or equivalently, if (4) is injective for any $\mu < 0$.*

The equivalence of the two versions of this definition is not immediate, since one could imagine the existence of Jacobi fields which decay polynomially rather than exponentially, but is true nonetheless. A duality argument then gives that if Σ is nondegenerate, (4) is surjective when $\mu > 0$, $\mu \notin \Lambda$. There is an important refinement of this surjectivity. Choose a cutoff function $\chi(s)$ which equals 1 for $s \geq 1$ and 0 for $s \leq 0$, and consider the $6k$-dimensional vector space W_Σ spanned by the elements

$$\Phi_{ij} = \chi(s_j)\phi_{i,\epsilon_j}(s_j, \theta_j), \ j = 1, \ldots, k, \ i = 1, \ldots, 6,$$

where $\{\phi_{i,\epsilon_j}\}$ are the 6 geometric Jacobi fields on D_{ϵ_j}. Thus we have transplanted these special Jacobi fields to each end of Σ. Because of the exponential decay of the ends to their Delaunay models, we have

$$L_\Sigma \phi = \mathcal{O}(e^{-cs_j}) \qquad \forall \phi \in W_\Sigma, \ j = 1, \ldots, k.$$

Here $c = \inf\{\gamma : 0 < \gamma \in \Lambda_\Sigma\} > 0$. Hence for $\mu > 0$ sufficiently small,

$$(5) \qquad L_\Sigma : W_\Sigma \oplus \mathcal{C}^{k+2,\alpha}_{-\mu}(\Sigma) \longrightarrow \mathcal{C}^{k,\alpha}_{-\mu}(\Sigma)$$

is well defined.

PROPOSITION 2. *If Σ is nondegenerate and $0 < \mu < \mu_0 = \inf\{|\lambda| : \lambda \in \Lambda_\Sigma\}$, then the mapping (5) is surjective.*

We call W_Σ the deficiency subspace of Σ. It also plays a role in the regularity of Jacobi fields.

PROPOSITION 3. *Let ϕ be any Jacobi field with at most polynomial growth on Σ. Then there exist functions $\Phi \in W_\Sigma$ and $\psi \in \mathcal{C}^{2,\alpha}_{-\mu}(\Sigma)$, $\mu > 0$, such that $\phi = \Phi + \psi$.*

The nonlinear mean curvature operator N does not preserve any of the spaces $\mathcal{C}^{k,\alpha}_\mu(\Sigma)$, $\mu > 0$, because the nonlinearity amplifies the exponential increase. On the other hand, while this nonlinear operator is defined on the spaces of exponentially decreasing functions $\mathcal{C}^{k,\alpha}_{-\mu}$, its linearization always has a cokernel there. We can extend the definition of N to W_Σ as follows: any $\Phi \in W_\Sigma$ corresponds to a one-parameter family of CMC deformations of the model Delaunay surface for that end, and so we can 'integrate' Φ by transplanting this to a one-parameter family of asymptotically CMC deformations localized to each of the ends of Σ. For example, if Φ corresponds to the rotation of a Delaunay surface, then we rotate the corresponding end, using some cutoff to join this to the identity on the rest of Σ. The constant mean curvature condition is destroyed only at the pivoting locus which is contained in a compact set. (The definition is slightly more complicated when Φ includes a component requiring a change of Delaunay parameter.) More generally, $N(\Phi)$ is just the mean curvature of this new surface, and this is either compactly supported or exponentially decreasing, so that

$$(6) \qquad N : W_\Sigma \oplus \mathcal{C}^{k+2,\alpha}_{-\mu}(\Sigma) \longrightarrow \mathcal{C}^{k,\alpha}_{-\mu}(\Sigma)$$

is well defined. When Σ is nondegenerate, this map has surjective linearization. Thus the addition of W_Σ provides an intermediate space where the conflicting requirements of well-definedness and surjectivity are balanced.

We conclude this section by noting that there is a more general way to define nondegeneracy. The crucial feature in the discussion above is that we were able to extend the definition of N to W_Σ because elements of the deficiency space correspond, at least asymptotically, to actual CMC deformations. Another way to say this is that every element of W_Σ asymptotically integrates to a one-parameter family of asymptotically CMC surfaces. Thus the 'correct' definition of nondegeneracy should be to require that all tempered Jacobi fields on Σ are similarly asymptotically integrable. The point here is that one should be able to use explicit geometric deformations of Σ to compensate for the cokernel of L_Σ. As a very simple but important example, the sphere $\mathbb{S}^2$ is degenerate in the more restricted sense; it has a three-dimensional space of Jacobi fields, which are the restrictions of the linear coordinates on $\mathbb{R}^3$ to $\mathbb{S}^2$. However, these Jacobi fields correspond to translations of the sphere, and it is precisely this which allowed Kapouleas to use them in his gluing construction.

The possible existence of degenerate CMC surfaces creates a lot of complications throughout this whole theory. However, even amongst immersed surfaces, there are no known examples which are degenerate in the sense of this broader definition. Indeed the only known examples of degenerate surfaces in the narrower sense are the sphere (or any compact immersed CMC surface) and certain immersed 'bubbleton' solutions which are globally cylindrically bounded and asymptotically cylindrical, so that translation along the cylinder's axis provides a decaying, but nevertheless 'integrable', Jacobi field. A major problem is to determine whether there ever exists degenerate CMC surfaces in the broader sense.

3. Gluing constructions

Up to this point the only examples of complete Alexandrov embedded CMC surfaces of finite topology we have seen are the Delaunay unduloids. In this section we discuss a number of closely related gluing constructions; taken together these show that CMC surfaces are quite malleable and exist in great profusion. As with all gluing constructions, the rough idea is that one starts with 'building blocks', pieces of surfaces which already are, or are near to, simpler CMC surfaces, and then pieces them together to form complete CMC surfaces. We use here three types of building blocks: Delaunay surfaces (or finite or semi-infinite segments of them), spheres and complete Alexandrov-embedded *minimal* surfaces with finite total curvature. It is necessary that all components be nondegenerate in the extended sense. Amongst the motivations for developing new analytic methods for these sorts of problems is that one wishes to show that the glued surfaces are again nondegenerate; this is important in the moduli space theory and also has the important consequence that these constructions can be iterated. In addition, until the advent of newer methods developed after Kapouleas' work, it was unclear whether there existed any nondegenerate surfaces beyond the unduloids.

I shall first give a rough overview of the type of argument needed here (or indeed any singular perturbation problem). After describing Kapouleas' construction I list the more general gluing constructions now known to work, and conclude by describing idea of Cauchy data matching which can be used to prove any of these. All of these surfaces have very small necksizes, or otherwise nearly degenerate (i.e. near to the boundary of Teichmüller space) conformal structures. It is an important philosophical point that all of the 'constructible' CMC surfaces are very nearly degenerate. Gluing constructions provide parametrizations of ends of the moduli spaces. The (presumably more commonplace) surfaces with larger necksizes are not 'reachable' by direct methods.

Thus, again at a rough level, these components are arranged, by rotating and translating them in space, so that they form a configuration which is an approximate solution. Such a configuration is either CMC or approximately CMC everywhere. There is always a parameter η involved, so that as $\eta \searrow 0$, the discrepancy of the configuration from being exactly CMC diminishes. Unfortunately, in this same limit, the geometry of the configuration degenerates. The main step involves using the Jacobi operator to perturb the approximate configuration to an exact CMC surface, but again this is an operator with 'degenerating coefficients'. Hence we must not only show that its Jacobi operator is surjective as in (5) when η is small, but also estimate the growth of the norm of a suitable right inverse as $\eta \searrow 0$. The game is to show that this norm does not blow up faster than the inverse of the size of the error term. There are many well-known constructions which follow these outlines, including Taubes' famous instanton patching, etc., and there are many different methods of carrying out the details. I shall sketch one, initially developed in [**21**], based on Cauchy data matching; this seems to be the most efficient and involves the fewest hard estimates.

The first construction of this type for CMC surfaces was accomplished by Kapouleas [**12**], inspired by a similar construction by Schoen [**34**] for the singular Yamabe problem. This establishes the existence of many complete CMC surfaces, but his method does not give a sufficiently fine understanding of their geometry; in particular, it seems very difficult to tell whether any of the surfaces he constructs are nondegenerate. (They are, in fact, nondegenerate, as follows from the alternate construction v) below.) I have already sketched his method in §2: one begins with a simplicial graph in $\mathbb{R}^3$ with k of the edges semi-infinite rays, and where each edge is labelled with a force vector. This graph must satisfy a certain balancing and flexibility condition. The approximate CMC surface is assembled by substituting for each vertex a sphere, for each finite edge a finite Delaunay segment (possibly just a single neck region), and for each ray a half Delaunay surface. The force vectors include information about the relative scaling of the Delaunay necksizes, and the parameter η gives the absolute size of these necks. Kapouleas joins these pieces with cutoff functions; the resulting error terms are large in L^∞, but are supported in small sets. A delicate analysis of the Jacobi operator (complicated by the fact that he works on a space where the Laplacian does not have closed range), allows him to show that the necessary perturbation can be made when η is small.

We now list and give brief descriptions of various geometric operations which, it has more recently been proved, may all be done in the (nondegenerate) CMC category. Afterwards we sketch a representative proof.

i) Adding Delaunay ends [**24**]: If Σ is any nondegenerate CMC surface and $p \in \Sigma$, then we may attach a half Delaunay surface D_ϵ^+ with small necksize ϵ to Σ at p. The axis of D_ϵ^+ is directed in the outward normal direction.

ii) Connected sums [**23**]: Let Σ_1 and Σ_2 be nondegenerate CMC surfaces and choose points $p_j \in \Sigma_j$. Rotate and translate these surfaces so that the tangent spaces coincide at these points, but with opposite orientation. Then there exists a new CMC surface $\Sigma_1 \#_\epsilon \Sigma_2$ obtained by 'bridging' these surfaces with a small Delaunay neck near these points. (One can also insert any finite Delaunay segment instead of a single neck.) To show that the resulting surface is nondegenerate requires minor restrictions concerning the locations of the points.

iii) Attaching Delaunay ends to minimal k-noids [**21**]: Let Σ be a complete nondegenerate *minimal* surface of finite total curvature with k ends (a so-called minimal k-noid), all of which are asymptotic to catenoids. Form a new surface with boundary Σ_ϵ by truncating the ends (at distance proportional to $-1/\log \epsilon$ on the catenoid) and then rescaling by a factor of ϵ. Then there is a CMC surface with k ends obtained by attaching half-Delaunay surfaces $D_{\epsilon_j}^+$ to these truncated catenoidal ends. The necksize parameters are $\epsilon a_j + \mathcal{O}(\epsilon^2)$, where $(a_1, \ldots, a_k)$ is a vector of relative dilation factors for the catenoids modelling the ends of Σ. Since many topologically complicated nondegenerate minimal k-noids are known to exist, this gives similarly complex CMC surfaces. It may seem that being *minimal* is very far away from having constant mean curvature 1, but this is where the scaling by ϵ becomes important. For, rescaling the ensemble by $1/\epsilon$, the construction is equivalent to attaching (albeit on a very large scale) Delaunay ends with mean curvature ϵ to a minimal surface, which seems more plausible.

iv) End-to-end gluing [**29**]: Let Σ_1 and Σ_2 be two nondegenerate CMC surfaces, each of which has an end with the same asymptotic necksize parameter (not necessarily small). Then we may translate and rotate these surfaces so that the axes of these particular ends are on the same line, but oppositely oriented. Truncating these ends very far out, it is possible to attach the surfaces to obtain a new CMC surface with a very long approximately unduloidal tube. This requires a small strengthening of the nondegeneracy condition. Notice that while these surfaces do not necessarily have small necksizes, they contain embedded essential annuli with large conformal modulus, hence still have nearly degenerate conformal structures.

v) Kapouleas-type constructions [**25**]: One may attach half Delaunay surfaces and finite Delaunay segments, arrayed along the rays and edges of a

suitable simplicial graph, with spheres at the vertices, to obtain a CMC
surface which is contained in a tubular neighbourhood of this graph.

vi) Attaching Delaunay cross-pieces [**25**]: Suppose Σ is a nondegenerate CMC
surface and $p_1, p_2 \in \Sigma$. Suppose that the tangent planes at these points
are parallel and oppositely oriented (so that the outward normal of $T_{p_1}\Sigma$
points towards the outward normal for $T_{p_2}\Sigma$), and the distance $|p_1 - p_2|$
is an even integer (hence essentially a multiple of a Delaunay period for
ϵ small). Then, subject to a flexibility condition on this configuration as
well as a slight strengthening of the nondegeneracy condition as in iv), one
may attach a Delaunay segment along the axis connecting these points.

For all of these constructions, with minor provisos concerning the locations of
the points where the gluings are done, the resulting surfaces are nondegenerate, and
hence these operations may be performed iteratively and in various combinations.
I have also omitted a detailed description of the slightly strengthened form of nondegeneracy which is needed in iv) and vi). In any case, using this, it is possible to
show, cf. [**24**], [**25**], that there exist nondegenerate elements in $\mathcal{M}_{g,k}$ for any $k \geq 3$
and $g \geq 0$.

As promised, we sketch the proof of i). Thus let Σ be a nondegenerate CMC
surface and fix $p \in \Sigma$. Excise a small ball $B_\zeta(p)$, and denote by $\Sigma(\zeta)$ the surface
with boundary, $\Sigma \setminus B_\zeta(p)$. Now let D'_ϵ be a half Delaunay surface, truncated near,
but not quite at the neck. More specifically, recalling the radial function $\rho_\epsilon(t)$ in the
original parametrization of D_ϵ, which attains its minimum value of ϵ at $t = 0$, we
let D'_ϵ denote that portion of the Delaunay surface corresponding to $t \geq t_\epsilon$ for some
small $t_\epsilon < 0$. We now have two surfaces with boundary, and the naive hope is that if
we have truncated them at the correct radii, they should fit together passably well.
This is not quite true without some minor modifications. To remedy this, we replace
$\Sigma(\zeta)$ by a normal graph over it, $\Sigma(\zeta)' = \Sigma(\zeta)_{\epsilon G}$, where G is the Green function
for the Jacobi operator with pole at p (this exists because of the nondegeneracy
assumption!). The radius ζ is chosen so that the mean curvature of this surface is
still bounded. The reason for doing this is that the shape of the graph of the Green
function, hence of this new surface, is approximately logarithmic, and this matches
the neck region of the Delaunay surface. At any rate, we have now obtained two
surfaces depending on a parameter which almost match at their boundaries; for
convenience we relabel them as Σ_1^ϵ and Σ_2^ϵ, respectively (so the first corresponds to
the original surface Σ, and the second to D_ϵ^+).

The second step of the construction is to consider over each of these surfaces
separately the space of *all* nearby CMC surfaces $(\Sigma_j^\epsilon)_\phi$ which are written as normal
perturbations. Here $\phi \in \mathcal{C}^{2,\alpha}(\Sigma_j^\epsilon)$ is small. This is an infinite dimensional space.
For Σ_1^ϵ it is parametrized by arbitrary (small) boundary data $\psi \in \mathcal{C}^{2,\alpha}(\partial\Sigma_1^\epsilon)$. For
Σ_2^ϵ almost the same is true, except that not every function ψ on the boundary of a
half Delaunay surface corresponds to a CMC normal perturbation. The functions
for which this fails lie in the span of the cross-sectional eigenmodes $e^{i\ell\theta}$, $\ell = 0, \pm 1$.
What saves the day is that these functions are precisely the boundary values of
the explicit geometric Jacobi fields on Σ_2^ϵ, and correspond to CMC surfaces which
are rotations or translations of the original Delaunay surface. Thus we can also
regard all small boundary values on $\partial\Sigma_2^\epsilon$ as corresponding to CMC deformations
of Σ_2^ϵ. The analysis required here is quite simple, and reduces to a straightforward
contraction mapping argument.

The final step is to consider the set of all Cauchy data of these infinite dimensional spaces of CMC deformations of the summands. The Cauchy data of a normal perturbation $(\Sigma_j^\epsilon)_\phi$, by definition, is the pair of functions (ψ_1, ψ_2), where ψ_1 is the restriction of ϕ to the boundary and $\psi_2 = \partial_\nu \phi$ is its normal derivative. The whole point of this argument is that if we can show that these Cauchy data subspaces intersect, then that point of intersection corresponds to CMC deformations of each surface such that these perturbed surfaces match up to second order along the interface. By elliptic regularity for the mean curvature equation, these surfaces must actually fit together smoothly, and we are done. The argument that these infinite dimensional submanifolds must intersect is also not too difficult. The tangent space of either submanifold at $\phi = 0$ is the graph of the Dirichlet-to-Neumann operator for $L_{\Sigma_j^\epsilon}$. One shows that the sum of the graphs of these two Dirichlet-to-Neumann operators spans the whole space when ϵ is small enough (by verifying that the same is true for the operators obtained in the limit at $\epsilon = 0$). Finally, the nonlinear Cauchy data submanifolds are graphs off of these linear subspaces, and some estimates are required to see that the neighbourhood in which these graphical representations are valid is large enough to contain the putative point of intersection.

The proofs of the remaining constructions ii) - vi) may all be done similarly. We note that in v) spheres are being used as summands, and these are only non-degenerate in the broader sense discussed at the very end of §2.

4. Moduli space theory

We have defined the spaces $\mathcal{M}_{g,k}$ as the set of all Alexandrov embedded complete CMC surfaces of genus g with k ends. Since we do not mod out by rigid motions of the ambient space or internal isometries, this might reasonably be called the premoduli space, but we shall simply call it the moduli space of CMC surfaces with given g and k.

We list a few facts about these moduli spaces which are well-known, or which follow immediately from the material in §2 and 3:

- $\mathcal{M}_{g,1}$ is empty for all $g \geq 0$, [**28**].
- $\mathcal{M}_{g,2}$ is empty when $g \geq 1$. $\mathcal{M}_{0,2}$ is equal to the set of all rotations and translations of Delaunay surfaces. These follow from Alexandrov reflection arguments, [**18**].
- $\mathcal{M}_{g,k}$ is nonempty for every $g \geq 0$ and $k \geq 3$. This was essentially proved by Kapouleas [**12**], but also follows from the gluing constructions of §3. We mention some special cases. Define $\mathbb{M}_{g,k} = \{(\Sigma, p) : \Sigma \in \mathcal{M}_{g,k}, \ p \in \Sigma\}$. Then there are continuous mappings

$$
\begin{aligned}
\mathbb{M}_{g,k} &\longrightarrow \mathcal{M}_{g,k+1} && \text{(from i))}, \\
\mathbb{M}_{g_1,k_1} \times \mathbb{M}_{g_2,k_2} &\longrightarrow \mathcal{M}_{g_1+g_2,k_1+k_2} && \text{(from ii))}, \\
\mathcal{M}_{g,k} &\longrightarrow \mathcal{M}_{2g,2k-2} && \text{(from iv))}.
\end{aligned}
$$

This final map, which comes from applying the end-to-end construction iv) to two copies of the same surface, thus doubling it across an end, is only defined on some open set in the moduli space where some nondegeneracy condition is satisfied. We mention finally that there are many known minimal k-noids of high genus, as catalogued in [**24**],and for each such g, k we obtain by construction iii) an element of $\mathcal{M}_{g,k}$.

The general structure of these moduli spaces is contained in the

THEOREM 1 ([**20**]). *For each g, k, the moduli space $\mathcal{M}_{g,k}$ is a finite dimensional real analytic variety in a neighbourhood of each of its points. If $\Sigma \in \mathcal{M}_{g,k}$ is nondegenerate, then there exists a neighbourhood $\Sigma \in \mathcal{U} \subset \mathcal{M}_{g,k}$ which is a real analytic manifold of dimension $3k$.*

The first assertion in this theorem means that if $\Sigma \in \mathcal{M}_{g,k}$, there is an open neighbourhood $\mathcal{V}$ of Σ in the space of *all* surfaces near to Σ (in the topology of an appropriate separable Banach space), and a real analytic diffeomorphism $\Phi : \mathcal{V} \to \mathcal{V}'$ such that $\Phi(\mathcal{V} \cap \mathcal{M}_{g,k})$ lies in a finite dimensional linear subspace B; furthermore, there exists a real analytic function F defined on $B \cap \mathcal{V}'$ such that $\Phi(\mathcal{M}_{g,k} \cap \mathcal{V}) = F^{-1}(0)$. Notice that the dimension of B may depend on Σ, and in general it is unclear if it is bounded as Σ varies over the moduli space.

When Σ is nondegenerate, the proof is a straightforward application of Proposition 2 and the implicit function theorem. To prove the more general assertion, one uses Proposition 1 and the 'Kuranishi trick' (Ljapunov-Schmidt reduction).

Note that the generic dimension of this moduli space only depends on the number of ends, but not on the genus g.

This result illuminates the importance of nondegeneracy in the moduli space theory. Some consequences of this theorem are that $\mathcal{M}_{g,k}$ is a real analytic stratified space, or more precisely, admits a locally finite decomposition into real analytic strata. If Σ is a nondegenerate element in some stratum, then that stratum is maximal, i.e. it is not contained in the closure of any other stratum, and has the predicted dimension $3k$. Unfortunately, this theorem says nothing about the structure of the moduli space near its boundary, nor does it limit the number of components it might have.

We explain the dimension $3k$ which appears here. Supposing that Σ is nondegenerate, the implicit function theorem applies to (6) since its linearization (5) is surjective. Hence a neighbourhood of Σ in $\mathcal{M}_{g,k}$ has a real analytic coordinate chart parametrized by a ball in the nullspace of (5), and so we must show that this nullspace is $3k$-dimensional. Although not stated explicitly in §2, this nullspace is precisely the same as the nullspace of L_Σ on $\mathcal{C}_\mu^{2,\alpha}(\Sigma)$ for small $\mu > 0$. This mapping is surjective, by nondegeneracy, and the usual integration by parts argument shows that its nullspace is identified with the cokernel of L_Σ on $\mathcal{C}_{-\mu}^{2,\alpha}$. On the other hand,

$$6k = \dim W_\Sigma = \dim \ker L_\Sigma\big|_{\mathcal{C}_\mu^{2,\alpha}} + \dim \operatorname{coker} L_\Sigma\big|_{\mathcal{C}_{-\mu}^{2,\alpha}}.$$

These facts together show that the dimension of the nullspace is $3k$, as claimed. In general, however, and as an alternate derivation of this formula, we have that

$$\ker L_\Sigma\big|_{\mathcal{C}_\mu^{2,\alpha}} = \operatorname{coker} L_\Sigma\big|_{\mathcal{C}_{-\mu}^{2,\alpha}}, \qquad \operatorname{coker} L_\Sigma\big|_{\mathcal{C}_\mu^{2,\alpha}} = \ker L_\Sigma\big|_{\mathcal{C}_{-\mu}^{2,\alpha}},$$

where each equality here connotes that the left and right sides are identified by the integration pairing. Hence

$$\operatorname{ind}\left(L_\Sigma : \mathcal{C}_\mu^{2,\alpha}(\Sigma) \to \mathcal{C}_\mu^{0,\alpha}(\Sigma)\right) =$$

$$\frac{1}{2}\left(\operatorname{ind}\left(L_\Sigma : \mathcal{C}_\mu^{2,\alpha}(\Sigma) \to \mathcal{C}_\mu^{0,\alpha}(\Sigma)\right) - \operatorname{ind}\left(L_\Sigma : \mathcal{C}_{-\mu}^{2,\alpha}(\Sigma) \to \mathcal{C}_{-\mu}^{0,\alpha}(\Sigma)\right)\right).$$

This last quantity can be computed by a relative index theorem [**20**]: the answer can be computed *locally* at infinity. The contribution from each end is 3, and the full relative index is the sum of these contributions over all k ends.

For any $\Sigma \in \mathcal{M}_{g,k}$, Proposition 3 gives a map

$$\iota : \ker L_\Sigma|_{\mathcal{C}^{2,\alpha}_\mu} \longrightarrow W_\Sigma;$$

each Jacobi field ϕ gets mapped to a $6k$-tuple consisting of the components of each of the 6 model geometric Jacobi fields on each of the k ends. If $\Sigma \in \mathcal{M}_{g,k}$ is nondegenerate, this map is injective, and hence we can regard $T_\Sigma \mathcal{M}_{g,k} = \ker L_\Sigma$ as a subspace of W_Σ.

THEOREM 2 ([**20**]). *There is a natural symplectic structure on W_Σ such that when Σ is nondegenerate, the $3k$-dimensional subspace $\iota(\ker L_\Sigma) \subset W_\Sigma$ is a Lagrangian subspace.*

This result is somewhat at odds with the geometric structures which exist on other standard moduli spaces. In particular, there does not seem to be a canonical Riemannian metric on these moduli spaces, and it remains unclear what this Lagrangian structure means, or how it may be used. Kusner [**19**] has ventured some interesting speculations about its interpretation.

At almost the same time these theorems were proved, Perez and Ros [**32**] discovered the analogous result for certain moduli spaces of complete minimal surfaces of finite total curvature. Their techniques were rooted in the underlying Riemann surface theory, in particular the Weierstraß representation.

In the past few years, some new advances have been made concerning these moduli spaces. The first is a striking result of Große-Brauckman, Kusner, and Sullivan.

THEOREM 3 ([**10**], [**11**]). *Let E_3 be the group of rigid motions of $\mathbb{R}^3$. Then $\mathcal{M}_{0,3}/E_3$ is homeomorphic to the 3-ball.*

The argument to prove this relies on a beautiful mix of a classical geometric construction relating CMC surfaces in $\mathbb{R}^3$ with minimal surfaces in $\mathbb{S}^3$ and (Dennis Sullivan's) $\mathbb{Z}_2$ degree theory for proper mappings in the real analytic category. This theorem requires the existence of a nondegenerate element in $\mathcal{M}_{0,3}$, which is known from [**21**]. Notice that the dimension count is correct: $\dim E_3 = 6$ and so the (pre)moduli space is $6 + 3 = 9$ dimensional, which agrees with the $3 \cdot 3$ dimensions predicted by Theorem 1.

It is not known whether the homeomorphism in Theorem 3 is a diffeomorphism, nor whether can exist any triunduloids, i.e. elements of $\mathcal{M}_{0,3}$, which are degenerate. This latter question is quite important, and if answered in the negative, would have several interesting implications.

To phrase the final results, we let $\mathcal{R}_{g,k}$ be the Riemann moduli space for surfaces of genus g with k punctures, and define the 'forgetful map':

$$\mathcal{F} : \mathcal{M}_{g,k} \longrightarrow \mathcal{R}_{g,k}.$$

Thus, to any CMC surface Σ we associate its marked conformal completion $[\overline{\Sigma}, p_1, \cdots, p_k]$. Thus punctured neighbourhoods around these points p_j correspond to (slightly perturbed) half Delaunay surfaces. We also let $\mathcal{P}_{g,k} = \mathcal{R}_{g,k} \times \mathbb{R}^k$ and define an enhanced forgetful map

$$\tilde{\mathcal{F}} : \mathcal{M}_{g,k} \longrightarrow \mathcal{P}_{g,k},$$

$$\tilde{\mathcal{F}}(\Sigma) = \big([\overline{\Sigma}, p_1, \cdots, p_k], \epsilon_1, \ldots, \epsilon_k\big).$$

In other words, this map also records the asymptotic necksizes of each end. By generalizing the arguments in [**18**], Kusner has recently proved the very useful

THEOREM 4 ([**19**]). *Up to rigid motions, this enhanced forgetful map $\tilde{\mathcal{F}}$ is proper.*

This result implies that if Σ_j is any sequence of elements in $\mathcal{M}_{g,k}$, then one of the following three possibilities must hold:

- There exists a sequence of rigid motions T_j such that $T_j(\Sigma_j)$ converges to an element $\Sigma \in \mathcal{M}_{g,k}$,
- The marked conformal structures of the sequence $([\overline{\Sigma}_j], p_{j,1}, \ldots, p_{j,k})$ must degenerate in $\mathcal{R}_{g,k}$, or
- At least one of the necksizes $\epsilon_{j,\ell}$ tends to zero as $j \to \infty$.

This gives a very clean and intuitive picture of the modes of possible degeneration of sequences of CMC surfaces. Kusner also shows that the $\mathbb{Z}_2$ degree of this map is zero, so there an even number of preimages of any regular value.

A topic which warrants much closer examination is the structure of the moduli spaces $\mathcal{M}_{g,k}$ near their ends. More specifically, it would be very interesting to prove that $\mathcal{M}_{g,k}$ has a tractable, e.g. (semi)analytic, compactification, presumably obtained by adding moduli spaces $\mathcal{M}_{g',k'}$ with $g' \leq g$, $k' \leq k$, with at least one of the inequalities strict.

It is also natural to try to characterize the image of the unenhanced forgetful map $\mathcal{F}$. One motivation for this is that the Teichmüller space $\mathcal{R}_{g,k}$ is known to have fairly complicated topology (its fundamental group is the Artin braid group on k generators of the topological surface $\overline{\Sigma}$), and so if the image of $\mathcal{F}$ is relatively large, then $\mathcal{M}_{g,k}$ itself must have lots of topology.

THEOREM 5 ([**24**]). *For all $g \geq 0$ and $k \geq 3$, the mapping $\mathcal{F}$ is real analytic, and its image in $\mathcal{R}_{g,k}$ is a semianalytic (hence stratified) variety $\mathcal{I}_{g,k}$. When $g = 0$, then for every $k \geq 3$, $\mathcal{F}$ is surjective. Likewise, if $g > 0$, then the codimension of the principal stratum of $\mathcal{I}_{g,k}$ in $\mathcal{R}_{g,k}$ is uniformly bounded as $k \to \infty$.*

The main tool in the proof is the half Delaunay addition construction from §3; the proof of the first analyticity assertion reduces to showing that the uniformizing map which takes Σ to the unique conformally related constant negative curvature surface is real analytic. The surjectivity statement when $g = 0$ was also obtained by Kusner [**19**], and a somewhat weaker statement can also be deduced from Ratzkin's end-to-end gluing construction [**29**].

There are many interesting questions raised by these theorems. Kusner [**19**] lists several important ones concerning the finer structure of the map $\tilde{\mathcal{F}}$, and in particular concerning the precise number of points in the preimages of regular values.

5. CMC surfaces in other asymptotically Euclidean 3-manifolds

Let (X, h) be a complete three dimensional manifold with asymptotically Euclidean ends. Recall that this means that the following hypotheses hold: there exists a compact set $K \subset X$ such that $X \setminus K$ is a disjoint union of ends, $F_1, \ldots, F_\ell$, each of which is equipped with a diffeomorphism $\Psi_j : \mathbb{R}^3 \setminus B(0, R_j) \to F_j$ such that the metric h on X pulls back via Ψ_j^* to a metric on the exterior region in $\mathbb{R}^3$ of the form $\delta + \mathcal{O}(r^{-1})$, with corresponding decay for its derivatives.

We shall consider the space $\mathcal{M}_{g,k}(X, h)$ consisting of all proper Alexandrov embedded CMC surfaces in X of genus g with k ends. If X has more than one end, then we could also consider a decomposition into moduli spaces which explicitly label which end of Σ tends to infinity in which end of X, but we shall not introduce special notation for this. As noted in §2, the asymptotics result of [18] most likely extends to this setting, but since this is not worked out anywhere, we circumvent this by assuming that this moduli space contains only CMC surfaces which are asymptotically Delaunay.

It is straightforward to check that the main results about the structure of the moduli space for CMC surfaces in $\mathbb{R}^3$ persist in this more general situation. In particular, we still have that for any (X,h), $\mathcal{M}_{g,k}(X,h)$ is a locally real analytic variety, and near nondegenerate points it has dimension $3k$. Note that we do not need to assume that X or h are real analytic because this theorem relies on the implicit function theorem in some function space on X, and the mean curvature functional depends real analytically on the metric and its derivatives.

Consider first the case where $X = \mathbb{R}^3$ and h is a small (decaying) perturbation of the standard metric δ. We prove below that $\mathcal{M}_{g,k}(\mathbb{R}^3,h)$ is a small perturbation of the 'standard' moduli space $\mathcal{M}_{g,k}(\mathbb{R}^3,\delta)$, as well as a standard 'generic regularity' theorem, that this perturbed moduli space is smooth for generic h; in other words, for most small alterations of the ambient metric there are no degenerate CMC surfaces.

As an aside, I have already noted that no degenerate CMC surfaces are known to exist, and it seems very unlikely that it will be possible to construct one directly. If they exist at all, I suspect they will only be found by indirect means, by showing that the projection mapping Π of the big moduli space $\mathcal{S}$ defined below must have fold singularities. It is completely unclear whether this happens for the standard (or any!) metric on $\mathbb{R}^3$.

Let $Z \equiv S_{-1}^{2,\alpha}(\mathbb{R}^3)$ denote the space of all symmetric 2-tensors with coefficients in the weighted Hölder space $\mathcal{C}_{-1}^{2,\alpha}(\mathbb{R}^3)$. Thus if x is the standard Cartesian coordinate chart on $\mathbb{R}^3$, then any $\gamma \in Z$ satisfies $|\partial_x^\beta \gamma_{ij}| \leq C|x|^{-1-|\beta|}$ as $|x| \to \infty$, $|\beta| \leq 2$, and with the Hölder quotients of the second derivatives decaying like $|x|^{-3-\alpha}$. Let $\mathcal{U}$ be an open ball in this space such that if $\gamma \in \mathcal{U}$, then $\delta + \gamma$ is everywhere positive definite, hence is an asymptotically Euclidean metric on $\mathbb{R}^3$.

THEOREM 6. *There exists a dense subset $\mathcal{U}' \subset \mathcal{U}$ such that when $\gamma \in \mathcal{U}'$, $\mathcal{M}_{g,k}(\mathbb{R}^3,\delta+\gamma)$ is a smooth analytic manifold of dimension $3k$.*

PROOF. Consider the Banach manifold $\mathcal{B}$ of all Alexandrov embedded $\mathcal{C}^{2,\alpha}$ surfaces in $\mathbb{R}^3$ of genus g with k asymptotically Delaunay ends. A coordinate chart on this manifold near any point Σ is given by a small ball in $W_\Sigma \oplus \mathcal{C}_{-\mu}^{2,\alpha}(\Sigma)$. Thus $w \in W_\Sigma$ induces a change of the ends, either by rigid motion or change of necksize, and $u \in \mathcal{C}_{-\mu}^{2,\alpha}(\Sigma)$ gives a normal perturbation. We write the resulting surfaces as $\Sigma_{w,u}$.

The first step is to show that the set

$$\mathcal{S} = \{(\Sigma,\gamma) \in \mathcal{B} \times Z : \Sigma \in \mathcal{M}_{g,k}(\mathbb{R}^3,\delta+\gamma)\}$$

has the structure of a Banach submanifold near $\gamma = 0$. This is a straightforward application of the implicit function theorem. Fix $\Sigma \in \mathcal{M}_{g,k}$ and a coordinate chart $\mathcal{V}$ on $\mathcal{B}$, and define the functional $N : \mathcal{V} \times Z \to \mathcal{C}_{-\mu}^{0,\alpha}(\Sigma)$ by setting $N(w,u,\gamma)$ equal to the mean curvature function on $\Sigma_{w,u}$ with respect to the metric $\delta + \gamma$. The differential of this mapping $D_{12}N$ in the first two slots, at $(w,u) = (0,0)$, is the Jacobi operator L_Σ, and by Proposition 2 this operator has closed range of finite codimension. We must show that the full differential DN at $(0, 0, 0)$ is actually surjective. Granting this for the moment, we then obtain an analytic mapping Ψ defined in a neighbourhood of 0 in Z to $\mathcal{V}$ such that all points in $\mathcal{S}$ near to Σ are of the form $(\Psi(\gamma), \gamma)$.

To prove the claim about surjectivity, we must show that

$$\mathrm{ran}\left(L_\Sigma\right) + \mathrm{ran}\left(D_3 N|_0\right) = \mathcal{C}^{0,\alpha}_{-\mu}(\Sigma).$$

It is necessary to reinterpret this slightly. First, the mapping (5) is the restriction of the mapping (4), i.e. from $\mathcal{C}^{2,\alpha}_\mu(\Sigma)$ to $W_\Sigma \oplus \mathcal{C}^{2,\alpha}_{-\mu}(\Sigma)$. By Proposition 3, the nullspaces of these mappings are the same, hence their cokernels also agree. As in §4, duality considerations show that

$$\mathcal{C}^{0,\alpha}_\mu = \mathrm{ran}\left(\left.L_\Sigma\right|_{\mathcal{C}^{2,\alpha}_\mu}\right) \oplus \ker \left.L_\Sigma\right|_{\mathcal{C}^{2,\alpha}_{-\mu}}.$$

Hence we must show that there does not exist any $\psi \in \mathcal{C}^{2,\alpha}_{-\mu}(\Sigma)$ such that $L_\Sigma \psi = 0$ and $\langle \psi, D_3 N(\gamma) \rangle = 0$ for all $\gamma \in Z$. In the end, we show that when γ ranges over all $\mathcal{C}^{2,\alpha}$ symmetric 2-tensors supported in some small ball $\mathcal{W}$ in Σ, the functions $D_3 N(\gamma)$ fill out $\mathcal{C}^{0,\alpha}(\mathcal{W})$, and this will prove the claim since if the orthogonality condition were to hold, then ψ would have to vanish in this set $\mathcal{W}$, which is impossible since it is in the nullspace of L_Σ.

By dilating the space substantially and using an approximation argument, it will be enough to consider the simpler problem where $\Sigma = \mathbb{R}^2 \subset \mathbb{R}^3$, and so we must compute the linearization $D_3 N$ which measures how the mean curvature of the flat plane $\{(x_1, x_2, 0)\}$ in $\mathbb{R}^3$ changes as we vary the metric. This computation is not too horrible, fortunately, and we obtain that

$$D_3 N(\gamma) = (\mathrm{div}\, \gamma)_3 + \frac{\partial}{\partial x_3} \frac{1}{2}(\mathrm{tr}\, \gamma);$$

here $\gamma = (\gamma_{ij})$ and the divergence and trace are computed only in the (x_1, x_2) directions. The subscript 3 in the first term on the right denotes the third component of $\mathrm{div}\, \gamma$, i.e. $\gamma_{i3;}{}^i$. Examining this expression, it is clear that we can specify it arbitrarily in the bounded set $\mathcal{W}$. This proves the claim.

Consider the projections $\Pi_1 : \mathcal{B} \times Z \to \mathcal{B}$ and $\Pi_2 : \mathcal{B} \times Z \to Z$. The restriction of Π_2 to $\mathcal{S}$ has surjective differential at Σ' if and only Σ' is nondegenerate with respect to the metric $\delta + \gamma$, $\gamma = \Pi_2(\Sigma')$. Hence by the Sard-Smale theorem, if $K \subset \mathcal{B}$ is a compact set containing Σ, then there is an open dense set $\mathcal{U}_K \subset Z$ of metrics near δ such that every surface in $\Pi_1^{-1}(K) \cap \mathcal{S} \cap \Pi_2^{-1}(\mathcal{U}_K)$ is nondegenerate (with respect to the appropriate metric).

To globalize this, cover $\mathcal{M}_{g,k}$ by a countable union of compact sets $\mathcal{K}_j$ and choose open dense neighbourhoods $\mathcal{U}_j$ in Z as above. Then the intersection of these sets $\mathcal{U}_j$ is still dense, by Baire's theorem, and has the desired property. $\square$

It is most likely possible to modify this proof to show something stronger and much more useful. Let $\mathcal{D}$ denote the set of 'bad' metrics, i.e. those for which the perturbed moduli space contains degenerate elements, then I expect it is true that $\mathcal{D}$ decomposes into a union of $\mathcal{D}' \cup \mathcal{D}''$ where $\mathcal{D}'$ is a codimension $2k$ submanifold in $\mathcal{U}$ and $\mathcal{D}''$ is quantitatively smaller. (In fact, probably $\mathcal{D}$ has a stratified structure with only strata of bounded codimension.) The reason for this conjecture is as follows. We may view this process of destroying degenerate surfaces as a problem in eigenvalue perturbation theory. As explained in §2, the essential spectrum of

L_Σ is a locally finite union of intervals I_j and 0 always lies on the left endpoint of one of these intervals. In addition, there is always a small interval $(-\epsilon, 0)$ which is disjoint from the spectrum; here ϵ depends on Σ. Σ is degenerate if and only if 0 is also in the point spectrum. There is a good procedure for tracking what happens to this L^2 eigenvalue as Σ' varies in $\mathcal{B}$. To explain it, consider (for any fixed Σ') the resolvent $R(\lambda) = (-L_{\Sigma'} - \lambda^2)^{-1}$; if we restrict λ^2 to lie in $B(0,\epsilon) \setminus [0,\epsilon)$, or equivalently, if λ lies in the half-ball $B(0, \sqrt{\epsilon}) \cap \{\mathrm{Im}\,\lambda < 0\}$, which is contained in the resolvent set of $-L_{\Sigma'}$, then $R(\lambda)$ is bounded on $L^2(\Sigma')$. Using the machinery developed in [**26**] it is quite straightforward to show that $R(\lambda)$ continues meromorphically to some neighbourhood of 0 in either the complex plane or the logarithmic complex plane (the latter being necessary if 0 is a nontrivial ramification point in this continuation). This continued resolvent has at most a double pole at the origin, and the coefficient of $1/\lambda^2$ is the projection onto the space of L^2 Jacobi fields. Thus we are interested in the dependence of this meromorphic structure as Σ' varies. If we were starting with some eigenvalue $\lambda_0^2 > 0$ (i.e. inside the continuous spectrum, but away from the 'threshold value' 0), then $R(\lambda)$ has only a simple pole at λ_0. As we deform Σ, this pole could move either to the left or right on the real axis or upwards, into the so-called nonphysical half-plane (where it would become a resonance, or scattering pole). This is predicted by 'Fermi's golden rule' [**33**]. Agmon and Herbst have studied persistence of embedded eigenvalues and have shown in as yet unpublished work, but cf. also [**5**], that for non-threshold eigenvalues, the set of perturbations which leave the eigenvalue fixed is a submanifold of codimension equal to the multiplicity of the continuous spectrum, which in this case equals $2k$. Unfortunately, their proof does not carry over in an obvious way to our case, where the pole is double and occurs at the ramification point. I hope that this will be possible, and if it is, full details will be given elsewhere.

This somewhat arcane issue is interesting for the following reason. We have shown that for most h near δ, $\mathcal{M}_{g,k}(\mathbb{R}^3, h)$ is a smooth (in fact, analytic) manifold of dimension $3k$. However, the topology of this manifold may depend on h. It would be quite interesting if there were a canonical smoothing of the moduli space $\mathcal{M}_{g,k}(\mathbb{R}^3, \delta)$, and to show this it would be sufficient to know that the set $\mathcal{D}$ of bad metrics has codimension at least two, so that there are no 'wall-crossing' transitions. This is precisely the point of the preceding discussion. We note that this discussion carries over to the more general setting of the moduli space of CMC surfaces in general asymptotically Euclidean manifolds (X, h), and so we might then expect to get canonical smooth moduli spaces which depend only on X, but not on h! A case of particular interest is when $X = \overline{X} \setminus \{p_1, \ldots, p_\ell\}$, the complement of an ordered finite set of points in a smooth compact manifold, endowed with a metric which is asymptotically Euclidean near these punctures. We would then obtain a canonical object $\mathcal{M}_{g,k}(\overline{X}, \{p_1, \ldots, p_\ell\})$ associated only to the smooth structure of $\overline{X}$ and the isotopy class of the marked set of points. As ventured earlier, one might hope that this CMC moduli space could reflect some of the topology of this data.

We conclude by discussing a final example: if $p \in \overline{X}$, and h is an asymptotically Euclidean metric on $X = \overline{X} \setminus \{p\}$, then $\mathcal{M}_{0,2}(X, \epsilon^{-2}h)$ contains many nontrivial elements when ϵ is small enough. The construction is quite simple. Let γ be a bi-infinite geodesic which is properly embedded in X; thus both ends converge to ∞ in the punctured ball around p. Choose Fermi coordinates $(s, r, \theta) \in \mathbb{R} \times (0, 1) \times S^1$ around γ corresponding to the metric $\epsilon^{-2}h$ and use these to define the approximate

Delaunay surface $r = \rho_\epsilon(s)$. The mean curvature of this is approximately 1, but is not exactly constant. A straightforward perturbation argument shows that there is a small perturbation of this surface which is CMC; we omit the details. Notice that there may be many nonisotopic geodesics, depending on the topology of X, and hence there may be many components in this moduli space. In fact, one expects that the components are in one-to-one correspondence with $\pi_1(X, p)$.

Once these Delaunay-type surfaces have been constructed, it is also not difficult to carry out the analogue of the construction i), of adding Delaunay ends, and most of the other constructions too.

While it is far from clear that this extension of the CMC moduli space theory will have any uses, these spaces are certainly natural geometric objects and I suspect that there is a lot more structure which has yet to be uncovered.

References

[1] A.D. Alexandrov, *A characteristic property of spheres*, Ann. Mat. Pura Appl. **58** (1962) 303-315.

[2] H. Brezis and J.-M. Coron, *Multiple solutions of H-systems and Rellich's conjecture*, Comm. Pure Appl. Math. **37** (1984), No. 2, 149-187.

[3] R. Bryant, *Surfaces of mean curvature one in hyperbolic space*, in *Théorie des variétes minimales et applications (Palaiseau, 1983-1984)*, Astérisque **154-155** (1987) 321-347.

[4] *Proceedings of the Clay Mathematics Institute Summer School on the Global Theory of Minimal Surfaces*. To appear.

[5] J. Cruz-Sampedro, I. Herbst and R. Martinez-Avendaño, *Perturbations of the Wigner-von Neumann potential leaving the embedded eigenvalue fixed*, Annales Henri Poincare **3** (2002), no. 2, 331-345.

[6] P. Collin, R. Kusner, W. Meeks III and H. Rosenberg, *The topology, geometry and conformal structure of properly embedded minimal surfaces*. Preprint (2001).

[7] C. Delaunay, *Sur la surface de revolution dont la courbure moyenne est constante*, Jour. de Mathématiques, **6** (1841) 309-320.

[8] G. Gregori and R. Mazzeo, *Constant mean curvature surfaces in hyperbolic space with prescribed asymptotic behaviour at infinity*, in preparation.

[9] K. Große-Brauckmann, *New surfaces of constant mean curvature*, Math. Z. **214** (1993) 527-565.

[10] K. Große-Brauckmann, R. Kusner and J. Sullivan, *Constant mean curvature surfaces with three ends*, Proc. Nat. Acad. Sci. USA **97** (2000) 14067-14068.

[11] K. Große-Brauckmann, R. Kusner and J. Sullivan, *Triunduloids: embedded constant mean curvatures surfaces with three ends and genus zero*, ArXiv:math.DG/0102183, To appear in J. Reine Angewandte Mathematik.

[12] N. Kapouleas, *Complete constant mean curvature surfaces in Euclidean three space*, Ann. of Math. (2) **131** (1990), 239-330.

[13] N. Kapouleas *Compact constant mean curvature surfaces in Euclidean three-space*, J. Diff. Geom. **33** (1991), 683–715.

[14] N. Kapouleas *Constant mean curvature surfaces constructed by fusing Wente tori*, Invent. Math. **119** (1995), 443–518.

[15] M. Kilian, I. McIntosh, N. Schmitt, *New constant mean curvature surfaces*, Experiment. Math. **9** (2000) 595-611.

[16] N. Korevaar and R. Kusner *The global structure of constant mean curvature surfaces*, Invent. Math. **114** (1993) 311-332.

[17] N. Korevaar, R. Kusner, W. Meeks III and B. Solomon, *Constant mean curvature surfaces in hyperbolic space*, Amer. Jour. Math. **114** (1992), 1-43.

[18] N. Korevaar, R. Kusner and B. Solomon, *The structure of complete embedded surfaces with constant mean curvature*, J. Differential Geometry 30 (1989) 465–503

[19] R. Kusner, *Conformal structures and necksizes of embedded constant mean curvature surfaces*, in [**4**]

[20] R. Kusner, R. Mazzeo and D. Pollack, *The moduli space of complete embedded constant mean curvature surfaces,* Geom. Funct. Anal. **6** (1996) 120–137.

[21] R. Mazzeo and F. Pacard, *Constant mean curvature surfaces with Delaunay ends,* Comm. Anal. Geom. **9** No. 1 (2001) 169–237.

[22] R. Mazzeo and F. Pacard, *Bifurcating Nodoids* in Topology and Geometry: Commemorating SISTAG, Contemporary Mathematics Vol. 314 (2002), 169–186.

[23] R. Mazzeo, F. Pacard and D. Pollack, *Connected Sums of constant mean curvature surfaces in Euclidean 3 space,* J. Reine Angew. Math. **536** (2001), 115–165.

[24] R. Mazzeo, F. Pacard and D. Pollack, *The conformal theory of Alexandrov embedded constant mean curvature surfaces in* $\mathbb{R}^3$, in [**4**].

[25] R. Mazzeo, F. Pacard, D. Pollack and J. Ratzkin, In Preparation.

[26] R. Mazzeo, D. Pollack and K. Uhlenbeck, *Moduli spaces of singular Yamabe metrics,* J. Amer. Math. Soc. **9** (1996), no. 2, 303–344.

[27] R. Mazzeo and D. Pollack, *Gluing and moduli for some noncompact geometric problems,* in *Geometric Theory of Singular Phenomena in Partial Differential Equations, Symposia Mathematica Vol XXXVIII,* Cambridge Univ. Press (1998) 17-51.

[28] W. Meeks III, *The topology and geometry of embedded surfaces of constant mean curvature,* J. Differential Geom. **27** (1988), no. 3, 539–552.

[29] J. Ratzkin, *An end-to-end gluing construction for surfaces of constant mean curvature,* PhD Thesis, University of Washington (2001).

[30] R. Osserman, *A survey of minimal surfaces.* Second Edition, Dover Publications, Inc., New York, 1986. vi + 207pp.

[31] F. Pacard and F. Pimentel, *Attaching handles to Bryant surfaces,* arXiv:DG/0112224.

[32] J. Perez and A. Ros, *The space of properly embedded minimal surfaces with finite total curvature,* Indiana Univ. Math. J. **45** (1996) 177-204.

[33] M. Reed and B. Simon, *Methods of Mathematical Physics, Vol. IV,* Academic Press, New York,

[34] R. Schoen, *The existence of weak solutions with prescribed singular behavior for a conformally invariant scalar equation,* Comm. Pure and Appl. Math. **XLI** (1988) 317-392.

[35] M. Struwe, *Plateau's problem and the calculus of variations,* Mathematical Notes **35**, Princeton University Press, Princeton, NJ (1988).

[36] H. Wente, *Counterexamples to a conjecture of H. Hopf,* Pacific J. Math **121** (1986) 193-243.

DEPARTMENT OF MATHEMATICS, STANFORD UNIVERSITY, STANFORD, CA 94305
E-mail address: mazzeo@math.stanford.edu

Contemporary Mathematics
Volume **350**, 2004

On Some Properties of S^1-valued Fractional Sobolev Spaces

Petru Mironescu

This paper dedicated to Haim Brezis and Felix Browder

ABSTRACT. We describe, for $0 < s < \infty, 1 \leq p < \infty$, the path-connected components of the space $W^{s,p}(\Omega; S^1)$, where $\Omega \subset \mathbb{R}^N$. We determine also all the values of s and p for which $C^\infty(\bar{\Omega}; S^1)$ is dense in $W^{s,p}(\Omega; S^1)$.

I. Introduction

We present below several results obtained with H. Brezis in [10] and with J. Bourgain and H. Brezis in [6] concerning some properties of the spaces $W^{s,p}(\Omega; S^1)$. Here, $0 < s < \infty$, $1 \leqslant p < \infty$, Ω is a smooth, bounded, connected open set in $\mathbb{R}^N$ and

$$W^{s,p}(\Omega; S^1) = \{ u \in W^{s,p}(\Omega; \mathbb{R}^2); \ |u| = 1 \text{ a.e.} \}.$$

The first topic concerns the description of the the homotopy classes (i.e., path-connected components) of $W^{s,p}(\Omega; S^1)$. Our main results in this direction are

THEOREM 1. *If $sp < 2$, then $W^{s,p}(\Omega; S^1)$ is path-connected.*

THEOREM 2. *If $sp \geqslant 2$, then $W^{s,p}(\Omega; S^1)$ and $C^0(\bar{\Omega}; S^1)$ have the same homotopy classes in the sense of [8]. More precisely:*

a) each $u \in W^{s,p}(\Omega; S^1)$ is $W^{s,p}$-homotopic to some $v \in C^\infty(\bar{\Omega}; S^1)$;

b) two maps $u, v \in C^\infty(\bar{\Omega}; S^1)$ are C^0-homotopic if and only if they are $W^{s,p}$-homotopic.

Here a simple consequence of the above results

1991 *Mathematics Subject Classification.* 46E35, 46E40.

Key words and phrases. S^1-valued maps, fractional Sobolev spaces, connected components, density of smooth maps.

COROLLARY. *If $0 < s < \infty, 1 \leqslant p < \infty$ and Ω is simply connected, then $W^{s,p}(\Omega; S^1)$ is path-connected.*

Indeed, when $sp < 2$ this is the content of Theorem 1. When $sp \geqslant 2$, we use a) of Theorem 2 to connect u_1, $u_2 \in W^{s,p}(\Omega; S^1)$ to v_1, $v_2 \in C^\infty(\bar{\Omega}; S^1)$; since Ω is simply connected, we may write $v_j = e^{\imath\varphi_j}$ for $\varphi_j \in C^\infty(\bar{\Omega}; \mathbb{R})$ and then we connect v_1 to v_2 via $e^{\imath[(1-t)\varphi_1 + t\varphi_2]}$.

When M is a compact connected manifold, the study of the topology of $W^{1,p}(\Omega; M)$ was initiated in Brezis - Li [8] (see also White [26] for some related questions). In particular, these authors proved Theorems 1 and 2 in the special case $s = 1$. The analysis of homotopy classes for an arbitrary manifold M and $s = 1$ was subsequently tackled by Hang - Lin [15]. The passage to $W^{s,p}$ introduces two additional difficulties:

a) when s is not an integer, the $W^{s,p}$ norm is not "local";

b) when $s > 1 + 1/p$, gluing two maps in $W^{s,p}$ does not yield a map in $W^{s,p}$.

In our proofs, we exploit in an essential way the fact that the target manifold is S^1. (The case of a general target is widely open.) In particular, we use the existence of a lifting of $W^{s,p}$ unimodular maps when $s \geqslant 1$ and $sp \geqslant 2$ (see Bourgain - Brezis - Mironescu [4]). Another important tool is the following

COMPOSITION THEOREM (BREZIS - MIRONESCU [10]). *If $f \in C^\infty(\mathbb{R}; \mathbb{R})$ has bounded derivatives and $s \geqslant 1$, then $\varphi \longmapsto f \circ \varphi$ is continuous from $W^{s,p} \cap W^{1,sp}$ into $W^{s,p}$.*

Remark 1. A very elegant and straightforward proof of the above result has been given by V.Maz'ya and T.Shaposhnikova [18]. In particular, their proof covers the case $p = 1$, which we were not able to treat in our original approach.

A related question is the description, when $sp \geqslant 2$, of the homotopy classes of $W^{s,p}(\Omega; S^1)$ in terms of lifting. Here are our results

THEOREM 3. *We have*

a) if $s \geqslant 1$, $N \geqslant 3$ and $2 \leqslant sp < N$, then

$$[u]_{s,p} = \{ue^{\imath\varphi}; \varphi \in W^{s,p}(\Omega; \mathbb{R}) \cap W^{1,sp}(\Omega; \mathbb{R})\};$$

b) if $sp \geqslant N$, then

$$[u]_{s,p} = \{ue^{\imath\varphi}; \varphi \in W^{s,p}(\Omega; \mathbb{R})\};$$

c) if $0 < s < 1$, $N \geqslant 3$ and $2 \leqslant sp < N$, then

$$[u]_{s,p} = \overline{\{ue^{\imath\varphi}; \varphi \in W^{s,p}(\Omega; \mathbb{R})\}}^{W^{s,p}}.$$

Theorem 3 is due to Rubinstein - Sternberg [21] in the special case where $s = 1$, $p = 2$ and Ω is the solid torus in $\mathbb{R}^3$. In case c), the answer is not completely satisfactory. However, using the "non-lifting" results in Bourgain - Brezis - Mironescu [4], it is easy to see that

$$[u]_{s,p} \neq \{e^{\imath\varphi}; \varphi \in W^{s,p}(\Omega; \mathbb{R})\}.$$

In a different but related direction, we investigate the local path-connectedness of $W^{s,p}(\Omega; S^1)$. Our main result is

THEOREM 4. *Let $0 < s < \infty, 1 \leqslant p < \infty$. Then $W^{s,p}(\Omega; S^1)$ is locally path-connected. Consequently, the homotophy classes coincide with the connected components and they are open and closed.*

The heart of the matter in the proof is the following

Claim. Let $0 < s < \infty, 1 \leqslant p < \infty$. Then there is some $\delta > 0$ such that, if $\|u - 1\|_{W^{s,p}} < \delta$, then u may be connected to 1 in $W^{s,p}$.

We suspect this result to hold in the case of general target manifolds:

Conjecture 1. Let Ω be a manifold with or without boundary. Then $W^{s,p}(\Omega; M)$ locally path-connected for every s, p and every compact manifold M.

The case $s = 1$ can be settled using the methods of Hang - Lin [15]. We will return to this question in a subsequent work; see Brezis - Mironescu [12].

The second topic concerns the density of $C^\infty(\bar\Omega; S^1)$ into $W^{s,p}(\Omega; S^1)$.

THEOREM 5. *We have, for $0 < s < \infty$, $1 < p < \infty$:*

$C^\infty(\bar\Omega; S^1)$ *is always dense in* $W^{s,p}(\Omega; S^1)$ *except when $N \geqslant 2$ and $1 \leqslant sp < 2$.*

Special cases of Theorem 5 were already known. Density when $sp < 1$ is due to Escobedo [14]; so is non-density when $N \geqslant 2$ and $1 \leqslant sp < 2$, but in this case the idea goes back to Schoen - Uhlenbeck [24] (see also Bourgain - Brezis - Mironescu [5]). For $s = 1$ and $p \geqslant N$, density is due to Schoen - Uhlenbeck [24]; their argument can be adapted to the general case (see, e.g., Brezis - Nirenberg [13] or Brezis - Li [8]). The only new result concerns the density when $N \geqslant 3$ and $2 \leqslant sp < N$. The proof relies heavily on the Composition Theorem, Theorems 2 and 3 and Theorem 6 below. We also mention that for $s = 1$ and $\Omega = B_1$, Theorem 4 was established by Bethuel - Zheng [3]. For a general compact connected manifold M and for $s = 1$, the question of density of $C^\infty(\bar\Omega; M)$ into $W^{1,p}(\Omega; M)$ was settled by Bethuel [1] and Hang - Lin [15].

When there is no density of smooth maps, i.e., when $N \geqslant 2$ and $1 \leqslant sp < 2$, we have instead the following

THEOREM 6. *Assume $N \geqslant 2$, $0 < s < \infty$, $1 \leqslant p < \infty$ and $1 \leqslant sp < N$. Set*

$\mathcal{R} = \{u \in W^{s,p}(\Omega; S^1); u$ *is smooth except a finite number of $N-[sp]-1$ manifolds*$\}$.

(Here, the number and location of singular manifolds is left free). Then $\mathcal{R}$ is dense in $W^{s,p}(\Omega; S^1)$.

Comment. $\mathcal{R}$ was already known to be dense in $W^{s,p}(\Omega; S^1)$ in many cases, e.g.:

a) $s = 1$ and $1 \leqslant p < 2$; see Bethuel-Zheng [3];

b) $s = 1 - 1/p$ and $2 < p < 3$; see Bethuel [2];

c) $s = 1/2$ and $p = 2$; see Rivière [20].

Remark 2. In Theorems 2 and 5, one may replace Ω by a manifold with or without boundary. The statements are unchanged. However, the argument in the proof of Theorem 1 does not quite go through to the case of a manifold without boundary. Nevertheless, we make the following

Conjecture 2. Let Ω be a manifold without boundary with $\dim \Omega \geqslant 2$. Then $W^{s,p}(\Omega; M)$ is path-connected for every $0 < s < \infty$, $1 \leqslant p < \infty$ with $sp < 2$, and for every compact connected manifold M.

Note that the condition $\dim \Omega \geqslant 2$ is necessary, since $W^{s,p}(S^1; S^1)$ is not path-connected when $sp \geqslant 1$.

II. Outline of the proofs

We present below the structure of the proofs of the main results. For the complete proofs, we refer to [6], [11].

Proof of Theorem 1. The main ingredient is the following

LEMMA 1. *Let* $0 < s < \infty$, $1 \leqslant p < \infty$, $sp < 2$. *Then each* $u \in W^{s,p}(\Omega; S^1)$ *has an extension* $v \in W^{s,p}_{\mathrm{loc}}(\mathbb{R}^N; S^1)$ *such that* $v(x) \equiv 1$ *for large* $|x|$.

The proof of Lemma 1 relies heavily on the lifting results in [4]. In some special cases, the above result is a consequence of the methods developed in [16]. However, the techniques of [16] do not seem to yield directly the full statement of Lemma 1.

Theorem 1 is an easy consequence of Lemma 1. Indeed, fix a vector $\vec{e} \in \mathbb{R}^N \setminus \{0\}$ and consider the map $t \to v_t = v(\cdot - t\vec{e})_{|\Omega}$. Then $v_0 = u$ and, for large t, $v_t \equiv 1$. Thus each map in $W^{s,p}(\Omega; S^1)$ is homotopic to the constant 1, which ends the proof.

Proof of Theorem 2 when $0 < s < 1$ **and** $2 \leqslant sp < N$. We adapt the techniques developed in [8] and subsequently used in [15]; see also [1]. Roughly speaking, we consider, for $u \in W^{s,p}(\Omega; S^1)$, its restriction to "generical" $[sp] - 1$ dimensional skeletons. This restriction is continuous and has an extension $v \in C^0(\overline{\Omega}; S^1) \cap W^{s,p}(\Omega; S^1)$. We next prove that two $W^{s,p}$-maps that coincide on a $[sp] - 1$ dimensional skeleton are homotopic. By mollifying v, we prove that any map in $W^{s,p}(\Omega; S^1)$ can be connected to a smooth map. As a consequence, the set of path-connnected components of $W^{s,p}(\Omega; S^1)$ is naturally included in the set of path-connected components of $C^\infty(\overline{\Omega}; S^1)$.

In order to prove that the path-connected components of $W^{s,p}(\Omega; S^1)$ are exactly the same as those of $C^\infty(\overline{\Omega}; S^1)$, we rely on the following easy consequence of the degree theory developed in [9] (see also [26]):

LEMMA 2. *Let* $u, v \in C^\infty(\overline{\Omega}; S^1)$ *and* $sp \geqslant 2$. *Then the following assertions are equivalent:*
a) $u \sim v$ *in* $W^{s,p}(\Omega; S^1)$;
b) the restrictions of u, v *on each closed curve in* Ω *have the same winding number;*
c) $u \sim v$ *in* $C^0(\overline{\Omega}; S^1)$.

Proof of Theorem 2 when $s \geqslant 1$. In this case, the proof relies essentially on lifting techniques. By adapting the techniques in [4], we obtain the following

LEMMA 3. *Let* $s \geqslant 1$, $1 \leqslant p < \infty$.
a) if $sp \geqslant N$, *then each* $u \in W^{s,p}(\Omega; S^1)$ *may be written as* $u = ve^{i\varphi}$ *for some* $\varphi \in W^{s,p}(\Omega; \mathbb{R})$ *and* $v \in C^\infty(\overline{\Omega}; S^1)$;
b) if $2 \leqslant sp < N$, *then each* $u \in W^{s,p}(\Omega; S^1)$ *may be written as* $u = ve^{i\varphi}$ *for some* $\varphi \in W^{s,p}(\Omega; \mathbb{R}) \cap W^{1,sp}(\Omega; S^1)$ *and* $v \in C^\infty(\overline{\Omega}; S^1)$.

When Ω is simply connected, these results were obtained in [4] (with $v \equiv 1$). Using Lemma 3 and the Composition Theorem, we may easily connect any u as above to the corresponding v. Finally, Lemma 2 implies that the path-connected components of $W^{s,p}(\Omega; S^1)$ and $C^\infty(\overline{\Omega}; S^1)$ are the same.

Proof of Theorem 4. By Theorem 1, it suffices to consider the case $sp \geqslant 2$. Using the following immediate equality

$$(1) \qquad\qquad [u]_{s,p} = u[1]_{s,p},$$

the proof amounts to proving that $[1]_{s,p}$ is a neighborhood of the constant 1, i.e., the Claim stated in the Introduction. The main ingredient in the proof of the Claim is the following

LEMMA 4. *Let* $1 \leqslant p < \infty$, $sp \geqslant 2$ *and* $u \in W^{s,p}(\Omega; S^1)$. *Then there is a* $\delta > 0$ *such that: if* $\|u - 1\|_{W^{s,p}} < \delta$, *then there is a generic 1-dimensional skeleton $\mathcal{C}$ such that u has degree* 0 *on each square of the skeleton.*

Note that, on generic 1-skeletons, u is in $W^{s,p}$ in one dimension, and thus continuous. Therefore, the above degrees make sense. For u as in Lemma 4, let $v \in C^\infty(\overline{\Omega}; S^1)$ be such that $u \sim v$ in $W^{s,p}$. By the degree theory in [9], the degree of v on each square of $\mathcal{C}$ is 0. Then, clearly, v is of the form $v = e^{i\varphi}$, for some $\varphi \in C^\infty(\overline{\Omega}; S^1)$. In particular, $v \sim 1$ in $W^{s,p}$, so that $u \sim 1$ in $W^{s,p}$. This proves the Claim.

Proof of Theorems 5 and 6. When $s \geqslant 1$ and $sp \geqslant 2$, Lemma 3 combined with the Composition Theorem yield immediately the density of smooth maps in $W^{s,p}(\Omega; S^1)$.

When $N \geqslant 2$, $s = 1$, $1 \leqslant p < 2$, Theorem 6 was proved in [3]. In the case $N \geqslant 2$, $s > 1$, $1 \leqslant sp < 2$, the proof of Theorem 6 relies on a technique developed in [16], that goes as follows: let $u \in W^{s,p}(\Omega; S^1)$ and consider a sequence $(v_n) \subset C^\infty(\Omega; \mathbb{R}^2)$ such that $v_n \to u$ in $W^{s,p}$. For $a \in \mathbb{C}$ with $|a| \leqslant 1/2$, consider the maps $v_{n,a} = (v_n - a)/|v_n - a|$ and $u_a = (u - a)/|u - a|$. Following [16], under our hypotheses on s and p, we may find a sequence (a_n) such that $v_{n,a_n} - u_{a_n} \to 0$ in $W^{s,p}$. If Φ_a is the inverse of the map

$$S^1 \ni z \mapsto (z - a)/|z - a| \in S^1,$$

a straightforward variant of the Composition Theorem implies that

$$u_n = \Phi_{a_n} \circ v_{n,a_n} \to u = \Phi_{a_n} \circ u_{a_n} \qquad \text{in } W^{s,p},$$

which ends the proof of Theorem 6 in this case, since $u_n \in \mathcal{R}$.

The real novelty concerns the case $0 < s < 1$, $1 \leqslant sp < N$. In this case, we consider a new technique for approximating general maps with maps having a simple structure. Here it is how it works: let $\varepsilon > 0$ and $a \in \mathbb{R}^N$. Consider the ε-grid $\mathcal{G} = \mathcal{G}_{\varepsilon,a}$ having a as one of the edges. For f defined in $\mathbb{R}^N$, let $g = g_{\varepsilon,a}$ be its restriction to the $[sp]$-dimensional skeleton of $\mathcal{G}$. We may extend g by 0-homogeneity , first to the $([sp] + 1)$-dimensional skeleton, next to the $([sp] + 2)$-dimensional one, and so on. Finally, we obtain a map $f_{\varepsilon,a}$ defined on $\mathbb{R}^N$. We have

LEMMA 5. *If $sp < N$, and $f \in W^{s,p}$, then we may find sequences $\varepsilon_n \to 0$ and $(a_n) \subset \mathbb{R}^N$ such that $f_{\varepsilon_n, a_n} \to f$ in $W^{s,p}$.*

Lemma 5 essentially implies Theorem 6. Indeed, the sequence of maps provided by Lemma 5 is not quite in $\mathcal{R}$, but almost: for any f_{ε_n, a_n}, its restriction to the corresponding $[sp]$-dimensional skeleton is VMO. Using standard mollification techniques (see, e.g., [8]), we may approximate these restrictions by smooth maps $g_{n,m}$. If we consider the homogeneous extension $f_{n,m}$ of $g_{n,m}$ to $\mathbb{R}^N$, it is easy to see, by further mollification, that each $f_{n,m}$ is in the closure of $\mathcal{R}$. A standard computation shows that each f_{ε_n, a_n} (and thus f) is in the closure of $\mathcal{R}$. This argument completes the proof of Theorem 6.

We note that, while the idea of considering homogeneous maps as above appeared already in the context of $W^{1,p}$-maps (see [1], [8], [15]), the conclusion of Lemma 6 **fails** when $s = 1$.

Going back to the proof of Theorem 5, the remaining case is $N \geqslant 2$, $0 < s < 1$, $2 \leqslant sp < N$. In this case, in view of Theorem 6, it suffices to prove that each map in $\mathcal{R}$ can be approximated with smooth maps. This was proved by Bethuel [1] when $s = 1$; his argument goes through the case $0 < s < 1$.

Proof of Theorem 3. In all the cases, the inclusion $[u]_{s,p} \supset \ldots$ is an easy consequence of the Composition Theorem (in case c), one has also to use Theorem 4). Conversely, cases a) and b) follow from Lemma 3. Indeed, if $u_1 \sim u_2$ in $W^{s,p}$, write $u_j = v_j e^{i\varphi_j}$, with $v_j \in C^\infty$ and φ_j as in Lemma 3. Using the Compsition Theorem, it is easy to see that $v_1 \sim v_2$ in $W^{s,p}$. By Lemma 2, it follows that $v_2 = e^{i\psi}$ for some $\psi \in C^\infty$. Thus $u_2 = u_1 e^{i\varphi}$, where $\varphi = \psi - \varphi_2 + \varphi_1$ has the regularity required by the conclusion of Theorem 3.

In case c), the argument combines Lemma 2 and Theorems 4 and 5. By (1), it suffices to prove c) when $u \equiv 1$. If $v \sim 1$ in $W^{s,p}$, consider a sequence $(v_n) \subset C^\infty(\overline{\Omega}; S^1)$ such that $v_n \to v$ in $W^{s,p}$. By Theorem 4, $v_n \sim v \sim 1$ for large n. By Lemma 2, we have, for large n, $v_n \sim 1$ in C^0, so that, for such n's, we have $v_n = e^{i\varphi_n}$ for some smooth φ_n's. This implies easily the desired result.

References

[1] **F. Bethuel**, *The approximation problem for Sobolev maps between two manifolds*, Acta Math., **167** (1991), 153-206.

[2] **F. Bethuel**, *Approximation in trace spaces defined between manifolds*, Nonlinear Anal. TMA, **24** (1994), 121-130.

[3] **F. Bethuel and X. Zheng**, *Density of smooth functions between two manifolds in Sobolev spaces*, J. Funct. Anal., **80** (1988), 60-75.

[4] **J. Bourgain, H. Brezis and P. Mironescu**, *Lifting in Sobolev spaces*, J. d'Analyse Mathématique, **80** (2000), 37-86.

[5] **J. Bourgain, H. Brezis and P. Mironescu**, *On the structure of the space $H^{1/2}$ with values into the circle*, C.R. Acad. Sci. Paris, **321** (2000), 119-124.

[6] **J. Bourgain, H. Brezis and P. Mironescu**, *Lifting, degree and minimal connections revisited*, (to appear).

[7] **A. Boutet de Monvel-Berthier, V. Georgescu and R. Purice**, *A boundary value problem related to the Ginzburg-Landau model*, Comm. Math. Phys., **142** (1991), 1-23.

[8] **H. Brezis and Y. Li**, *Topology and Sobolev spaces*, J. Funct.Anal., (to appear).

[9] **H. Brezis, Y. Li, P. Mironescu and L. Nirenberg**, *Degree and Sobolev spaces*, Top. Meth. in Nonlinear Anal,. **13** (1999), 181-190.

[10] **H. Brezis and P. Mironescu**, *Gagliardo-Nirenberg, composition and products in fractional Sobolev spaces*, Journal of Evolution Equations, **1** (2001), 387-404.

[11] **H. Brezis and P. Mironescu**, *On some questions of topology for S^1-valued fractional Sobolev spaces*, Rev. R. Acad. Cien. Serie A Mat. **95** (2001), 121-143.

[12] **H. Brezis and P. Mironescu**, in preparation.

[13] **H. Brezis and L. Nirenberg**, *Degree Theory and BMO, Part I: Compact manifolds without boundaries*, Selecta Math., **1** (1996), 197-263; *Part II: Compact manifolds with boundaries*, Selecta Math., **2** (1996), 309-368.

[14] **M. Escobedo**, *Some remarks on the density of regular mappings in Sobolev classes of S^M-valued functions*, Rev. Mat. Univ. Complut. Madrid, **1** (1988), 127-144.

[15] **F. B. Hang and F. H. Lin**, *Topology of Sobolev mappings II,*(to appear).

[16] **R. Hardt, D. Kinderlehrer and F. H. Lin**, *Stable defects of minimizers of constrained variational principles*, Ann. IHP, Anal. Nonlinéaire, **5** (1988), 297-322.

[17] **J. L. Lions and E. Magenes**, Problèmes aux limites non homogènes, Dunod, 1968; English translation, Springer, 1972.

[18] **V. Maz'ya and T. Shaposhnikova**, *An elementary proof of the Brezis and Mironescu theorem on the composition operator in fractional Sobolev spaces*, (to appear).

[19] **J.Peetre**, *Interpolation of Lipschitz operators and metric spaces*, Mathematica (Cluj), **12** (1970), 1-20.

[20] **T. Rivière**, *Dense subsets of $H^{1/2}(S^2;S^1)$*, Ann. of Global Anal. and Geom., (to appear).

[21] **J. Rubinstein and P. Sternberg**, *Homotopy classification of minimizers of the Ginzurg-Landau energy and the existence of permanent currents*, Comm. Math. Phys., **179** (1996), 257-263.

[22] **T. Runst**, *Mapping properties of nonlinear operators in spaces of Triebel-Lizorkin and Besov type*, Analysis Mathematica, **12** (1986), 313-346.

[23] **T. Runst and W. Sickel**, Sobolev spaces of fractional order, Nemytskij operators, and nonlinear partial differential equations, Walter de Gruyter, Berlin and New York, 1996.

[24] **R. Schoen and K. Uhlenbeck**, *Boundary regularity and the Dirichlet problem for harmonic maps*, J. Diff. Geom., **18** (1983), 253-268.

[25] **H. Triebel**, Theory of function spaces, Birkhäuser, Basel and Boston, 1983.

[26] **B. White**, *Homotopy clases in Sobolev spaces and the existence of energy minimizing maps*, Acta Math., **160** (1988), 1-17.

DÉPARTEMENT DE MATHÉMATIQUES, UNIVERSITÉ PARIS-SUD, 91405 ORSAY, FRANCE
E-mail address: `Petru.Mironescu@math.u-psud.fr`

Contemporary Mathematics
Volume **350**, 2004

Homoclinics for a Semilinear Elliptic PDE

Paul H. Rabinowitz

Dedicated with best wishes to my good friends
Haïm Brezis and Felix Browder

ABSTRACT. This paper establishes the existence of solutions of a semilinear elliptic PDE that are spatially heteroclinic in a cylindrical domain to a pair of periodic solutions. The existence of more complex spatial 'multibump' solutions is also discussed. Constrained minimization arguments are used to obtain the solutions.

1. Introduction

In earlier papers, [1]–[3], the existence of solutions spatially heteroclinic to a pair of periodic solutions was established for two families of semilinear elliptic partial differential equations. The goal of the current paper is to show how to construct more complex solutions which are near formal chains obtained by gluing together basic heteroclinics. The simplest case is that of spatial homoclinics and it will be studied in detail here. We confine ourselves to the first model case treated in [1]–[3]. The same ideas can be used in the setting of the second case which contains some water wave models treated by Kirchgässner [4], Turner [5], and Mielke [6].

To be more precise, the setting of [1]–[3] will be recalled. The partial differential equation to be studied here is

(PDE) $$-\Delta u = g(x, u), \quad x \in \Omega$$

where Ω is a cylindrical domain in $\mathbb{R}^n$, i.e. $\Omega = \mathbb{R} \times \mathcal{D}$ with $\mathcal{D}$ a bounded domain in $\mathbb{R}^{n-1}$ whose boundary is a smooth manifold. Thus if $x \in \Omega$, $x = (x_1, y)$ with $x_1 \in \mathbb{R}$ and $y \in \mathcal{D}$. (PDE) will be considered together with the Neumann boundary conditions:

(BC) $$\frac{\partial u}{\partial \nu}(x_1, y) = 0, \quad y \in \partial\mathcal{D}$$

where $\nu = \nu(y)$ denotes the outward pointing normal derivative to $\partial\mathcal{D}$.

1991 *Mathematics Subject Classification.* 35J60, 49J99, 58E30.

Key words and phrases. Spatial homoclinics; constrained minimization.

This research was supported in part by NSF grant #MCS-8110556. Any reproduction for the purpose of the U.S. Government is permitted.

Suppose g satisfies:

(g_1) $\qquad\qquad\qquad g(x,z) \in C^1(\overline{\Omega} \times \mathbb{R}, \mathbb{R})$ and is 1-periodic in x_1

and

(g_2) $\qquad\qquad\qquad G(x,z) = \int_0^z g(x,t)dt$ is 1-periodic in z.

As will be recalled more fully in §2, (PDE)–(BC) possesses an ordered family, $\mathcal{M}_0$, of solutions which are 1-periodic in x_1. Moreover if $v < w$ are adjacent members of $\mathcal{M}_0$, then there exist a pair of solutions, U, V, of (PDE)–(BC), U being heteroclinic in x_1 from v to w and V heteroclinic from w to v. Minimization arguments were used to prove these results. In §2, it will be shown that the sets of minimizers, $\mathcal{M}(v,w), \mathcal{M}(w,v)$ to which U and V belong are each ordered sets. Under a further assumption on these sets, namely that there exist adjacent minimizers in each set, it will be shown in §3 that there exist infinitely many distinct solutions of (PDE)–(BC) which are homoclinic in x_1 from v to v that are near the 2-chains of solutions obtained by formally concatenating phase shifts in x_1 of U and V. Results of this type are familiar in the theory of dynamical systems but the technicalities of obtaining such solutions are much greater in a PDE setting. A constrained minimization argument will be used to find these homoclinic solutions of (PDE)–(BC). The work of [1]–[3] will play a strong role but we also rely crucially on some new ideas such as the analysis of certain limit problems at infinity that are being treated in on-going work on Allen-Cahn model equations [7]–[8].

The methods used in §2–3 can be applied to construct more complicated heteroclinic and homoclinic solutions of (PDE)–(BC) that are near formal concatenations of k-chains of phase shifts of U and V. This will be discussed in §4.

2. Some preliminaries

In this section, a more precise statement of the results of [1]–[3] will be given and some extensions of this earlier work will be made. In addition some further results which will play a role in the comparison arguments and constructions of §3 will be presented.

To begin, set

$$L(u) = \frac{1}{2}|\nabla u|^2 - G(x,u),$$

the Lagrangian associated with (PDE). For $j \in \mathbb{Z}$, let $\Omega_j = [j, j+1] \times \mathcal{D}$. Set

$$(2.1) \qquad\qquad I(u) = \int_{\Omega_0} L(u)dx$$

and let

$$E = \{u \in W^{1,2}(\Omega_0) \mid u \text{ is 1-periodic in } x_1\}.$$

Then as was shown in [1], minimizing I over E produces solutions of (PDE) that are 1-periodic in x_1. Set

$$(2.2) \qquad\qquad c_0 = \inf_E I$$

and

$$\mathcal{M}_0 = \{v \in E \mid I(v) = c_0\}.$$

By (g_2) and (BC), if $v \in \mathcal{M}_0$, so is $v + j$ for all $j \in \mathbb{Z}$. It was proved in [2] that:

PROPOSITION 2.3. *$\mathcal{M}_0$ is an ordered set, i.e. if $v, w \in \mathcal{M}_0$, then either $v \equiv w$, $v < w$, or $v > w$.*

In the nicest case, $\mathcal{M}_0$ simply consists of $\{v + j \mid j \in \mathbb{Z}\}$ for some fixed $v \in E$. However it is possible that $\mathcal{M}_0$ is a continuum of solutions of (PDE) and (BC). This is so if e.g. g is independent of u. To avoid this degenerate setting, we assume:

(∗) There are adjacent $v, w \in \mathcal{M}_0$ with $v < w$.

Then as was shown in [2]–[3], there is a solution U, of (PDE) and (BC) with u heteroclinic in x_1 from v to w (and likewise a solution V, of (PDE) and (BC) heteroclinic in x_1 from w to v). Moreover for $j \in \mathbb{Z}$, setting $\tau_j U(x) = U(x_1 - j, y)$, U is monotone in the sense that

$$(2.4) \qquad \tau_{-1} U(x) > U(x), \quad x \in \Omega.$$

A minimization argument was used to find U. It is not sufficient to simply minimize

$$(2.5) \qquad \int_\Omega L(U) dx$$

over the class of candidates for a solution of (PDE)–(BC), namely functions asymptotic to v and w as $x_1 \to -\infty, \infty$ respectively, because the functional in (2.5) will generally be infinite on this class. Therefore a "renormalized" functional was introduced in [2] to avoid this difficulty. For $p \in \mathbb{Z}$, set

$$a_p(u) = \int_{\Omega_p} L(u) dx - c_0$$

and for $\ell, m \in \mathbb{Z}$ with $\ell \leq m$, set

$$J_{\ell,m}(u) = \sum_\ell^m a_p(u).$$

Define

$$(2.6) \qquad J(u) = \lim_{\ell \to -\infty} J_{\ell,0}(u) + \lim_{m \to \infty} J_{1,m}(u)$$

and

$$
\begin{aligned}
\Gamma(v, w) \;=\; & \big\{ u \in W^{1,2}_{\text{loc}}(\Omega) \mid v \leq u \leq \tau_{-1} u \leq w, \\
& \|u - v\|_{L^2(\Omega_j)} \to 0 \text{ as } j \to -\infty, \text{ and} \\
& \|u - w\|_{L^2(\Omega_j)} \to 0 \text{ as } j \to \infty \big\}.
\end{aligned}
$$

Then it was shown in [2] that J possesses the following properties:

PROPOSITION 2.7. *There is a constant $K \geq 0$ such that for all $\ell \leq m$ in $\mathbb{Z}$ and $u \in \Gamma(v, w)$,*

1° $-K \leq J_{\ell,m}(u) \leq J(u) + 2K$.
2° *If $J(u) < \infty$, the $\underline{\lim}$'s in (2.6) are limits, i.e. $\sum_{-\infty}^\infty a_p(u)$ converges. Moreover $\|u - v\|_{W^{1,2}(\Omega_j)} \to 0$ as $j \to -\infty$ and $\|u - w\|_{W^{1,2}(\Omega_j)} \to 0$ as $j \to \infty$.*

The existence of a solution of (PDE)–(BC) heteroclinic in x_1 and satisfying (2.4) then follows from:

PROPOSITION 2.8. *Let g satisfy (g_1)–(g_2) and let $(*)$ hold. Define*

$$(2.9) \qquad c(v,w) = \inf_{u \in \Gamma(v,w)} J(u).$$

Then there exists $U \in \Gamma(v,w)$ such that $J(U) = c(v,w)$. Moreover U is a classical solution of (PDE)–(BC) with $\tau_{-1} U > U$.

Solutions of (PDE)–(BC) near a formal gluing of U and its analogue $V \in \Gamma(w,v)$ are certainly not monotone. Therefore one needs another characterization of U that does not involve (2.4). This was given in [**3**]. Set

$$\widehat{\Gamma}(v,w) \;=\; \big\{ w \in W^{1,2}_{\mathrm{loc}}(\Omega) \mid v \le u \le w,$$
$$\|u - v\|_{W^{1,2}(\Omega_j)} \to 0 \;\text{ as } j \to -\infty, \text{ and}$$
$$\|u - w\|_{W^{1,2}(\Omega_j)} \to 0 \;\text{ as } j \to \infty \big\}$$

and

$$(2.10) \qquad \widehat{c}(v,w) = \inf_{u \in \widehat{\Gamma}(v,w)} J(u)$$

Then it was proved in [**3**] that:

PROPOSITION 2.11. $\widehat{c}(v,w) = c(v,w)$.

Thus U of Proposition 2.8 minimizes J over $\widehat{\Gamma}(v,w)$.

The above preliminaries form a starting point to construct more complex homoclinic and heteroclinic solutions of (PDE)–(BC). A few further remarks, results, and techniques that will be helpful in this task conclude this section.

Remark 2.12. An examination of the proof of Proposition 2.7 — see Propositions 2.4 and 2.6 and Lemma 2.5 in [**2**] — shows its conclusions hold if $\Gamma(v,w)$ is replaced by

$$\Gamma^*(v,w) \;=\; \big\{ u \in W^{1,2}_{\mathrm{loc}}(\Omega) \mid v \le u \le w,$$
$$\|u - v\|_{L^2(\Omega_j)} \to 0 \;\text{ as } j \to -\infty \text{ and}$$
$$\|u - w\|_{L^2(\Omega_j)} \to 0 \;\text{ as } j \to \infty \big\}$$

Since $\widehat{\Gamma}(v,w) \subset \Gamma^*(v,w)$, a fortiori Proposition 2.7 applies to $\widehat{\Gamma}(v,w)$.

Set

$$\mathcal{M}(v,w) = \{ u \in \Gamma(v,w) \mid J(u) = c(v,w) \}$$

and

$$\widehat{\mathcal{M}}(v,w) = \{ u \in \widehat{\Gamma}(v,w) \mid J(u) = \widehat{c}(v,w) \}.$$

By Proposition 2.11, $\mathcal{M}(v,w) \subset \widehat{\mathcal{M}}(v,w)$. The next result shows these sets are the same.

THEOREM 2.13. 1° $u \in \widehat{\mathcal{M}}(v,w)$ *implies u is a classical solution of (PDE)–(BC).*

2° $\widehat{\mathcal{M}}(v,w)$ *is an ordered set.*

3° $\mathcal{M}(v,w) = \widehat{\mathcal{M}}(v,w)$.

Proof. 1° is a consequence of a local minimization property possessed by members of $\widehat{\mathcal{M}}(v,w)$. Following arguments from [**2**] or [**7**], for $r > 0$ and $z \in \mathbb{R}^n$, set

$$B_r(z) = \{ x \in \mathbb{R}^n \mid |x - z| < r \}.$$

Choose $\xi \in \Omega$ and $r = r(\xi)$ such that $B_r(\xi) \subset \Omega$. Set

$$(2.14) \qquad \widehat{Z}(B_r(\xi)) = \{ \varphi \in W^{1,2}_{\mathrm{loc}}(\Omega) \mid \varphi = u \text{ in } \Omega \setminus B_r(\xi) \}$$

and for $\varphi \in \widehat{Z}(B_r(\xi))$, define

$$(2.15) \qquad \Phi_{r,\xi}(\varphi) = \int_{B_r(\xi)} L(\varphi)dx$$

and

$$(2.16) \qquad \widehat{c}(B_r(\xi)) = \inf_{\varphi \in \widehat{Z}(B_r(\xi))} \Phi_{r,\xi}(\varphi).$$

Since $\widehat{Z}(B_r(\xi))$ is closed and convex and $\Phi_{r,\xi}$ is weakly lower semicontinuous on $\widehat{Z}(B_r(\xi))$, there is a $\widehat{u} \in \widehat{Z}(B_r(\xi))$ such that $\Phi_{r,\xi}(\widehat{u}) = \widehat{c}(B_r(\xi))$. Standard regularity arguments show any minimizer of (2.16) is a solution of (PDE) in $B_r(\xi)$. The given $u \in \widehat{\mathcal{M}}(v,w)$ belongs to $\widehat{Z}(B_r(\xi))$. We claim

$$(2.17) \qquad \Phi_{r,\xi}(u) = \widehat{c}(B_r(\xi)),$$

i.e. u is a minimizer of (2.16) and hence a solution of (PDE). Since $\widehat{u}$ is a minimizer of (2.16), $\widehat{u} \in \widehat{\Gamma}(v,w)$ and

$$(2.18) \qquad J(\widehat{u}) \leq \widehat{c}(v,w).$$

Thus there must be equality in (2.18) and

$$(2.19) \qquad \Phi_{r,\xi}(u) = \Phi_{r,\xi}(\widehat{u}) = \widehat{c}(B_r(\xi)).$$

Similar arguments show u satisfies (BC) — see e.g. [2] — and $1°$ is proved.

To verify $2°$, suppose that $v, w \in \widehat{\mathcal{M}}(v,w)$. Set $\varphi = \max(V,W)$ and $\psi = \min(V,W)$. Then $\varphi, \psi \in \widehat{\Gamma}(v,w)$ so

$$(2.20) \qquad \widehat{c}(v,w) \leq J(\varphi), J(\psi).$$

On the other hand, for all $k \in \mathbb{Z}$,

$$(2.21) \qquad \int_{\Omega_k} (L(\varphi) + L(\psi))dx = \int_{\Omega_k} (L(V) + L(W))dx.$$

Therefore with the aid of $2°$ of Proposition 2.7 and Remark 2.12,

$$(2.22) \qquad 2\widehat{c}(v,w) = J(\varphi) + J(\psi) = J(V) + J(W)$$

so

$$(2.23) \qquad J(\varphi) = \widehat{c}(v,w) = J(\psi).$$

Hence $\varphi, \psi \in \widehat{\mathcal{M}}(v,w)$ and by $1°$ are solutions of (PDE)–(BC).

Suppose that V and W are not ordered. Then it can be assumed that there are points $\zeta, \eta \in \Omega$ such that $V(\zeta) = W(\zeta)$ and $V(\eta) > W(\eta)$. Therefore $\varphi(\zeta) = W(\zeta)$ and $\varphi(\eta) > W(\eta)$. Set $f = \varphi - W$. Then f satisfies the linear elliptic PDE:

$$(2.24) \qquad -\Delta f - b^- f = b^+ f \quad \text{in } \Omega$$

where

$$\begin{aligned} b(x) &= \frac{g(x,\varphi(x)) - g(x,W(x))}{\varphi(x) - W(x)} && \text{if } \varphi(x) > W(x), \\ &= g_u(x, \varphi(x)) && \text{if } \varphi(x) = W(x) \end{aligned}$$

and $b, b^+ = \max(b,0)$, and $b^- = \min(b,0)$ are continuous with $b^- \leq 0 \leq b^+$. Note that $f \geq 0$ in Ω and $f(\zeta) = 0$. By the Maximum Principle, $f \equiv 0$ in Ω. But $f(\eta) > 0$. This contradiction shows V and W are ordered.

Lastly to prove $3°$, note that U given by Proposition 2.8 belongs to $\widehat{\mathcal{M}}(v,w)$ and this implies $\tau_{-j}U \in \widehat{M}(v,w)$ for all $j \in \mathbb{Z}$. Let $u \in \widehat{\mathcal{M}}(v,w) \setminus \{\tau_{-j}U \mid j \in \mathbb{Z}\}$. By $2°$, there is a $k \in \mathbb{Z}$ such that for all $x \in \Omega$,

$$(2.25) \qquad \tau_{-k}U(\tau) < u(x) < \tau_{-k-1}U(x).$$

Replacing x by $(x_1 + 1, y)$, (2.25) shows

$$(2.26) \qquad \tau_{-k-1}U(x) < \tau_{-1}u(x)$$

so $u < \tau_{-1}u$ via (2.25)–(2.26), i.e. $u \in \mathcal{M}(v,w)$.

Remark 2.27. The above arguments also provide an analogue of Theorem 2.13 for $\widehat{\mathcal{M}}(w,v)$.

Next define

$$\Gamma(v,v) = \{u \in W^{1,2}_{\text{loc}}(\Omega) \mid \|u - v\|_{L^2(\Omega_j)} \to 0 \ \text{ as } |j| \to \infty\}.$$

To be consistent with the notation used earlier in this section, superscript $*$'s should be used here, but to simplify matters we drop the $*$'s.

Set

$$(2.28) \qquad c(v,v) = \inf_{u \in \Gamma(v,v)} J(u).$$

and

$$\mathcal{M}(v,v) = \{u \in \Gamma(v,v) \mid J(u) = c(v,v)\}.$$

As in Remark 2.12, the arguments of [**2**] show $1°$, $2°$ of Proposition 2.7 hold for $u \in \Gamma(v,v)$. The next result provides some useful facts about $c(v,v)$ and $\mathcal{M}(v,v)$.

THEOREM 2.29. $1°$ $c(v,v) = 0$,
      $2°$ *If* $u \in \mathcal{M}(v,v)$, *then* u *is a solution of (PDE) and (BC).*
      $3°$ $\mathcal{M}(v,v) = \{v\}$.

Proof: Since $v \in \Gamma(v,v)$ and $J(v) = 0$,

$$(2.30) \qquad c(v,v) \le 0.$$

Let $u \in \Gamma(v,v)$ and $k \in \mathbb{N}$. Define

$$
\begin{aligned}
(2.31) \qquad u_k \ &= \ v, \quad x_1 \le -k \\
&= \ (x_1 + k)u + (-k + 1 - x_1)v, \quad -k \le x_1 \le -k+1 \\
&= \ u, \quad -k+1 \le x_1 \le k-1 \\
&= \ (x_1 - k + 1)v + (k - x_1)u, \quad k-1 \le x_1 \le k \\
&= \ v, \quad k \le x_1.
\end{aligned}
$$

Then $u_k \in \Gamma(v,v)$ and since u_k restricted to $[-k,k] \times \mathcal{D}$ extends naturally as a $W^{1,2}$ $2k$-periodic function of x_1, by Lemma 2.10 of [**2**],

$$(2.32) \qquad J_{-k,k-1}(u_k) \ge 0.$$

By $2°$ of Proposition 2.7, as $|j| \to \infty$,

$$(2.33) \qquad \|u - v\|_{W^{1,2}(\Omega_j)} \to 0.$$

A straightforward computation then shows

$$(2.34) \qquad |a_{-k}(u_k)|, \ |a_{k-1}(u_k)| \to 0$$

as $k \to \infty$ and by (2.33),

$$(2.35) \qquad |a_{-k}(u)|, \ |a_k(u)| \to 0.$$

Moreover by $2°$ of Proposition 2.7 again,

$$(2.36) \qquad J_{-\infty,-k-1}(u), \ \ J_{k,\infty}(u) \to 0$$

as $k \to \infty$. Consequently

$$
\begin{aligned}
0 \ &\leq \ J_{-k,k-1}(u_k) = J(u_k) = J(u) + a_{-k}(u_k) - a_{-k}(u) \\
&\qquad + a_{k-1}(u_k) - a_{k-1}(u) - J_{-\infty,-k-1}(u) - J_{k,\infty}(u) \\
&\equiv \ J(u) - R_k(u)
\end{aligned}
$$

or

$$(2.37) \qquad\qquad R_k(u) \leq J(u).$$

Thus letting $k \to \infty$, by (2.34)–(2.37),

$$(2.38) \qquad\qquad 0 \leq J(u)$$

for all $u \in \Gamma(v,v)$. Consequently (2.30) and (2.38) yield $1°$.

Next $2°$ follows exactly as in the proof of $1°$ of Theorem 2.13. Lastly to get $3°$, note that the argument of Theorem 2.13 shows $\mathcal{M}(v,v)$ is an ordered set. Also, $u \in \mathcal{M}(v,v)$ implies $\tau_{-j}u \in \mathcal{M}(v,v)$ for all $j \in \mathbb{Z}$. Suppose e.g.

$$(2.39) \qquad\qquad u < \tau_{-1}u.$$

The argument is essentially the same if inequality (2.39) is reversed. By (2.39), for any $j \in \mathbb{N}$,

$$\tau_j u < u < \tau_{-j}u$$

so letting $j \to \infty$ shows

$$(2.40) \qquad\qquad v(x) \leq u(x) \leq v(x)$$

for all $x \in \Omega$. Thus $u \equiv v$ and $3°$ is proved.

Remark 2.41. The same theorem holds if v is replaced by w.

CCOROLLARY 2.42. $c(v,w) + c(w,v) > 0$.

Proof: Let $U \in \mathcal{M}(v,w)$ and $V \in \mathcal{M}(w,v)$. Set $\varphi = \max(U,V)$ and $\psi = \min(U,V)$. Then $\varphi \in \Gamma(w,w)$ and $\psi \in \Gamma(v,v)$. As in (2.21)–(2.22),

$$(2.43) \qquad J(\varphi) + J(\psi) = J(U) + J(V) = c(v,w) + c(w,v).$$

Since $\psi = U$ for x_1 near $-\infty$, $\psi \neq v$ so by $1°$ and $3°$ of Theorem 2.29, $J(\psi) > 0$. Similarly $J(\varphi) > 0$ and Corollary 2.42 follows.

The final result in this section provides an estimate that will be required in §3.

PPROPOSITION 2.44. *Suppose $u \in \Gamma(v,v)$, $v \leq u \leq w$, and there is a $\gamma > 0$ such that*

$$(2.45) \qquad\qquad \|u - v\|_{L^1(\Omega_0)} \geq \gamma.$$

Then there is a $\beta = \beta(\gamma) > 0$ and independent of u such that $J(u) \geq \beta$.

Remark 2.46. The arguments used to prove Proposition 2.44 also give the analogous result for $\Gamma(w,w)$.

The proof of Proposition 2.44 requires several steps. Since two of them will be used repeatedly in §3, they will be formulated below as separate propositions so that they can be referred to easily later.

Proof of Proposition 2.44: An indirect argument will be used. If the result is not true, there is a sequence $(u_k) \subset \Gamma(v, v)$ satisfying (2.45) with $v \leq u_k \leq w$ and as $k \to \infty$,

$$(2.47) \qquad\qquad J(u_k) \to 0.$$

Thus (u_k) is a minimizing sequence for (2.28). By 1° of Proposition 2.7, there is an $M > 0$ such that

$$(2.48) \qquad\qquad J_{\ell,m}(u_k) \leq M + 2K$$

for all $\ell \leq m$ in $\mathbb{Z}$ and $k \in \mathbb{N}$. The form of $J_{\ell,m}$ then implies (u_k) is bounded in $W_{\mathrm{loc}}^{1,2}(\Omega)$. Therefore there is a $P \in W_{\mathrm{loc}}^{1,2}(\Omega)$ and a subsequence of (u_k) which converges weakly in $W_{\mathrm{loc}}^{1,2}(\Omega)$ to P, strongly in $L_{\mathrm{loc}}^2(\Omega)$, and pointwise a.e. Consequently

$$(2.49) \qquad\qquad v \leq P \leq w$$

and by (2.45),

$$(2.50) \qquad\qquad \|P - v\|_{L^1(\Omega_0)} \geq \gamma.$$

The weak lower semicontinuity of $J_{\ell,m}$ and (2.48) show

$$(2.51) \qquad\qquad J_{\ell,m}(P) \leq M + 2K.$$

Thus letting $\ell \to -\infty$ and $m \to \infty$ gives

$$(2.52) \qquad\qquad J(P) \leq M + 2K.$$

Now we need

PROPOSITION 2.53. *A subsequence of (u_k) can be chosen so that it converges to P in $W_{\mathrm{loc}}^{1,2}(\Omega)$.*

The proof of Proposition 2.53 will be postponed for now. The next step is

PROPOSITION 2.54. *P is a solution of (PDE)–(BC).*

The proof of this result will also be postponed for the moment and the proof of Proposition 2.44 will be completed. By Proposition 2.53 it can be assumed that $u_k \to P$ in $W_{\mathrm{loc}}^{1,2}(\Omega)$. Set $\varphi_k = \max(u_k, \tau_{-1}u_k)$ and $\psi_k = \min(u_k, \tau_{-1}u_k)$. Then $\varphi_k, \psi_k \in \Gamma(v, v)$ so

$$(2.55) \qquad\qquad J(\varphi_k),\ J(\psi_k) \geq 0$$

and as in (2.22),

$$(2.56) \qquad J(\varphi_k) + J(\psi_k) = J(u_k) + J(\tau_{-1}u_k) = 2J(u_k) \to 0$$

as $k \to \infty$ via (2.47). Thus by (2.55)–(2.56), (φ_k) and (ψ_k) are also minimizing sequences for (2.28). The proofs of Propositions 2.53 and 2.54 show it can be assumed that φ_k, ψ_k converge in $W_{\mathrm{loc}}^{1,2}(\Omega)$ to solutions $\varphi = \max(P, \tau_{-1}P)$ and $\psi = \min(P, \tau_{-1}P)$ of (PDE)–(BC). Moreover $\varphi \geq \psi$. Two cases arise: (a) $\varphi(z) = \psi(z)$ for some $z \in \Omega$ or (b) $\varphi(z) > \psi(z)$ for all $z \in \Omega$.

If (a) occurs, the Maximum Principle argument centered around (2.24) shows $\varphi \equiv \psi$, i.e. $P = \tau_{-1}P$. Thus P is 1-periodic in x_1. This fact and (2.52) imply $J(P) = 0$ and $P \in \mathcal{M}_0$. But then (2.49)–(2.50) and $(*)$ show $P = w$.

If (b) occurs, either $P > \tau_{-1}P$ or $\tau_{-1}P > P$. The argument is the same in either event so suppose that

$$(2.57) \qquad\qquad \tau_{-1}P > P.$$

Set $P_j = \tau_{-j} P|_{\Omega_0}$. Since P is a bounded (in $C^{2,\alpha}(\Omega)$ for any $\alpha \in (0,1)$) solution of (PDE)–(BC) and $P_{j+1} > P_j$ via (2.57), as $j \to \pm\infty$, P_j converges in C^2 to $Q^\pm$, a pair of sol of (PDE)–(BC) in Ω_0. By (2.57) again, $\tau_{-1} Q^\pm = Q^\pm$ so $Q^\pm$ are 1-periodic in x_1 and $v \leq Q^- < Q^+ \leq w$. If $Q^+ \neq w$, $I(Q^+) > c_0$ so there is an $\alpha > 0$ such that $a_j(P) \geq \alpha$ for all large j and $J(P) = \infty$ contrary to (2.52). Thus $Q^+ = w$ and similarly, $Q^- = v$. In particular as $j \to \infty$,

$$(2.58) \qquad \|P - w\|_{C^2(\Omega_j)} \to 0$$

which is also the case for (a).

Now it will be shown that (2.58) is impossible. Define

$$
(2.59) \qquad
\begin{aligned}
f_k &= u_k, \quad x_1 \leq q \\
&= (x_1 - q)w + (q + 1 - x_1)u_k, \quad q \leq x_1 \leq q + 1 \\
&= w, \quad q + 1 \leq x_1 \leq p \\
&= (x_1 - p)u_k + (p + 1 - x_1)w, \quad p \leq x_1 \leq p + 1 \\
&= u_k, \quad p + 1 \leq x_1.
\end{aligned}
$$

Then $f_k \in \Gamma(v, v)$ and

$$(2.60) \qquad J(u_k) - J(f_k) = J_{q,p}(u_k) - a_q(f_k) - a_p(f_k).$$

Since

$$(2.61) \qquad \|u_k - w\|_{W^{1,2}(\Omega_j)} \leq \|u_k - P\|_{W^{1,2}(\Omega_j)} + \|P - w\|_{W^{1,2}(\Omega_j)},$$

given any $\sigma > 0$, by (2.58), for j large the second term on the right in (2.61) can be made $\leq \sigma/2$. The $W^{1,2}_{\mathrm{loc}}$ convergence of u_k to P shows the first term on the right in (2.61) can also be made $\leq \sigma/2$ for k large. Thus for such j and k,

$$(2.62) \qquad \|u_k - w\|_{W^{1,2}(\Omega_j)} \leq \sigma.$$

Taking p large, $q = p - 2$, and $k = k(p)$ large, as in (2.33)–(2.35), it can be assumed that

$$(2.63) \qquad |a_q(u_k)|, \; |a_q(f_k)|, \; |a_p(u_k)|, \; |a_p(f_k)| \leq \kappa(\sigma)$$

where $\kappa(\sigma) \to 0$ as $\gamma \to 0$. Thus

$$(2.64) \qquad J(f_k) \leq J(u_k) + 4\kappa(\sigma).$$

Note that if

$$
(2.65) \qquad
\begin{aligned}
g_k &= f_k, \quad x_1 \leq p \\
&= w, \quad x_1 \geq p
\end{aligned}
$$

and

$$
(2.66) \qquad
\begin{aligned}
h_k &= w, \quad x_1 \leq p \\
&= f_k, \quad x_1 \geq p,
\end{aligned}
$$

then

$$(2.67) \qquad J(f_k) = J(g_k) + J(h_k).$$

But $g_k \in \widehat{\Gamma}(v, w)$, $h_k \in \widehat{\Gamma}(w, v)$ so by (2.64),

$$(2.68) \qquad c(v, w) + c(w, v) \leq J(u_k) + 4\kappa(\sigma).$$

Choose σ so small that

$$(2.69) \qquad 8\kappa(\sigma) < c(v, w) + c(w, v).$$

Then (2.69) and (2.47) show (2.68) is not possible for large k. Thus (2.58) cannot occur and Proposition 2.44 is proved.

It remains to verify Propositions 2.53 and 2.54.

Proof of Proposition 2.53: Adapting arguments from [**2**], [**7**], [**8**], consider

$$(2.70) \qquad \delta_i^2 \equiv \lim_{k \to \infty} a_i(u_k) - a_i(P) \geq 0.$$

By $1°$ of Proposition 2.7, (2.70), and (2.48),

$$(2.71) \qquad -K \ \leq \ J_{\ell,m}(P) = \sum_{\ell}^{m} \left(\lim_{k \to \infty} a_i(u_k) - \delta_i^2 \right)$$

$$\leq \ \lim_{k \to \infty} J_{\ell,m}(u_k) - \sum_{\ell}^{m} \delta_i^2$$

$$\leq \ c(v,v) + 2K - \sum_{\ell}^{m} \delta_i^2.$$

This implies

$$(2.72) \qquad \sum_{i \in \mathbb{Z}} \delta_i^2 \leq 3K$$

and therefore

$$(2.73) \qquad \delta_i \to 0, \ \ |i| \to \infty.$$

We claim (2.73) implies

$$(2.74) \qquad \lim_{k \to \infty} \|P - u_k\|_{W^{1,2}(\Omega_i)} \leq \sqrt{2}\delta_i$$

for all $i \in \mathbb{Z}$. To verify (2.74), observe that by (2.70),

$$(2.75) \qquad \lim_{k \to \infty} \int_{\Omega_i} \frac{1}{2}|\nabla u_k|^2 dx + \lim_{k \to \infty} \int_{\Omega_i} -G(x,u_k)dx$$

$$\leq \lim_{k \to \infty} \int_{\Omega_i} L(u_k)dx = \int_{\Omega_i} L(P)dx + \delta_i^2$$

$$= \int_{\Omega_i} \frac{1}{2}|\nabla P|^2 dx + \lim_{k \to \infty} \int_{\Omega_i} -G(x,u_k)dx + \delta_i^2$$

since it can be assumed that

$$\int_{\Omega_i} G(x,P)dx = \lim_{k \to \infty} \int_{\Omega_i} G(x,u_k)dx.$$

Therefore

$$(2.76) \qquad \lim_{k \to \infty} \int_{\Omega_i} |\nabla u_k|^2 dx \leq \int_{\Omega_i} |\nabla P|^2 dx + 2\delta_i^2.$$

Since

$$|\nabla(u_k - P)|^2 = |\nabla u_k|^2 - 2\nabla u_k \cdot \nabla P + |\nabla P|^2,$$

by (2.76),

$$(2.77) \qquad \lim_{k \to \infty} \int_{\Omega_i} \frac{1}{2}|\nabla(u_k - P)|^2 dx \leq \delta_i^2.$$

It can be assumed that $u_k \to P$ in $L^2_{\mathrm{loc}}(\Omega_1)$ so

$$(2.78) \qquad \lim_{k \to \infty} \|P - u_k\|_{L^2(\Omega_i)} = 0.$$

Combining (2.77)–(2.78) gives (2.74). In fact by passing to a subsequence of u_k, which shall be done for what follows, it can be assumed that (2.74) holds with $\underline{\lim}$ replaced by lim.

Next we claim

$$(2.79) \qquad \delta_i = 0, \quad i \in \mathbb{Z}.$$

Assuming (2.79) for the moment, passing to an appropriate subsequence of u_k then yields Proposition 2.53. To verify (2.79), let f_k be as in (2.59) with $q = -p$ and w replaced by P. Then $f_k \in \Gamma(v, v)$ and

$$(2.80) \qquad \begin{aligned} 0 \;=\;& c(v, v) \le J(f_k) = J_{-\infty, -p}(u_k) + J_{-p+1, p-1}(P) \\ &+ J_{p, \infty}(u_k) + a_{-p}(f_k) - a_{-p}(u_k) + a_p(f_k) - a_p(u_k). \end{aligned}$$

By (2.74) (with lim), there exists a $\gamma_p \to 0$ as $p \to \infty$ such that

$$(2.81) \qquad |a_{-p}(f_k) - a_{-p}(u_k)| + |a_p(f_k) - a_p(u_k)| \le \gamma_p.$$

Therefore for large p for which (2.81) holds, by (2.80)–(2.81) and (2.70),

$$(2.82) \qquad \begin{aligned} 0 \;=\;& c(v, v) \le J(u_k) + \lim_{s \to \infty} J_{-p+1, p-1}(u_s) \\ & -J_{-p+1, p-1}(u_k) - \sum_{-p+1}^{p-1} \delta_i^2 + \gamma_p. \end{aligned}$$

Letting $k \to \infty$ and recalling (2.47), (2.82) shows

$$(2.83) \qquad \sum_{-p+1}^{p-1} \delta_i^2 \le \gamma_p.$$

Hence letting $p \to \infty$ yields (2.79).

Finally this section concludes with

Proof of Proposition 2.54: An argument in the spirit of $1°$ of Theorem 2.13 will be employed. First to show that P satisfies (PDE), it suffices to show that for each $z \in \Omega$ and $r = r(z)$ such that $B_r(z) \subset \Omega$,

$$(2.84) \qquad J(P) \le J(P + t\varphi)$$

for all smooth φ with support in $B_r(z)$ and t near 0. Indeed (2.84) implies P is a weak solution of (PDE) in $B_r(z)$ and standard regularity arguments then show P is a classical solution of (PDE) in $B_r(z)$. To obtain (2.84), set $f_k = \max(u_k + t\varphi, w)$, $g_k = \min(u_k + t\varphi, w)$. Then as in (2.22),

$$(2.85) \qquad J(f_k) + J(g_k) = J(u_k + t\varphi).$$

For $|x_1|$ large, $\varphi = 0$ so $u_k \le w$ implies $f_k \in \Gamma(w, w)$. Therefore by Theorem 2.29 and Remark 2.41,

$$(2.86) \qquad J(f_k) \ge 0.$$

Thus (2.85) and (2.86) imply

$$(2.87) \qquad J(g_k) \le J(u_k + f\varphi).$$

Next define $\chi_k = \max(g_k, v)$, $\psi_k = \min(g_k, v)$. Then

$$(2.88) \qquad J(\chi_k) + J(\psi_k) = J(g_k).$$

Since $u_k \geq v$, for $|x_1|$ large, $\psi_k = v$ so $\psi_k \in \Gamma(v, v)$ and by Theorem 2.29 again,

$$(2.89) \qquad J(\psi_k) \geq 0.$$

Hence by (2.87)–(2.89),

$$(2.90) \qquad J(\chi_k) \leq J(u_k + t\varphi).$$

But $\chi_k \in \Gamma(v, v)$ so

$$(2.91) \qquad J(\chi_k) \geq 0.$$

Thus

$$(2.92) \qquad J(u_k) \leq J(u_k) + J(\chi_k) \leq J(u_k) + J(u_k + t\varphi).$$

Suppose $B_r(z) \subset [p, p+s] \times \mathcal{D}$ where $p \in \mathbb{Z}$ and $s \in \mathbb{N}$. Then by (2.92),

$$(2.93) \qquad J_{p,p+s-1}(u_k) \leq J(u_k) + J_{p,p+s-1}(u_k + t\varphi).$$

Letting $k \to \infty$ and using (2.47) and Proposition 2.53 yields

$$(2.94) \qquad J_{p,p+s-1}(P) \leq J_{p,p+s-1}(P + t\varphi)$$

from which (2.94) follows.

The verification of (BC) is similar to the above argument — see e.g. Proposition 2.9 of [**2**] — and will be omitted.

3. Homoclinic solutions of (PDE)

Whenever one has basic heteroclinic solutions of (PDE)–(BC) such as those provided by Proposition 2.8 and its companion result for $\Gamma(w, v)$, by analogy with related situations, in the theory of dynamical systems, one can hope to find solutions homoclinic in x_1 from v to v or w to w near formal concatenations of elements of $\mathcal{M}(v, w)$ and $\mathcal{M}(w, v)$. That is the goal of this section. Such solutions will be obtained using a constrained variational problem. The following analogue of $(*)$ is required.

$(**)$ $\mathcal{M}(v, w)$ and $\mathcal{M}(w, v)$ are not continua.

For $u \in W^{1,2}_{\mathrm{loc}}(\Omega)$, define

$$(3.1) \qquad \rho(u) = \int_{\Omega_0} u\,dx.$$

Then by Theorem 2.13, ρ is strictly monotone on $\mathcal{M}(v, w)$ and on $\mathcal{M}(w, v)$, i.e. if $u_1 < u_2$, $\rho(u_1) < \rho(u_2)$. Now ρ_i can be chosen so that

$$(3.2) \qquad \begin{cases} \rho_i \in (\rho(v), \rho(w)), & 1 \leq i \leq 4, \\ \rho_i \notin \rho(\mathcal{M}(v, w)), & i = 1, 2 \\ \rho_i \notin \rho(\mathcal{M}(w, v)), & i = 3, 4. \end{cases}$$

To define the class of admissible homoclinic-like functions that will be used in the variational problem that will be formulated here, let $\ell \in \mathbb{N}$ and $m \in \mathbb{Z}^4$ with

$$(3.3) \qquad m_1 < m_2 < m_2 + 2\ell < m_3 < m_4.$$

Define

$$(3.4) \qquad \Gamma_m = \{u \in W^{1,2}_{\mathrm{loc}}(\Omega) \mid u \text{ satisfies } (3.5)\text{–}(3.7)\}$$

where

$$(3.5) \qquad v \le u \le w,$$

$$(3.6) \qquad \begin{cases} \text{(i)} & \rho(\tau_{-i}u) \le \rho_1, & i = m_1 - \ell, \dots, m_1 - 1, \\ \text{(ii)} & \rho_2 \le \rho(\tau_{-i}u), & i = m_2, \dots, m_2 - 1, \\ \text{(iii)} & \rho_3 \le \rho(\tau_{-i}u), & i = m_3 - \ell, \dots, m_3 - 1, \\ \text{(iv)} & \rho(\tau_{-i}u) \le \rho_4, & i = m_4, \dots, m_4 - 1, \end{cases}$$

$$(3.7) \qquad \|u - v\|_{L^2(\Omega_j)} \to 0, \quad |j| \to \infty.$$

With Γ_m so defined, set

$$(3.8) \qquad c_m = \inf_{\varphi \in \Gamma_m(v,v)} J(u).$$

The main result of this section is:

THEOREM 3.9. *Let g satisfy (g_2)–(g_2) and suppose that $(*)$ and $(**)$ hold. Then if ℓ is sufficiently large, there is a $U_m \in \Gamma_m(v,v)$ such that $J(U_m) = c_m(v,v)$. Moreover if $m_2 - m_1$, $m_4 - m_3$ are also sufficiently large, U_m is a solution of (PDE) and (BC) and*

$$(3.10) \qquad \|U_m - v\|_{C^2(\Omega_j)} \to 0, \quad |j| \to \infty.$$

Remark 3.11. Theorem 3.9 yields infinitely many such heteroclinic solutions of (PDE)–(BC). Once it has been established that $U_m \in \Gamma_m(v,v)$ and satisfies (PDE)–(BC), (3.10) follows from standard interpolation inequalities via (3.7) and the fact that solutions of (PDE)–(BC) in Ω are bounded in $C^{2,\alpha}(\Omega)$ for any $\alpha \in (0,1)$. It will also be shown that U_m is C^2 close to w in $[m_2, m_3] \times \mathcal{D}$. A sharper result than Theorem 3.9 would be to show that there are solutions of (PDE)–(BC) between prescribed pairs of adjacent solutions (as provided by $(**)$) in $\mathcal{M}(v,w)$ and $\mathcal{M}(w,v)$.

Proof of Theorem 3.9: The proof requires several steps. First the proof of Proposition 2.7 — see [2] — shows it holds for $u \in \Gamma_m(v,v)$. Therefore choosing a minimizing sequence, (u_k), for (3.8), there is an $M > 0$ such that

$$(3.12) \qquad J(u_k) \le M$$

for all $k \in \mathbb{N}$. Hence by 1° of Proposition 2.7, (u_k) is bounded in $W^{1,2}_{\text{loc}}(\Omega)$ and along a subsequence converges to $U_m \in W^{1,2}_{\text{loc}}(\Omega)$ weakly in $W^{1,2}_{\text{loc}}(\Omega)$, strongly in $L^2_{\text{loc}}(\Omega)$, and pointwise a.e. Thus U_m satisfies (3.5)–(3.6). Moreover by Proposition 2.7 and the weak lower semicontinuity of $J_{p,q}$,

$$(3.13) \qquad J_{p,q}(U_m) \le \varliminf_{k \to \infty} J_{p,q}(u_k) \le M + 2K$$

so letting $p \to -\infty$, $q \to \infty$ gives

$$(3.14) \qquad J(U_m) \le M + 2K.$$

The remainder of the proof will be accomplished in 5 steps: (A) U_m is a solution of (PDE)–(BC) outside of the constraint regions; (B) For $i \in \mathbb{Z}$, set $X_i = \Omega_{i-1} \cup \Omega_i \cup \Omega_{i+1}$. For ℓ large, there is an X_i in each constraint region such that U_m is a solution of (PDE)–(BC) in X_i; (C) U_m satisfies (3.7) and therefore $U_m \in \Gamma_m(v,v)$; (D) $J(U_m) = c_m(v,v)$. This completes the first assertion of Theorem 3.9. Finally, (E) if $m_2 - m_1$, $m_4 - m_3$ are sufficiently large, U_m is a solution of (PDE)–(BC) in the constraint regions.

Before presenting the proofs of (A)–(E), the following analogue of Proposition 2.53 is required.

PROPOSITION 3.15. *A subsequence of (u_k) can be chosen so that for each $k \in \mathbb{Z}$,* $\|u_k - U_m\|_{W^{1,2}(\Omega_j)} \to 0$ *as $j \to \infty$.*

Proof: The proof of Proposition 3.15 is identical to that of Proposition 2.53 except that $c(v,v)$ is replaced by $c_m(v,v)$. Therefore in the current setting, (2.82) becomes

$$(3.16) \qquad c_m(v,v) \;\leq\; J(u_k) + \varliminf_{s \to \infty} J_{-p+1,p-1}(u_s)$$

$$-J_{-p+1,p-1}(u_k) - \sum_{-p+1}^{p-1} \delta_i^2 + \gamma_p.$$

Since $J(u_k) \to c_m(v,v)$ as $k \to \infty$, (2.83) and Proposition 3.15 then follow.

Now (A)–(E) will be proved.

Proof of (A): Let $\mathcal{R}$ denote the constraint region. To verify that U_m satisfies (PDE)–(BC) in $\Omega \setminus \mathcal{R}$, suppose that $z \in \Omega \setminus \mathcal{R}$. Then there is an $r = r(z) > 0$ such that $B_r(z) \subset \Omega \setminus \mathcal{R}$. As in the proof of Proposition 2.54, it suffices to show

$$(3.17) \qquad J(U_m) \leq J(U_m + t\varphi)$$

for all smooth φ with support in $B_r(z)$. Follow the proof of Proposition 2.54 through (2.90). Since as $k \to \infty$, $J(u_k) \to c_m(v,v)$,

$$(3.18) \qquad c_m(v,v) \leq J(u_k) = c_m(v,v) + r_k$$

where $r_k \to 0$ as $k \to \infty$. Observe that $\chi_k \in \Gamma_m(v,v)$. Therefore by (3.18) and (2.90),

$$(3.19) \qquad J(u_k) \leq J(\chi_k) + r_k \leq J(u_k + t\varphi) + r_k$$

which replaces (2.92). The remainder of the proof of (A) follows that of Proposition 2.54.

Proof of (B): Set $\mathcal{R} = \bigcup_{j=1}^{4} \mathcal{R}_j$ where $\mathcal{R}_1 = [m_1 - \ell, m_1] \times \mathcal{D}$, etc. Suppose $X_i \subset \mathcal{R}_1$ and

$$(3.20) \qquad \int_{\Omega_{i+s}} U_m \, dx < \rho_1, \quad s = 0, \pm 1.$$

Then for $z \in X_i$ and smooth φ with support in $B_r(z) \subset X_i$, for $|t|$ sufficiently small, $U_m + t\varphi$ will satisfy (3.20). Therefore the argument of Part (A) shows U_m is a solution of (PDE)–(BC) in X_i. Similar reasoning applies if $X_i \subset \mathcal{R}_k$, $k = 2, 3, 4$. Thus to prove (B), it suffices to verify (3.20) and its analogues for $\mathcal{R}_k$, $k = 2, 3, 4$. This is a consequence of taking

$$\sigma < \frac{1}{2} \min\{|\rho_i - \rho(\varphi)| \mid 1 \leq i \leq 4, \varphi = v, w\}$$

in the following more general result:

PROPOSITION 3.21. *For any $\sigma > 0$, there is an $\ell_0 = \ell_0(\sigma)$ such that whenever $\ell \geq \ell_0$, $\ell \in \mathbb{N}$, $u \in W^{1,2}_{\mathrm{loc}}(\Omega)$ with $v \leq u \leq w$, $J(u) \leq M$, and u satisfies (3.6)(j), $j \in \{1, \ldots, 4\}$, then*

$$(3.22) \qquad \|u - \varphi\|_{L^1(X_i)} \leq \sigma$$

for some $X_i \subset \mathcal{R}_j$ where $\varphi = v$ if $j = 1$ or 4 and $\varphi = w$ if $j = 2$ or 3.

Proof: If not, there is a sequence $(u_k) \subset W^{1,2}_{\text{loc}}(\Omega)$ such that $v \leq u_k \leq w$, $J(u_k) \leq M$, (u_k) satisfies (3.6)(j), and

$$(3.23) \qquad \|u_k - \varphi\|_{L^1(X_i)} \geq \sigma$$

for all $X_i \subset \mathcal{R}_j$. As in (A), the functions (u_k) are bounded in $W^{1,2}_{\text{loc}}(\Omega)$ as are $(\tau_{-k} u_k)$. Therefore there is a $U^* \in W^{1,2}_{\text{loc}}(\Omega)$ such that along a subsequence, $\tau_{-k} u_k \to U^*$ weakly in $W^{1,2}_{\text{loc}}(\Omega)$, strongly in $L^2_{\text{loc}}(\Omega)$, and pointwise a.e. Hence $v \leq U^* \leq w$ and

$$(3.24) \qquad J(U^*) \leq M + 2K.$$

By (3.6)(j),

$$(3.25) \qquad \|U^* - \varphi\|_{L^1(\Omega_i)} \leq |\rho_j - \rho(\varphi)|$$

for all $i \in \mathbb{Z}$ and by (3.23),

$$(3.26) \qquad \|U^* - \varphi\|_{L^1(X_i)} \geq \sigma$$

for all $i \in \mathbb{Z}$. Consequently if $\psi \in \{v, w\} \setminus \{\varphi\}$,

$$\|U^* - \psi\|_{L^1(\Omega_i)} \geq \|\psi - \varphi\|_{L^1(\Omega_i)} - \|U^* - \varphi\|_{L^1(\Omega_i)}$$

for all $i \in \mathbb{Z}$ so by (3.2), there is an $\alpha > 0$ such that

$$(3.27) \qquad \|U^* - \psi\|_{L^1(\Omega_i)} \geq \frac{\alpha}{3}$$

and

$$(3.28) \qquad \|U^* - \psi\|_{L^1(X_i)} \geq \alpha$$

for all $i \in \mathbb{Z}$. Set

$$(3.29) \qquad \mu = \min(\alpha, \sigma).$$

It will be shown next that the existence of such a U^* is not possible. Define

$$\Gamma(v, U^*) = \{u \in W^{1,2}_{\text{loc}}(\Omega) \mid v \leq u \leq w, \|v - u\|_{L^2(\Omega_i)} \to 0 \text{ as } i \to -\infty,$$
$$\|v - U^*\|_{L^2(\Omega_i)} \to 0 \text{ as } i \to \infty\}$$

and

$$(3.30) \qquad c(v, U^*) = \inf_{u \in \Gamma(v, U^*)} J(u).$$

Note that $\Gamma(v, U^*) \neq \phi$, e.g. if $u = v$, $x_1 \leq 0$, $u = U^*$, $x_1 \geq 1$; and u is defined in the usual fashion in between, then $u \in \Gamma(v, U^*)$. Moreover

$$c(v, U^*) \leq J(u) < \infty.$$

Let (φ_k) be a minimizing sequence for (3.30). By the definition of $\Gamma(v, U^*)$ and with μ as in (3.29), there is an $i_k \in \mathbb{N}$ such that for $i \geq i_k$,

$$(3.31) \qquad \|\varphi_k - U^*\|_{L^1(X_i)} \leq \mu/3.$$

Since $\tau_{-i_k} \varphi_k \in \Gamma(v, U^*)$ and $J(\tau_{-i_k} \varphi_k) = J(\varphi_k)$, $\tau_{-i_k} \varphi_k$ is also a minimizing sequence for (3.30). Hence by a familiar argument, there is a $\Phi \in W^{1,2}_{\text{loc}}(\Omega)$ such that along a subsequence, $\tau_{-i_k} \varphi_k \to \Phi$ weakly in $W^{1,2}_{\text{loc}}(\Omega)$. Furthermore the proof of Proposition 2.53 with $c(v, U^*)$ replacing $c(v, v)$ shows it can be assumed that $\tau_{-i_k} \varphi_k \to \Phi$ in $W^{1,2}_{\text{loc}}(\Omega)$.

224   PAUL H. RABINOWITZ

By (3.31), (3.26), and (3.28), for $i \geq 0$,

$$(3.32) \qquad \int_{X_i} (\tau_{-i_k} \varphi_k - v) dx = \int_{X_{i+i_k}} (\varphi_k - v) dx$$

$$= \int_{X_{i+i_k}} (U^* - v) dx - \int_{X_{i+i_k}} |\varphi_k - U^*| dx \geq \frac{2}{3}\mu.$$

Thus letting $k \to \infty$,

$$(3.33) \qquad \int_{X_i} (\Phi - v) dx \geq \frac{2}{3}\mu, \quad i \geq 0.$$

Similarly

$$(3.34) \qquad \int_{X_i} (w - \Phi) dx \geq \frac{2}{3}\mu, \quad i \geq 0.$$

An examination of the argument of (A) with $c(v, U^*)$ replacing $c_m(v, v)$ shows Φ is a solution of (PDE)–(BC) in Ω. A second such solution will be produced next. By Proposition 2.8, there is a $U \in \mathcal{M}(v, w)$ such that

$$(3.35) \qquad \int_{X_0} (U - v) dx < \mu/3.$$

Set $\chi_k = \min(\tau_{-i_k} \varphi_k, U)$ and $\psi_k = \max(\tau_{-i_k} \varphi_k, U)$. Then $\chi_k \in \Gamma(v, U^*)$, $\psi_k \in \widehat{\Gamma}(v, w)$, and

$$(3.36) \qquad J(\chi_k) + J(\psi_k) = J(U) + J(\varphi_k).$$

Since $U \in \mathcal{M}(v, w)$,

$$(3.37) \qquad J(\psi_k) \geq J(U).$$

Thus by (3.36)–(3.37),

$$(3.38) \qquad J(\chi_k) \leq J(\varphi_k).$$

Hence the choice of φ_k shows as $k \to \infty$,

$$(3.39) \qquad J(\chi_k) \to c(v, U^*).$$

Since $\tau_{-i_k} \varphi_k \to \Phi$ in $W^{1,2}_{\text{loc}}(\Omega)$ as $k \to \infty$, by the continuity of the map $u \to \min(u, U)$ in $W^{1,2}_{\text{loc}}(\Omega)$,

$$(3.40) \qquad \chi_k \to \chi = \min(\Phi, U)$$

in $W^{1,2}_{\text{loc}}(\Omega)$ as $k \to \infty$. Another application of the argument of (A) shows χ is a solution of (PDE)–(BC) in Ω. Set $\Psi = \Phi - \chi$. Then $\Psi \geq 0$ in Ω and as in 1° of the proof of Theorem 2.13, Ψ satisfies a linear elliptic PDE like (2.24) to which the Maximum Principle applies. Thus Ψ cannot both have a zero in Ω and be $\not\equiv 0$. Since $\|U - w\|_{C^2(X_i)} \to 0$ as $i \to \infty$, by (3.34) there are points in X_i for large i where $\chi = \Phi$. Therefore Ψ has zeroes in X_i for large i. On the other hand by (3.33) and (3.35), there are points in X_0 where $\Phi > U$. Thus $\Psi \not\equiv 0$. Consequently we have a contradiction and there must exist an $X_i \subset \mathcal{R}_j$ for which (3.22) holds, establishing Proposition 3.21.

Proof of (C): To verify that U_m satisfies (3.7), it suffices to do so for $i \to \infty$; the proof for $i \to -\infty$ is the same. First it will be shown that

$$(3.41) \qquad \|U_m - \varphi_{j_i}\|_{L^1(X_{j_i})} \to 0$$

as $i \to \infty$ where $j_i \to \infty$ as $i \to \infty$ and $\varphi_{j_i} \in \{v, w\}$. Indeed if (3.41) is false, then there is a $\sigma > 0$ such that

$$\|U_m - v\|_{L^1(X_j)}, \quad \|U_m - w\|_{L^1(X_j)} \geq \sigma \tag{3.42}$$

for all large $j \in \mathbb{N}$. Consider $\{\tau_{-j} U_m \mid j \in \mathbb{N}, \ j \geq m_4 + \ell\}$. Since U_m is a bounded solution of (PDE)–(BC) for $x_1 > m_4 + \ell$, a subsequence of $\tau_{-j} U_m$ converges to Φ, a solution of (PDE)–(BC) in Ω with Φ satisfying $v \leq \Phi \leq w$ and

$$\|\Phi - v\|_{L^1(X_i)}, \quad \|\Phi - w\|_{L^1(X_i)} \geq \sigma \tag{3.43}$$

for all $i \in \mathbb{Z}$. But then we are in the setting of (A) with Φ replacing U^*. Thus (3.42) is impossible and (3.41) must hold. Since U_m, v, w are bounded in C^2 for $x_1 > m_4 + \ell$, (3.41) and an interpolation argument shows

$$\|U_m - \varphi_{j_i}\|_{C^1(X_{j_i})} \to 0, \quad i \to \infty. \tag{3.44}$$

Next we claim $\varphi_{j_i} = v$ for all large i. Arguing indirectly note first that for $\ell \geq \ell_0(\sigma)$, by Proposition 3.21 there is an $X_t \subset \mathcal{R}_4$ such that U_m is a solution of (PDE)–(BC) in X_t and

$$\|U_m - v\|_{L^1(X_t)} \leq \sigma. \tag{3.45}$$

As for (3.44), (3.45) implies

$$\|U_m - v\|_{C^1(X_t)} \leq \psi(\sigma) \tag{3.46}$$

where $\psi(\sigma) \to 0$ as $\sigma \to 0$. By (3.44), i can be chosen so large that

$$\|U_m - w\|_{C^1(X_{j_i})} \leq \psi(\sigma). \tag{3.47}$$

Now define

$$\begin{aligned}
f_k \ &= \ u_k, \quad && x_1 \leq t - 1 \\
&= \ v, \quad && t \leq x_1 \leq t + 1 \\
&= \ u_k, \quad && t + 2 \leq x_1 \leq j_i - 1 \\
&= \ w, \quad && j_i \leq x_1 \leq j_i + 1 \\
&= \ u_k, \quad && j_i + 2 \leq x_1
\end{aligned} \tag{3.48}$$

where f_k is defined in the intermediate intervals as e.g. in (2.59). By Proposition 3.15, $u_k \to U_m$ in $W^{1,2}_{\text{loc}}(\Omega)$. Hence for $k = k(\sigma)$ sufficiently large, as in (2.61)–(2.63),

$$|a_p(f_k)|, |a_p(u_k)| \leq \kappa(\psi(\sigma)) \equiv \omega(\sigma) \tag{3.49}$$

for each intermediate interval. Thus

$$|J(u_k) - J(f_k)| \leq 8\omega(\sigma). \tag{3.50}$$

Set

$$\begin{aligned}
g_k \ &= \ f_k, \quad && x_1 \leq j_i - 1 \\
&= \ u_k, \quad && x_1 \geq j_i - 1
\end{aligned} \tag{3.51}$$

and

$$\begin{aligned}
h_k \ &= \ u_k, \quad && x_1 \leq j_i - 1 \\
&= \ f_k, \quad && x_1 \geq j_i - 1
\end{aligned} \tag{3.52}$$

so

$$J(f_k) = J(g_k) + J(h_k). \tag{3.53}$$

But $h_k \in \Gamma(v, v)$ and $\tau_{-j_i} h_k$ satisfies (2.45) with

$$\gamma = \int_{\Omega_0} (w - v)dx.$$

Hence by Proposition 2.44,

$$(3.54) \qquad J(\tau_{-j_i} h_k) = J(h_k) \geq \beta(\gamma).$$

Consequently (3.49), (3.53)–(3.54) show

$$(3.55) \qquad J(u_k) \geq J(g_k) + \beta(\gamma) - 8\omega(\sigma).$$

Choosing σ so small that

$$(3.56) \qquad 16\omega(\sigma) < \beta(\gamma)$$

yields

$$(3.57) \qquad J(u_k) \geq J(g_k) + \frac{1}{2}\beta(\gamma).$$

But $g_k \in \Gamma_m(v, v)$ so (3.57) is contrary to (u_k) being a minimizing sequence for (3.8). Thus $\varphi_{j_i} = v$ for large i. Moreover $j_i = i$ for such i. Otherwise there is a $\gamma > 0$ and a subsequence $s_i \to \infty$ as $i \to \infty$ such that

$$(3.58) \qquad \|\sigma_m - v\|_{W^{1,2}(X_{s_i})} \geq \gamma.$$

Therefore the above argument with w in (3.48) replaced by v and j_i by s_i provides a contradiction. The proof of (C) is complete.

Proof of (D): Since by (C), $U_m \in \Gamma_m(v, v)$,

$$(3.59) \qquad J(U_m) \geq c_m(v, v).$$

Let $\varepsilon > 0$. By (C) again, p can be chosen so that

$$(3.60) \qquad \|U_m - v\|_{W^{1,2}(X_i)} \leq \varepsilon/2, \quad |i| \geq p.$$

Since u_k converges to U_m in $W^{1,2}_{\text{loc}}(\Omega)$,

$$(3.61) \qquad \|u_k - U_m\|_{W^{1,2}(\pm p)} \leq \varepsilon/2$$

for $k \geq k_0(p)$. Thus (3.60)–(3.61) imply

$$(3.62) \qquad \|u_k - v\|_{W^{1,2}(X_{\pm p})} \leq \varepsilon$$

for $k \geq k_0(p)$. For any fixed k, since $u_k \in \Gamma_m(v, v)$,

$$(3.63) \qquad \|u_k - v\|_{W^{1,2}(X_q)} \to 0, \quad |q| \to \infty.$$

For $q > p$, define f_k as in (3.48) with t replaced by p, j_i by q, and w by v. Then

$$(3.64) \qquad J_{1,\infty}(u_k) = J_{1,p-1}(u_k) + J_{p,q}(u_k) + J_{q,\infty}(u_k).$$

As for (3.49)–(3.50),

$$(3.65) \qquad |J_{p,q}(u_k) - J_{p,q}(f_k)| \leq 4\kappa(\varepsilon) + 4\kappa(\|u_k - v\|_{W^{1,2}(X_q)}).$$

Therefore

$$(3.66) \qquad J_{p,q}(u_k) \geq J_{p,q}(f_k) - 4\kappa(\varepsilon) - 4\kappa(\|u_k - v\|_{W^{1,2}(X_q)}).$$

Since $f_k|_{[p,q+1] \times \mathcal{D}}$ extends naturally as a $W^{1,2}_{\text{loc}}\, q + 1 - p$ periodic function in x_1, as in (2.32),

$$(3.67) \qquad J_{p,q}(f_k) \geq 0.$$

Note that

$$(3.68) \qquad J_{q,\infty}(u_k) \to 0$$

as $q \to \infty$ via Remark 2.12. Thus letting $q \to \infty$ in (3.64), (3.66)–(3.68) shows

$$(3.69) \qquad J_{1,\infty}(u_k) \geq J_{1,p-1}(u_k) - 4\kappa(\varepsilon).$$

Combining (3.69) with its counterpart for $J_{-\infty,0}(u_k)$ yields

$$(3.70) \qquad J(u_k) \geq J_{-p+1,p-1}(u_k) - 8\kappa(\varepsilon).$$

Letting $k \to \infty$ and using the $W^{1,2}_{\mathrm{loc}}(\Omega)$ convergence of u_k to U_m shows

$$(3.71) \qquad c_m(v,v) = \lim_{k\to\infty} J(u_k) \geq J_{-p+1,p-1}(U_m) - 8\kappa(\varepsilon).$$

Letting $p \to \infty$ gives

$$(3.72) \qquad c_m(v,v) \geq J(U_m) - 8\kappa(\varepsilon).$$

Finally since (3.72) is true for all $\varepsilon > 0$,

$$(3.73) \qquad c_m(v,v) \geq J(U_m)$$

which combined with (3.59) establishes (D).

Proof of (E): It remains to show that U_m is a solution of (PDE)–(BC) in $\overline{\mathcal{R}}$. If there is never equality in the constraints, (3.6), with $u = U_m$, the arguments of (A)–(B) apply equally well here. It will be shown that equality cannot occur for (3.6)(i)–(ii). The same argument gives the remaining cases. A comparison function argument involving a new variational problem will be used. To introduce the comparison value $d(v,w)$, define

$$(3.74) \qquad \Lambda(v,w) = \{u \in \Gamma^*(v,w) \mid (3.75) \text{ holds}\}$$

and

$$(3.75) \qquad \int_{\Omega_0} u \, dx = \rho_i$$

where $i = 1$ or 2. Set

$$(3.76) \qquad d(v,w) = \inf_{u \in \Lambda(v,w)} J(u).$$

The next proposition provides a crucial estimate for $d(v,w)$.

PROPOSITION 3.77. $d(v,w) > c(v,w)$.

Proof: Since $\Lambda(v,w) \subset \Gamma^*(v,w)$,

$$(3.78) \qquad d(v,w) \geq c(v,w).$$

To exclude equality, let (u_k) be a minimizing sequence for (3.76). As in earlier cases, it can be assumed that (u_k) converges in $W^{1,2}_{\mathrm{loc}}(\Omega)$ to P with P a solution of (PDE)–(BC) for $x_1 \notin [0,1]$ and P satisfies (3.75). Now arguments in the spirit of (C) will be used to obtain the Proposition.

We claim that as $i \to \infty$,

$$(3.79) \qquad \|P - v\|_{L^1(X_{j_i})} \to 0$$

or

$$(3.80) \qquad \|P - w\|_{L^1(X_{j_i})} \to 0$$

where $j_i \to -\infty$. Indeed as in the proof of (3.41),

$$(3.81) \qquad \|P - \varphi_{j_i}\|_{L^1(X_{j_i})} \to 0$$

along such a subsequence and therefore

$$(3.82) \qquad \|P - \varphi_{j_i}\|_{C^1(X_{j_i})} \to 0$$

as in (3.44). If $\varphi_{j_i} = w$ along a subsequence, then given any $\sigma > 0$, for j_i large enough,

$$(3.83) \qquad \|P - w\|_{C^1(X_{j_i})} \leq \sigma/2.$$

Fix such a j_i. Since $u_k \to P$ in $W^{1,2}_{\text{loc}}(\Omega)$, for all $k \geq k_0(\sigma)$,

$$(3.84) \qquad \|u_k - w\|_{W^{1,2}(X_{j_i})} \leq \sigma.$$

For each such k, a $t = t(k) < j_i$ can be chosen so that

$$(3.85) \qquad \|u_k - v\|_{W^{1,2}(X_t)} \leq \sigma.$$

Now define f_k as in (3.48). Then (3.49) is replaced by

$$(3.86) \qquad |a_p(f_k)|, \ |a_p(u_k)| \leq \kappa(\sigma)$$

for the intermediate intervals and

$$(3.87) \qquad |J(u_k) - J(f_k)| \leq 8\kappa(\sigma).$$

Define g_k and h_k as in (3.51)–(3.52) so (3.53) holds for the current setting. Now $g_k \in \Gamma^*(v, w)$ and $h_k \in \widehat{\Gamma}(v, w)$ and also satisfies (3.75). Hence as in (3.57), by Proposition 2.44 (for w),

$$(3.88) \qquad J(f_k) \geq c(v, w) + \min(\beta(\rho_1), \beta(\rho_2)).$$

Choosing σ so small that

$$(3.89) \qquad 16\kappa(\sigma) < \min(\beta(\rho_1), \beta(\rho_2)) \equiv \alpha,$$

$$(3.90) \qquad J(u_k) \geq c(v, w) + \frac{\alpha}{2}.$$

Thus letting $k \to \infty$ yields Proposition 3.77 for this case.

Next suppose that $\varphi_{j_i} = v$ for all i near ∞. Then $j_i = -i$ for otherwise there is a $\gamma > 0$ and a sequence $s_i \to -\infty$ as $i \to \infty$ such that

$$(3.91) \qquad \|P - v\|_{W^{1,2}(X_{s_i})} \geq \gamma.$$

The remark following (3.58) thus provides a contradiction. Thus

$$(3.92) \qquad \|P - v\|_{W^{1,2}(X_i)} \to 0$$

as $i \to -\infty$. If also

$$(3.93) \qquad \|P - w\|_{W^{1,2}(X_i)} \to 0$$

as $i \to \infty$, then $P \in \Lambda(v, w)$ and a slight modification of the proof of (D) yields

$$(3.94) \qquad J(P) = d(v, w).$$

Assuming (3.93), if there is equality in (3.78), the regularity results of [1] imply P is a solution of (PDE)–(BC) in Ω. But (3.75) and the choice of ρ_1, ρ_2 show this is impossible. Thus $d(v, w) > c(v, w)$ and Proposition 3.77 is also proved for this case. Finally if (3.93) fails to hold, by the reasoning of (3.79)–(3.89),

$$(3.95) \qquad \|P - v\|_{W^{1,2}(X_{j_i})} \to 0$$

for some sequence $j_i \to \infty$ and with v now playing the role of w, (3.90) holds with ρ_1, ρ_2 replaced by ρ_3, ρ_4.

Remark 3.96. Define $\Lambda(w, v)$ and $d(w, v)$ in the obvious way with $i = 3, 4$ in (3.75), it similarly follows that $d(w, v) > c(w, v)$.

Next to complete the proof of (E), additional comparison arguments will be used to show that equality cannot occur in (3.6)(i)–(ii).

By Proposition 3.21 again, for any $\sigma > 0$ and $\ell \geq \ell_0(\sigma)$,

$$(3.97) \qquad \|U_m - v\|_{L^1(X_t)} \leq \sigma$$

for some $X_t \subset \mathcal{R}_1$ and

$$(3.98) \qquad \|U_m - v\|_{L^1(X_s)} \leq \sigma$$

for some $X_s \subset \mathcal{R}_2$. By (B), U_m is a solution of (PDE)–(BC) in these special domains. As in (3.46),

$$(3.99) \qquad \|U_m - \varphi\|_{C^1(X_j)} \leq \psi(\sigma), \quad j = s, t$$

and $\varphi = v$ or w as appropriate. Suppose that U_m satisfies (3.6) with equality for some interval $\Omega_i \subset \mathcal{R}_1 \cup \mathcal{R}_2$. Two cases arise:

(E$_1$) Ω_i lies betwen X_t and X_s:

(E$_2$) Ω_i does not lie between X_t and X_s.

A different argument is used in each case.

Suppose (E$_1$) occurs. Define

$$(3.100) \qquad \begin{aligned} \Psi_m \;&=\; U_m, \quad x_1 \notin X_t \cup X_s \\ &=\; v, \qquad x_1 \in \Omega_t \\ &=\; w, \qquad x_1 \in \Omega_s \end{aligned}$$

with the usual interpolation in between. Then as e.g. in (3.49)–(3.50),

$$(3.101) \qquad |J_{t,s}(U_m) - J_{t,s}(\Psi_m)| \leq 4\omega(\sigma).$$

Choose σ so small that

$$(3.102) \qquad 12\omega(\sigma) < \min(d(v, w) - c(v, w), d(w, v) - c(w, v)) \equiv \delta.$$

Set

$$(3.103) \qquad \begin{aligned} V_m \;&=\; v, \qquad x_1 \leq t \\ &=\; \Psi_m, \quad t \leq x_1 \leq s + 1 \\ &=\; w, \qquad s + 1 \leq x_1. \end{aligned}$$

Then $V_m \in \Lambda(v, w)$ so

$$(3.104) \qquad J(V_m) = J_{t,s}(V_m) = J_{t,s}(\Psi_m) \geq d(v, w).$$

Thus (3.101) and (3.104) imply

$$(3.105) \qquad J_{t,s}(U_m) \geq d(v, w) - 4\omega(\sigma).$$

Set

$$(3.106) \qquad \begin{aligned} \Phi_m \;&=\; U_m, \qquad x_1 \leq t - 1 \text{ and } x_1 \geq s + 2 \\ &=\; t_{-q}U, \quad t \leq x_1 \leq s + 1 \end{aligned}$$

with the usual interpolation in between. Take $U \in \mathcal{M}(v, w)$. Choose ε so that

$$(3.107) \qquad 0 < \varepsilon < \delta/3.$$

For $m_2 - m_1$ sufficiently large, $q \in \mathbb{Z}$ can be chosen so that $\Phi_m \in \Gamma_m(v, v)$,

$$(3.108) \qquad J_{t,s}(\Phi_m) \leq c(v, w) + \varepsilon,$$

and

$$(3.109) \qquad \|\tau_{-q}U - v\|_{C^1(X_t)}, \ \|\tau_{-q}U - w\|_{C^1(X_s)} \leq \kappa(\sigma).$$

Since $\Phi_m \in \Gamma_m(v, v)$,

$$(3.110) \qquad J(U_m) = c_m(v, v) \leq J(\Phi_m) = J_{-\infty, t-2}(U_m) + J_{t-1, s+1}(\Phi_m) + J_{s+2, \infty}(U_m).$$

Therefore by (3.108)–(3.110)

$$(3.111) \qquad J_{t,s}(U_m) \leq c(v, w) + \varepsilon + 4\omega(\sigma).$$

Combining (3.105) and (3.111) yields

$$(3.112) \qquad d(v, w) - c(v, w) \leq \varepsilon + 8\omega(\sigma).$$

But (3.112) is contrary to (3.102) and (3.107). This (E_1) cannot occur.

Next suppose that (E_2) occurs. Then either $i < t$ or $i > s$. A similar argument can be employed for either alternative so suppose $i > s$. Invoking Proposition 3.21 again, for $\ell \geq \ell_0(\sigma)$ it can be assumed that there is an $X_q \subset \mathcal{R}_3$ such that

$$(3.113) \qquad \|U_m - w\|_{L^1(X_q)} \leq \sigma$$

and

$$(3.114) \qquad \|U_m - w\|_{C^1(X_q)} \leq \psi(\sigma).$$

Define

$$(3.115) \qquad \begin{aligned} F_m \ &= \ U_m, \quad x_1 \notin X_s \cup X_q \\ &= \ w, \quad x_1 \in \Omega_s \cup \Omega_q \end{aligned}$$

with the usual interpolation for the remaining 4 intervals. Then as earlier,

$$(3.116) \qquad |J(U_m) - J(F_m)| \leq 4\omega(\sigma).$$

Set

$$(3.117) \qquad \begin{aligned} G_m \ &= \ w, \quad x_1 \leq s \\ &= \ F_m, \quad s \leq x_1 \leq q+1 \\ &= \ w, \quad q+1 \leq x_1 \end{aligned}$$

and

$$(3.118) \qquad \begin{aligned} H_m \ &= \ F_m, \quad x_1 \leq s \\ &= \ w, \quad s \leq x_1 \leq q+1 \\ &= \ F_m, \quad q+1 \leq x_1. \end{aligned}$$

Hence

$$(3.119) \qquad J(F_m) = J(G_m) + J(H_m).$$

Observe that $G_m \in \Gamma(w, w)$ and

$$(3.120) \qquad \int_{\Omega_i} G_m \, dx = \int_{\Omega_i} U_m \, dx = \rho_2.$$

Therefore by Proposition 2.44,

$$(3.121) \qquad J(G_m) \geq \beta(\rho_2).$$

Since $H_m \in \Gamma_m(v, v)$, (3.116), (3.119)–(3.121) imply

$$(3.122) \qquad J(U_m) = c_m(v, v) \geq \beta(\rho_2) + c_m(v, v) - 4\omega(\sigma).$$

Further requiring that

$$(3.123) \qquad 8\omega(\sigma) < \min_{1 \leq j \leq 4} \beta(\rho_j)$$

shows (E$_2$) is not possible.

This completes the proof of (E) and of Theorem 3.9.

Remark 3.124. The argument of (E$_2$) can also be used to provide estimates for $U_m - w$ in $[m_2 + \ell, m_3 - \ell] \times \mathcal{D}$. Let X_s and X_q be as in case (E$_2$). Set

$$(3.125) \qquad \gamma = \min_{j \in [s+1, q] \cap \mathbb{Z}} \int_{\Omega_j} U_m dx.$$

Then by (3.116), (3.125), (3.121), and (3.122),

$$(3.126) \qquad c_m(v, v) \geq \beta(\gamma) + c_m(v, v) - 4\omega(\sigma)$$

so

$$(3.127) \qquad 4\omega(\sigma) \geq \beta(\gamma).$$

Thus σ small implies U_m is L^1 and therefore C^2 close to w in $[m_2 + \ell, m_3 - \ell]$.

4. More complex solutions

In this section a few remarks will be made about more complex solutions of (PDE)–(BC). The solutions constructed in §3 are the simplest type that can be found near a heteroclinic 2-chain on x_1 obtained by formally gluing members of $\mathcal{M}(v, w)$ and $\mathcal{M}(w, v)$. More generally one can seek solutions that approximate a formal k-chain of members of $\mathcal{M}(v, w)$, $\mathcal{M}(w, v)$. Such results are familiar in dynamical systems — see e.g. Mather [9] or [10] — and can also be obtained here using the approach of §3.

By way of illustration, to construct a heteroclinic solution of 3-chain type, begin by choosing $m \in \mathbb{Z}^6$. Now (3.3) is supplemented by

$$(4.1) \qquad m_4 + \ell < m_5 - \ell < m_5 < m_6.$$

Two additional constraints are required in (3.6):

$$(4.2) \qquad \begin{aligned} \rho(\tau_{-i}u) &\leq \rho_5, & i &= m_5 - \ell, \ldots, m_5 - 1 \\ \rho_6 &\leq \rho(\tau_{-i}u), & i &= m_6, \ldots, m_6 - 1 \end{aligned}$$

with ρ_5, ρ_6 chosen like ρ_1, ρ_2. The asymptotic condition as $x_1 \to \infty$ in (3.7) is replaced by

$$(4.3) \qquad \|u - w\|_{L^2(\Omega_j)} \to 0, \quad j \to \infty.$$

Then assuming also that $m_6 - m_5$ is sufficiently large, the analogue of Theorem 3.9 is valid where now

$$(4.4) \qquad \begin{cases} \|U_m - v\|_{C^2(\Omega_j)} \to 0, & j \to -\infty \\ \|U_m - w\|_{C^2(\Omega_j)} \to 0, & j \to \infty. \end{cases}$$

The proof follows steps (A)–(E). The only change required is in (E). The same comparison ideas are used but the estimates become a bit more complicated.

If the 4 values ρ_i, $i = 1, \ldots, 4$ are used for ρ_j, $j \geq 5$, then the requirements that ℓ and $m_{2p} - m_{2p-1}$ be large can be satisfied for a solution of k-chain type independently of k so a limit argument yields infinite chain solutions of (PDE)–(BC).

References

[1] Rabinowitz, P. H., Solutions of heteroclinic type for some classes of semilinear elliptic partial differential equations, J. Fac. Sci. Univ. Tokyo, **1**, (1994), 525–550.

[2] Rabinowitz, P. H., Spatially heteroclinic solutions for a semilinear elliptic PDE, to appear, ESAIM: Control, Optimization and Calculus of Variations.

[3] Rabinowitz, P. H., A new variational characterization of spatially heteroclinic solutions of a semilinear elliptic PDE, to appear Discrete and Continuous Dynamical Systems.

[4] Kirchgässner, K., Wave-solutions of reversible systems and applications, J. Diff. Eq., **45**, (1982), 113–127.

[5] Turner, R. E. L., Internal waves in fluids with rapidly varying density, Ann. Scuola. Norm. Pisa Cl. Sc. Ser. 4, (1981), 513–573.

[6] Mielke, A., Reduction of quasilinear elliptic equations in cylindrical domains with applications, Math. Meth. Appl. Sci., **10**, (1988) 51–66.

[7] Rabinowitz, P. H. and E. Stredulinsky, Mixed states for an Allen-Cahn type equation, to appear in Comm. Pure Appl. Math.

[8] Rabinowitz, P. H. and E. Stredulinsky, in progress.

[9] Mather, J. N., Variational construction of connecting orbits, Ann. Inst. Fourier (Grenoble), **43**, (1993), 1349–1386.

[10] Rabinowitz, P. H., On a theorem of Strobel, Calc. Var. Partial Differential Equations, **12**, (2001), 399–415.

DEPARTMENT OF MATHEMATICS, UNIVERSITY OF WISCONSIN–MADISON, MADISON, WI 53706
E-mail address: rabinowi@math.wisc.edu

Contemporary Mathematics
Volume **350**, 2004

Vortices for Ginzburg-Landau Equations : With Magnetic Field Versus Without

Sylvia Serfaty and Etienne Sandier

This paper is dedicated to H. Brezis and F. Browder

ABSTRACT. We present a review and comparison of results on the static Ginzburg-Landau models with and without magnetic field.

I. Presentation of the models

I.1. Vortices in the Ginzburg-Landau framework. Ginzburg-Landau type equations have been the subject of intense mathematical study over the past ten years (as a proof, one may see the list of names below). Here, we examine static problems and we distinguish between the model with magnetic field and the model without, and try to compare some results on both. Mathematical literature on the Ginzburg-Landau equation without magnetic field (in 2D or higher dimensions) include works by: Bethuel-Brezis-Hélein, Struwe, Mironescu, Comte, Sandier, Shafrir, F.H. Lin, Rivière, Pacard, Almeida, Jerrard-Soner, Alberti-Baldo-Orlandi... On the Ginzburg-Landau equations with magnetic field: Jaffe-Taubes, Berger-Chen, Bethuel-Rivière, Rubinstein, Chapman, Sternberg, Felmer, Del Pino, Bauman, Phillips, Giorgi, Alama, Bronsard, Du, Lin, Pan, Helffer, Almog, Aftalion, Dancer, Gustafson-Sigal, Bonnet-Monneau...

Let us start with a simple (naive) approach : try to minimize the Dirichlet energy of a map from a domain Ω, bounded, regular and *simply connected* in $\mathbb{R}^2$, to the unit-circle S^1. Let u denote such a map. The *topological degree* of u on $\partial\Omega$ is the integer number

$$\frac{1}{2\pi}\int_{\partial\Omega} u \times \frac{\partial u}{\partial \tau},$$

1991 *Mathematics Subject Classification.* 82D55; 35B40; 35B25; 35J60; 35J20; 35Q99; 58E50.

Key words and phrases. superconductivity, second critical field, phase-transitions, asymptotic analysis .

Supported by the C.N.R.S.

($\times$ denoting the vector product) or, if one writes $u = e^{i\varphi}$ with φ a local lifting of u (or multi-valued phase function, defined modulo $2\pi\mathbb{Z}$) the degree is

$$\frac{1}{2\pi} \int_{\partial\Omega} \frac{\partial\varphi}{\partial\tau}.$$

If we try to find harmonic S^1-valued maps i.e. ask u to minimize the Dirichlet energy

$$\int_\Omega |\nabla u|^2$$

over all S^1-valued maps with degree d on the boundary of Ω, then the answer is :
- if $d = 0$, there are minimizers
- if $d \neq 0$ there is no map of finite Dirichlet energy hence the problem has no meaning.

One can thus think of considering a regularization of this problem, dropping the constraint $|u| = 1$ and replacing it by a penalization (or singular perturbation): minimize over $u \in H^1(\Omega, \mathbb{C})$ with a boundary degree d,

$$(I.1) \qquad F_\varepsilon(u) = \frac{1}{2} \int_\Omega |\nabla u|^2 + \frac{1}{2\varepsilon^2}(1 - |u|^2)^2.$$

Then, u is no longer S^1 valued, but as $\varepsilon \to 0$, its modulus is forced to tend to 1. Moreover, the problem of minimizing F_ε does have test functions of finite energy. This is the simplest regularization of the original problem one can think of, and was studied extensively in the book by Bethuel, Brezis and Hélein [**BBH**]. This is one possible motivation for this problem, another one being, as we will see, the study of the Ginzburg-Landau model of superconductivity, a third one being an $\mathbb{R}^2$-valued generalization of the scalar phase-transition problem (studied by Modica-Mortola, Sternberg...)

From now on, we will call F_ε the "simple" Ginzburg-Landau functional. Bethuel, Brezis and Hélein studied it under the Dirichlet boundary condition $u = g$ on $\partial\Omega$ where g is a smooth S^1-valued map of degree $d > 0$. The Euler equation associated to the functional is

$$(I.2) \qquad -\Delta u = \frac{u}{\varepsilon^2}(1 - |u|^2).$$

When the degree of u at the boundary is nonzero, for the same topological reason as before, u has to vanish in Ω (otherwise $u/|u|$ would be S^1-valued and H^1 of degree d, a contradiction). Minimizers of the energy are almost everywhere of unit norm except on small regions of characteristic size ε around their zeroes. A zero of a solution of (I.2), or vortex of degree d looks like

$$\rho(r)e^{id\theta}$$

in polar coordinates, where ρ is a certain "profile" vanishing at 0 and close to 1 away from a ball of size ε. Thus,

$$\frac{1}{2} \int_\Omega |\nabla u|^2 \geq \frac{1}{2} \int_{|x|\geq\varepsilon} |\nabla\varphi|^2 = \frac{1}{2} \int_{|x|\geq\varepsilon} \frac{d^2}{|x|^2} \simeq \pi d^2 \log\frac{1}{\varepsilon}$$

is the approximate energy carried by such a vortex. It diverges as $\varepsilon \to 0$. Also notice that, contrarily to the case of the real-valued problem, it is the *phase* of u (and not the modulus ρ) which carries the main energy, and that the energy cost is not carried in the vortex-"core" $B(0,\varepsilon)$ where $|u|$ is small, but rather in an annulus of size $\gg \varepsilon$ around that core (sometimes called "neck region").

I.2. Vortex-balls constructions. In order to fully justify the rough picture of vortices we just presented, it is necessary to design a method to isolate the vortices and estimate their individual cost (in $\pi d^2 |\log \varepsilon|$) as well as their possible interactions. Bethuel, Brezis and Hélein used, for critical points of F_ε, a finite covering argument, which allowed them to isolate *disjoint* balls of radius $\lambda\varepsilon$, outside of which $|u| \geq \frac{1}{2}$ (i.e. that contain all the zeroes) and to evaluate precisely the cost of the vortices they contain according to their degree and the interaction with other vortices. The fact that they had the fixed Dirichlet boundary condition $u = g$, allowed, (using Pohozaev's identity) to have an a priori bound independent of ε on the number of these balls.

The main facts proved in [**BBH**] for energy-minimizers are
- the energetic cost of a vortex a degree d is $\pi d^2 |\log \varepsilon|$.
- vortices have a coulombian interaction through a "renormalized energy"
$W_g(a_i, \cdots, a_n) \simeq -\pi \sum_{i,j} d_i d_j \log |a_i - a_j|$ depending on their locations and degrees (and g).

THEOREM 1 ([**BBH**]). *Let u_ε be a minimizer of F for the boundary data g. From any sequence $\varepsilon_n \to 0$, one may extract a subsequence u_{ε_n} converging to*

$$u_*(x) = \frac{x - a_1}{|x - a_1|} \cdots \frac{x - a_d}{|x - a_d|} e^{ih(x)},$$

where $a_1, \cdots a_d$ are d distinct points of Ω and h is a harmonic function, the convergence taking place in $C_{loc}^k(\Omega \backslash \{a_1, \ldots, a_d\})$. Moreover, $(a_1, \cdots, a_d)$ is a minimizer of W_g and

$$F_{\varepsilon_n}(u_{\varepsilon_n}) \sim \pi d |\log \varepsilon| + W_g(a_1, \cdots, a_d) + o(1) \quad as \ n \to \infty.$$

Later on, Almeida and Bethuel [**AB**], proposed a method which allowed to define "vortex-balls" (in bounded number again) for any u such that $F_\varepsilon(u) \leq C |\log \varepsilon|$, without the assumption that u is a critical point of F_ε.

Then, Jerrard [**J**] and Sandier [**Sa**], proposed independently a method which allows to isolate the zeros or vortices of any reasonable map u in *disjoint* small balls with optimal lower bounds on the energy they contain. This allowed to consider possibly unbounded numbers of vortices (i.e. which blow up as $\varepsilon \to 0$) independently of their mutual interactions.

I.3. Ginzburg-Landau with magnetic field. The Ginzburg-Landau functional which comes from physics is

$$(\text{I.3}) \qquad J(u, A) = \frac{1}{2} \int_\Omega |(\nabla - iA)u|^2 + |h - h_{\text{ex}}|^2 + \frac{1}{2\varepsilon^2}(1 - |u|^2)^2$$

where u is coupled to a magnetic potential A. This functional describes the cross-section of a superconductor submitted to an applied magnetic field of intensity h_{ex}. $h = \text{curl}\, A$ is the magnetic field induced in the sample. ε corresponds to a material constant. Here, u is the "order-parameter", describing the local state of the material, its modulus $|u|$ measures the density of superconducting electrons, where $|u|$ is close to 1, it is the superconducting phase, whereas where $|u|$ is close to 0, it is the normal phase (often under the form of vortices). The associated "full" Ginzburg-Landau system is

$$(\mathrm{I.4})\qquad\begin{cases} -(\nabla - iA)^2 u = \kappa^2 u(1 - |u|^2) & \text{in } \Omega \\[2mm] -\nabla^\perp h = (iu, (\nabla - iA)u) & \text{in } \Omega \\[2mm] \dfrac{\partial u}{\partial n} - i(A.n)u = 0 & \text{on } \partial\Omega \\[2mm] h = h_{\mathrm{ex}} & \text{on } \partial\Omega. \end{cases}$$

Here $(.,.)$ denotes the scalar product in $\mathbb{C}$ identified with $\mathbb{R}^2$ and $\nabla^\perp$ denotes $(-\partial_2, \partial_1)$. Configurations are only defined up to a gauge-transformation :

$$\begin{cases} u \to ue^{i\Phi} \\ A \to A + \nabla\Phi \end{cases}$$

I.4. First correspondence. Consider $u = \rho e^{i\varphi}$ a solution of (I.2). Taking the scalar product of the equation with iu, one finds

$$\mathrm{div}\,(\rho^2 \nabla\varphi) = 0.$$

Then, using Poincaré's lemma, one can write

$$(\mathrm{I.5})\qquad \rho^2 \nabla\varphi = -\nabla^\perp H$$

for some real-valued function H. Taking the curl of this equation, one obtains

$$(\mathrm{I.6})\qquad -\Delta H = \mathrm{curl}\,(\rho^2 \nabla\varphi) = \mathrm{curl}\,(iu, \nabla u).$$

A crucial ingredient in the analysis of Ginzburg-Landau models in the small ε asymptotics is to prove the following estimate on the Jacobian determinant of u, or $\mathrm{curl}\,(u \times \nabla u) = u_{x_1} \times u_{x_2}$,

$$(\mathrm{I.7})\qquad \mathrm{curl}\,(iu, \nabla u) \sim 2\pi \sum_i d_i \delta_{a_i} \quad \text{as } \varepsilon \to 0.$$

Such relations have been established rigorously in variously strong forms either in the bounded vortex-number or the unbounded vortex-number settings, by Bethuel-Brezis-Helein [**BBH**], Bethuel-Rivière [**BR1, BR2**], Alberti-Baldo Orlandi [**ABO**], Jerrard-Soner [**JS1**], Sandier-Serfaty [**SS1, SS3, SS5**].

In the case of the "full" Ginzburg-Landau equation, one has a natural analogue of (I.5): the second Ginzburg-Landau equation

$$-\nabla^\perp h = \rho^2(\nabla\varphi - A)$$

so that, taking the curl

$$(\mathrm{I.8})\qquad -\Delta h + h \sim 2\pi \sum_i d_i \delta_{a_i}$$

This is the "London equation" in the physics of superconductivity. h can be considered as playing the same role as H for (I.2). In view of (I.5), one has

$$|\nabla H|^2 = \rho^4 |\nabla\varphi|^2 \sim |\nabla\varphi|^2$$

in the first case, and

$$(\mathrm{I.9})\qquad |\nabla h|^2 = \rho^4 |\nabla\varphi - A|^2 \sim |\nabla\varphi - A|^2$$

in the case with magnetic field. Since we saw the energy of a vortex is essentially carried by the phase, it will be carried by $\int |\nabla H|^2$ or $\int |\nabla h|^2$ and equal to $\pi d^2 |\log \varepsilon|$

per vortex.

On the other hand, going back to (I.3), one sees that if the applied field h_{ex} is large enough (typically bigger than the order of $|\log \varepsilon|$), then the term $|h - h_{\mathrm{ex}}|^2$ forces h to be close to the constant h_{ex}, and plugging this back into (I.8), one expects $2\pi \sum d_i \delta_{a_i} \sim h_{\mathrm{ex}}$ or $\frac{2\pi \sum d_i \delta_{a_i}}{h_{\mathrm{ex}}} \sim 1$, hence the number of vortices should blow up like h_{ex}. In the case of the minimization of (I.1), the vortices are forced into the system by the degree of the Dirichlet boundary condition g, and if one keeps this g fixed, the number of vortices remains bounded as $\varepsilon \to 0$. This is in contrast with the case of (I.3) where the presence of vortices is forced not through a boundary condition but through the intensity of the external field, and blows up for large external fields. (Notice that the functional (I.3) with a (fixed) Dirichlet boundary condition instead of the natural Neumann condition was studied in [**BR1**], with results somehow analogous to those of [**BBH**].)

II. Results on the case with magnetic field

II.1. Around the first critical field. Let us first present the energy-splitting argument of Bethuel and Rivière [**BR2**], which was carried out in more details in [**S1**]. In order to freeze the gauge-invariance, one may fix a gauge by setting div $A = 0$. Since the domain is simply connected, one can then write $A = \nabla^{\perp}\xi$ (and $h = \Delta\xi$), for some function ξ vanishing on $\partial\Omega$. Then,

$$(\mathrm{II.1}) \qquad \int_{\Omega} |(\nabla - iA)u|^2 \simeq \int_{\Omega} |\nabla\varphi - \nabla^{\perp}\xi|^2 = \int_{\Omega} |\nabla\varphi|^2 + |\nabla\xi|^2 - 2\int_{\Omega} \nabla\varphi \cdot \nabla^{\perp}\xi.$$

The last term is really equal to $\int_{\Omega} \xi\,\mathrm{curl}\,(iu, \nabla u)$ and can be rewritten $2\pi \sum d_i \xi(a_i)$ in view of (I.7). Thus, it will be possible to split the energy J as

$$(\mathrm{II.2}) \qquad J(u, A) \simeq F_{\varepsilon}(u) - 2\pi \sum d_i \xi(a_i) + \int_{\Omega} |\nabla\xi|^2 + |\Delta\xi - h_{\mathrm{ex}}|^2,$$

where F_{ε}, the functional of [**BBH**], naturally appears. We also see how the addition of a magnetic potential couples F with a magnetic effect term (in ξ), coupled with u through the vortices (a_i, d_i). Let us then look at how vortices may become energetically favorable. When it occurs, the external field is relatively large (of order $|\log \varepsilon|$), while the number of vortices is small. Thus, one can approximate (I.8) by the simplest London equation

$$(\mathrm{II.3}) \qquad \begin{cases} -\Delta h + h = 0 & \text{dans } \Omega \\ h = h_{\mathrm{ex}} & \text{sur } \partial\Omega. \end{cases}$$

Thus $h = h_{\mathrm{ex}} h_0$ where we denote by h_0 the solution of

$$(\mathrm{II.4}) \qquad \begin{cases} -\Delta h_0 + h_0 = 0 & \text{dans } \Omega \\ h_0 = 1 & \text{sur } \partial\Omega. \end{cases}$$

Then, it is justified to approximate ξ by $h_{\mathrm{ex}}(h_0 - 1)$. By "plugging" the analysis of [**BBH**] or [**AB**] into the F term in (II.2), one is led to

$$(\mathrm{II.5}) \qquad J(u, A) \geq \pi \sum d_i^2 |\log \varepsilon| - 2\pi h_{\mathrm{ex}} \sum d_i (h_0 - 1)(a_i) + h_{\mathrm{ex}}^2 J_0,$$

where $J_0 = \int_\Omega |\nabla h_0|^2 + |h_0 - 1|^2$. We can deduce that vortices become favorable when $J < h_{\mathrm{ex}}^2 J_0$ i.e. when

$$h_{\mathrm{ex}} > \frac{|\log \varepsilon|}{2 \max |h_0 - 1|} := H_{c_1}$$

for some vortices of degree $+1$, placed at the point(s) of minimum of $h_0 - 1$ (which is a negative function, depending only on Ω). In physics, this corresponds to a first-order phase-transition, and this corresponding field is called the "first critical field" H_{c_1}.

In order to make all this analysis rigorous, one has to justify that the number of vortices remains bounded for global minimizers in this field-regime. But the a priori energy bound is $J(u, A) \leq h_{\mathrm{ex}}^2 J_0 \leq O(|\log \varepsilon|^2)$. So a priori we only have $F(u) \leq C|\log \varepsilon|^2$, which, from the Bethuel-Brezis-Hélein analysis, could only provide an unsufficient bound in $|\log \varepsilon|$ on the total number of vortices. In [**SS5**], we were however able to prove with E. Sandier the following result

THEOREM 2 (Sandier-Serfaty [**SS5**]). *For all $M > 0$, there exist ε_0 and C such that $\forall \varepsilon < \varepsilon_0$, if $h_{\mathrm{ex}} \leq H_{c_1} + C \log |\log \varepsilon|$ and (u, A) is a minimizer of J, then (up to a gauge-transformation),*

$$F(u) \leq M|\log \varepsilon|.$$

This provides a handy upper bound on the number of vortices as long as the field remains close enough to H_{c_1}.

THEOREM 3 ([**S1, SS1, SS5**]). *For ε small enough,*
- if $h_{\mathrm{ex}} < H_{c_1} - C$, the minimizer of the energy J is unique, and has no vortex $(\inf_\Omega |u| \to 1$ as $\varepsilon \to 0)$
- if $h_{\mathrm{ex}} \geq H_{c_1} + C$, minimizers have vortices, of degree 1, tending to the point(s) of minimum of h_0.

For h_{ex} slightly larger than H_{c_1}, the behavior of minimizers was described for the case of the ball:

THEOREM 4 ([**S2, S3**]). *For $\Omega = B(0, R)$, $n \leq M$, and $\varepsilon < \varepsilon_0(M)$, if*

$$H_{c_1} + (n-1) \log |\log \varepsilon| + o(\log |\log \varepsilon|) \leq h_{\mathrm{ex}} \leq H_{c_1} + n \log |\log \varepsilon| + o(\log |\log \varepsilon|)$$

minimizers of J have exactly n vortices of degree 1, located at points a_i, and rescaling around the origin: $\tilde{a}_i = \sqrt{h_{\mathrm{ex}}} a_i$, the $\tilde{a}_i$'s converge to a minimizer of

$$(\mathrm{II.6}) \qquad w(x_1, \cdots, x_n) = -\pi \sum_{i \neq j} \log |x_i - x_i| + C \sum_i |x_i|^2.$$

The vortices are thus placed around the origin (at a distance $1/\sqrt{|\log \varepsilon|}$) and arranged in regular patterns minimizing w. w can be seen as a "renormalized energy" which takes into account the repulsion between the vortices and the confinement effect due to the magnetic field, and determines their location; thus an analogue of the renormalized energy of [**BBH**].

Theorem 4 provides branches of minimizing solutions with n vortices in a certain range of applied field. They can actually be extended to branches of *locally minimizing* solutions with the same number of vortices (same shapes...) to a much wider interval of h_{ex}. These are obtained by minimizing the energy over the set

$$U_n = \left\{ (u, A), \pi n |\log \varepsilon| < F(u) < \pi \left(n + \frac{1}{2} \right) |\log \varepsilon|, \ \mathrm{div}\, A = 0 \right\},$$

which, from the ideas of [**BBH, AB**], is roughly the set of configurations with n vortices.

II.2. The obstacle problem. Being interested in the case of global minimizers for larger fields (for which one has to expect a large number of vortices), one looks for the asymptotic limit μ_* of the rescaled vorticity measure :

$$(\text{II.7}) \qquad \mu_\varepsilon := \frac{2\pi \sum_i d_i \delta_{a_i}}{h_{\text{ex}}},$$

where the vortex-points are given by the ball-construction method mentioned in Section I.2; and h_*, that of the rescaled fields h/h_{ex} (bounded in H^1). Because of (I.8), one will have at the limit the relation

$$(\text{II.8}) \qquad -\Delta h_* + h_* = \mu_*.$$

The method is that of Γ-convergence. Bounding the energy from below relies on the construction of balls B_i mentioned in Section I.2 and expressing the energy in terms of h (see (I.9))

$$(\text{II.9}) \qquad J(u, A) \geq \frac{1}{2} \int_{\cup_i B_i} |\nabla h|^2 + \int_{\Omega \backslash \cup B_i} |\nabla h|^2 + \int_\Omega |h - h_{\text{ex}}|^2.$$

Dividing this relation by h_{ex}^2, the first term is bounded below by the ball construction, and for the last two, one may use the weak H^1 convergence of h/h_{ex}, thus obtaining

$$\begin{aligned} \frac{J(u, A)}{h_{\text{ex}}^2} &\geq \pi \sum |d_i| \frac{|\log \varepsilon|}{h_{\text{ex}}^2} + \frac{1}{2} \int_\Omega |\nabla h_*|^2 + |h_* - 1|^2 \\ &\geq \frac{1}{2} \frac{|\log \varepsilon|}{h_{\text{ex}}} \int_\Omega |\mu_*| + \frac{1}{2} \int_\Omega |\nabla h_*|^2 + |h_* - 1|^2. \end{aligned}$$

A matching upper bound can be obtained via the construction of a test-configuration (for all μ_*, h_* satisfying (II.8)).

THEOREM 5 ([**SS3**]). *Let $(u_\varepsilon, A_\varepsilon)$ be minimizers of J, $h_\varepsilon = \text{curl } A_\varepsilon$, then*

$$\frac{h_n}{h_{\text{ex}}} \rightharpoonup h_* \quad \text{weakly in } H^1(\Omega) \text{ strongly in } W^{1,p}(\Omega), \forall p < 2,$$

$$\mu_n \rightharpoonup \mu_* = -\Delta h_* + h_* \text{ in the weak sense of measures}$$

$$\lim_{n \to \infty} \frac{J(u_n, A_n)}{h_{\text{ex}}^2} = E(h_*) = \min E,$$

where h_* is the unique minimizer of

$$E(h) = \frac{1}{2\lambda} \int_\Omega |-\Delta h + h| + \frac{1}{2} \int_\Omega |\nabla h|^2 + |h - 1|^2 \qquad \lambda = \lim_{\varepsilon \to 0} \frac{h_{\text{ex}}}{|\log \varepsilon|}.$$

The problem of minimizing E turns out to be the dual (in the sense of convex duality) of

$$\min_{\substack{h \geq 1 - \frac{1}{2\lambda} \\ h = 1 \text{ on } \partial\Omega}} \frac{1}{2} \int_\Omega |\nabla h|^2 + h^2,$$

which is a standard obstacle problem, with obstacle $1 - \frac{1}{2\lambda}$. Thus the limiting field h_* is solution of a free-boundary problem. The free-boundary is the boundary of the coincidence set $\omega_\lambda = \{h_* = 1 - \frac{1}{2\lambda}\}$ and the limiting vorticity measure is

$$\mu_* = -\Delta h_* + h_* = (1 - 1/(2\lambda))\chi_{\omega_\lambda}$$

thus a uniform density supported on the set ω_λ (which depends on λ and thus on the field strength h_{ex}). ω_λ is nonempty if and only if $h_{\mathrm{ex}} > H_{c_1}$; for higher fields ω_λ inflates as h_{ex} increases, until $\omega_\lambda = \Omega$ for $h_{\mathrm{ex}} \gg |\log \varepsilon|$ (or $\lambda = +\infty$).

This theorem can also be formulated as a Gamma-convergence result for J/h_{ex}^2. The same type of result of Gamma-convergence has been written by Jerrard and Soner [**JS2**] for the functional F.

II.3. The case of critical points. In [**BBH**], Bethuel, Brezis and Hélein studied the asymptotics of minimizers of F under the boundary condition g, but also of critical points, not necessarily minimizing. They obtained the following result :

THEOREM 6 ([**BBH**]). *Let u_{ε_n} be a sequence of critical points of F_{ε_n} for the boundary data g with $F_{\varepsilon_n}(u_{\varepsilon_n}) \leq C|\log \varepsilon|$. Up to extraction, u_{ε_n} converges to a limiting function*

$$u_*(x) = \left(\frac{x - a_1}{|x - a_1|} \right)^{d_1} \cdots \left(\frac{x - a_k}{|x - a_k|} \right)^{d_k} e^{ih(x)}$$

where $d_1 + \cdots + d_k = d$ and h is a harmonic function. Moreover, the configuration of points+degrees $(a_1, \cdots, a_k)$ is a critical point of W_g and, writing

$$u_*(x) = \left(\frac{x - a_j}{|x - a_j|} \right)^{d_j} e^{i\varphi_j(x)},$$

one has

$$(\mathrm{II.10}) \qquad \forall j, \quad \nabla \varphi_j(a_j) = 0.$$

This last condition was called the "vanishing gradient property". It expresses the fact that the vortex-locations are stationary. It was obtained through a relation on the Hopf-differential of u (see [**BBH**]).

For (I.4), we obtained a corresponding result, starting from the fact that a critical (u, A) is stationary with respect to variations in the domain space, i.e that the "energy-momentum tensor"

$$T_{ij} = \left(\frac{h^2}{2h_{\mathrm{ex}}^2} - \frac{1}{4h_{\mathrm{ex}}^2 \varepsilon^2}(1 - |u|^2)^2 \right) \delta_{ij}$$
$$+ \frac{1}{2h_{\mathrm{ex}}^2} \left(\begin{array}{cc} |\partial_1^A u|^2 - |\partial_2^A u|^2 & 2(\partial_1^A u, \partial_2^A u) \\ 2(\partial_1^A u, \partial_2^A u) & |\partial_2^A u|^2 - |\partial_1^A u|^2 \end{array} \right),$$

is divergence-free. This is a particular case of Noether's theorem, and also the analogue of the relation on the Hopf differential in [**BBH**]. Considering critical points (u, A) such that $J(u, A) \leq Ch_{\mathrm{ex}}^2$ and $h_{\mathrm{ex}} \leq C|\log \varepsilon|$, one finds that h/h_{ex} and μ_ε (see (II.7)) remain bounded and have weak limits h_∞ and μ_∞. One would like to pass to the limit in T_{ij}/h_{ex}^2 (which is bounded L^1) in order to obtain necessary conditions on h_∞ and μ_∞. The formal limit of $T_{ij}h_{\mathrm{ex}}^2$ is

$$(\mathrm{II.11}) \quad \mathrm{div}\ L_{ij} = 0\,, i = 1, 2\,,$$
$$L_{ij} = \frac{h_\infty^2}{2}\delta_{ij} + \frac{1}{2} \left(\begin{array}{cc} (\partial_2 h_\infty)^2 - (\partial_1 h_\infty)^2 & -2\partial_1 h_\infty \partial_2 h_\infty \\ -2\partial_1 h_\infty \partial_2 h_\infty & (\partial_1 h_\infty)^2 - (\partial_2 h_\infty)^2 \end{array} \right).$$

However, passing to the limit in T_{ij}/h_{ex}^2 is problematic, since $(\nabla - iA)u$ does not converge strongly in L^2 (there is a lack of compactness in the vortices). Nevertheless, we were able to go around this problem and obtain

THEOREM 7 ([**SS4**]). *Assume (u_n, A_n) is a sequence of critical points of J for $\varepsilon_n \to 0$ with $J(u_n, A_n) \le Ch_{\mathrm{ex}}^2$ and $h_{\mathrm{ex}} \le C|\log \varepsilon|$. Up to extraction*

$$\frac{h_n}{h_{\mathrm{ex}}} \rightharpoonup h_\infty \quad in \ H^1(\Omega)$$

$$\mu_n = \frac{2\pi \sum d_i \delta_{a_i}}{h_{\mathrm{ex}}} \rightharpoonup \mu_\infty = -\Delta h_\infty + h_\infty$$

and

$$\mathrm{div}\ L_{ij} = 0.$$

If ∇h_∞ is continuous and BV (for example if $\mu_\infty \in L^p$, $p > 2$) then

$$(\mathrm{II}.12) \qquad \mathrm{div}\ L_{ij} = 0 \iff \mu_\infty \nabla h_\infty = 0, \qquad thus \ \mu_\infty = h_\infty \chi_{|\nabla h_\infty|=0}$$

With the same method we obtained the following result on the case of fewer vortices.

THEOREM 8 ([**SS4**]). *Under the same hypotheses, if $N_n := 2\pi \sum |d_i| \ll h_{\mathrm{ex}}$ then*

$$\frac{h_n}{h_{\mathrm{ex}}} \rightharpoonup h_0 \quad weakly \ in \ H^1$$

(where h_0 is the solution of (II.4)).

$$\nu_n = \frac{2\pi \sum_i d_i \delta_{a_i}}{N_n} \rightharpoonup \nu_\infty \quad as \ measures$$

where

$$(\mathrm{II}.13) \qquad \nu_\infty \nabla h_0 = 0.$$

Therefore, in the case where the number of vortices $(\sim N_n)$ is much smaller than h_{ex}, the limiting vorticity ν_∞ is supported in the set $\{\nabla h_0 = 0\}$ which is a finite set of points of Ω (reduced to one point if Ω is convex), and ν_∞ is a finite combination of Dirac masses at these points.

This method naturally also applies to solutions of (I.2) with divergent number of vortices. One just needs to replace h by H (see (I.5)).

THEOREM 9. *Let u_ε be a family of solutions of (I.2) such that $F(u_\varepsilon) \le \frac{1}{\varepsilon^\beta}$ ($\beta > 0$ small) and $\|\nabla u_\varepsilon\|_{L^\infty(\Omega)} \le \frac{C}{\varepsilon}$, as $\varepsilon \to 0$, if $N_\varepsilon \sim C\sqrt{F_\varepsilon(u_\varepsilon)}(C > 0)$, up to extraction,*

$$\frac{H_\varepsilon}{\sqrt{F_\varepsilon(u_\varepsilon)}} \rightharpoonup H_0 \quad weakly \ in \ H^1(\Omega) \ strongly \ in \ W_{loc}^{1,p}(\Omega), \ p < 2,$$

$$\frac{\mu_\varepsilon}{\sqrt{F_\varepsilon(u_\varepsilon)}} \rightharpoonup \mu_0 = \Delta H_0 \in \mathcal{M}(\Omega) \cap H^{-1}(\Omega) \quad in \ the \ weak \ sense \ of \ measures$$

and the Hopf differential of H_0

$$(\mathrm{II}.14) \qquad (\partial_1 H_0)^2 - (\partial_2 H_0)^2 - 2i\partial_1 H_0 \partial_2 H_0$$

is holomorphic in Ω. Therefore, if $\Omega' \subset \Omega$ and $\mu \in L^p(\Omega'), p > 2$, then $\mu = 0$ in Ω'.

This theorem only gives a result for $F_\varepsilon(u_\varepsilon)$ of order $|\log \varepsilon|^2$ and N_ε of order $|\log \varepsilon|$. It asserts that either the limiting measure is concentrated, or it is 0 (it is 0 wherever it does not concentrate). This is to be compared with a result of Sandier and Soret [**SaSo**], which indicates that for minimizers of F_ε with a boundary condition of degree $d \to \infty$, the vortices, because they repel each other, all go to the boundary of Ω as $\varepsilon \to 0$ (thus the limiting vorticity measure is 0 in Ω). This is in contrast with the case of J where the magnetic field has the effect of *confining the vortices* in the domain, even when their number gets large.

From Theorem 9, we deduce that there cannot be a limiting L^∞ density of vortices for large number of vortices in (I.2). For example, the model without magnetic field does not allow for lattices of vortices (of period going to 0) which are expected and observed in the case with magnetic field. This is another major difference.

III. Conclusion

In the small ε asymptotics, the behavior of solutions of Ginzburg-Landau is driven either by the Dirichlet boundary condition in the case without magnetic field, or by the parameter h_{ex} in the case with magnetic field. Thus, in this latter case, there are several *critical fields* $H_{c_1}, H_{c_2}, H_{c_3}$ for which phase transitions occur, which cannot be seen in the framework without magnetic field.

However, we have seen how the models for Ginzburg-Landau with and without magnetic field are related and how the analysis carried out by Bethuel, Brezis and Hélein could be used to obtain results for the full model, in particular when the number of vortices is small (see Section II.1). We have also seen how the analytical problem of constructing vortex-balls and evaluating their energy is crucial, especially when the number of vortices is unbounded. In Section II.2 and II.3 we saw how the number of vortices, contrarily to the model of [**BBH**], can blow up in the case with magnetic field and how useful it then becomes to replace the study of individual vortices by the study of the rescaled vorticity measure, thus deriving limiting vortex-densities as $\varepsilon \to 0$. This led to an obstacle problem for energy minimizers and to a stationarity condition comparable to a result of [**BBH**] for critical points. We finally saw how the magnetic field has a confinement effect on the vortices which allows to stabilize large numbers of them, and how this is not the case if one omits the magnetic coupling (in particular lattices of vortices cannot be observed and vortices would escape to the boundary).

In the following table, we tried to sum up the comparison between the model without magnetic field and the one with.

Ginzburg-Landau BBH	Ginzburg-Landau with magnetic field
Dirichlet boundary condition $u = g$ on $\partial\Omega$ imposes number of vortices	Neumann boundary condition
	the value of h_{ex} imposes number of vortices
a priori bounded number of vortices $F(u) \leq C|\log \varepsilon|$	a priori $J(u, A) \leq C h_{\mathrm{ex}}^2$, possibly unbounded number of vortices $(\leq C \frac{h_{\mathrm{ex}}^2}{|\log \varepsilon|})$ $F(u) \leq C|\log \varepsilon|$ for $h_{\mathrm{ex}} \leq H_{c_1} + O(\log |\log \varepsilon|)$
isolating vortex-balls of size $O(\varepsilon)$ energy study : cost of a vortex $= \pi d_i^2 |\log \varepsilon|$	isolating vortex-balls of size $O(1/|\log \varepsilon|)$ ball construction Sandier/Jerrard $+$ cost of a vortex $\geq \pi |d_i| |\log \varepsilon|$
$+$ *repulsion* between vortices	vortices submitted to *repulsion* $+$ *confinement*
renormalized energy $W \sim -\pi \sum_{i\neq j} d_i d_j \log |a_i - a_j| + ...$ *distances* between vortices $\sim O(1)$	case $h_{\mathrm{ex}} \sim H_{c_1} + O(\log |\log \varepsilon|)$ $w = -\pi \sum_{i\neq j} \log |a_i - a_j| + C \sum_i |a_i|^2$ *distances* $O(\frac{1}{\sqrt{h_{\mathrm{ex}}}}) \ll 1$
energy concentrated around the vortices	$h_{\mathrm{ex}} > H_{c_1}$ energy in vortex-balls $\sim$ energy outside of the balls $\rightarrow$ obstacle problem
discrete problems on point locations (cf other type of problems like critical exponent)	from discrete to continuous vorticity-measure $\mu_\varepsilon = \frac{2\pi \sum_i d_i \delta_{a_i}}{h_{\mathrm{ex}}} \rightarrow$ continuous measure (uniform densities)
time-dependent case $\rightarrow$ ODE on vortex location	time-dependent case $\rightarrow$ Euler equation on vortex-density (?)
critical points "vanishing gradient property" $\nabla \varphi_i(a_i) = 0 \; \forall i$	critical points $\nabla h_\infty \mu_\infty = 0$ ∇h_∞ vanishes on support of μ_∞
critical point of W with deg ± 1 $\Longrightarrow \exists$ a solution of (I.2) ([**Li, PR**]) $\rightarrow$ multiplicity of solutions (cf also [**AB**])	critical point of w $\Longrightarrow \exists$ a solution of (I.4)? (likely) $\nabla h_\infty \mu_\infty = 0 \Longrightarrow \exists$ a solution of (I.4)? multiplicity of locally minimizing solutions [**S3**]

References

[ABO] G. Alberti, S. Baldo and G. Orlandi, Functions with prescribed singularities, preprint.

[AB] L. Almeida and F. Bethuel, Topological Methods for the Ginzburg-Landau Equations, *J. Math. Pures Appl., 77*, (1998), 1-49.

[BBH] F. Bethuel, H. Brezis and F. Hélein, *Ginzburg-Landau Vortices*, Birkhäuser, (1994).

[Be] F. Bethuel, Vortices in Ginzburg-Landau equations. Proceedings of the International Congress of Mathematicians, Vol. III (Berlin, 1998). Doc. Math. 1998, Extra Vol. III, 11–19

[BR1] F. Bethuel and T. Rivière, Vortices for a Variational Problem Related to Superconductivity, *Annales IHP, Analyse non linéaire, 12*, (1995), 243-303.

[BR2] F. Bethuel and T. Rivière, Vorticité dans les modèles de Ginzburg-Landau pour la supraconductivité, *Séminaire E.D.P de l'École Polytechnique*, exposé XVI, (1994).

[J] R. Jerrard, Lower Bounds for Generalized Ginzburg-Landau Functionals, *SIAM J. Math. Anal. 30*, (1999), no. 4, 721–746 .

[JS1] R.L. Jerrard and H.M. Soner, The Jacobian and the Ginzburg-Landau functional, *Calc. Var., 14*, (2002), No 2. 151–191.

[JS2] R.L. Jerrard et H.M. Soner, Limiting behavior of the Ginzburg-Landau functional, *J. Funct. Analysis, 192*, (2002), No 2. 542–561.

[Li] F. H. Lin, Solutions of Ginzburg-Landau equations and critical points of the renormalized energy, *Ann. Inst. H. Poincaré Anal. Non Linéaire 12*, (1995), no. 5, 599–622.

[PR] F. Pacard and T. Rivière, *Linear and Non-linear Aspects of Vortices*, Birkhäuser, (2000).

[Sa] E. Sandier, Lower Bounds for the Energy of Unit Vector Fields and Applications, *J. Functional Analysis, 152, No 2*, (1998), 379-403.

[SS1] E. Sandier and S. Serfaty, Global Minimizers for the Ginzburg-Landau Functional below the First Critical Magnetic Field, *Annales IHP, Analyse non linéaire. 17*, 1, (2000), 119-145.

[SS2] E. Sandier and S. Serfaty, On the Energy of Type-II Superconductors in the Mixed Phase, *Reviews in Math. Phys., 12, No 9*, (2000), 1219-1257.

[SS3] E. Sandier and S. Serfaty, A Rigorous Derivation of a Free-Boundary Problem Arising in Superconductivity, *Annales Sci. Ecole Normale Supérieure*, 4e série, 33, (2000), 561-592.

[SS4] E. Sandier and S. Serfaty, Limiting vorticities for the Ginzburg-Landau equations, to appear in *Duke Math. J.*, (2001).

[SS5] E. Sandier and S. Serfaty, Ginzburg-Landau Minimizers Near the First Critical Field Have Bounded Vorticity, *Calc. Var. PDE, 17, No 1*, (2003), 17–28.

[SS6] E. Sandier and S. Serfaty, The decrease of bulk-superconductivity close to the second critical field in the Ginzburg-Landau model, *SIAM J. Anal, 34, No 4*, (2003), 939–956.

[SaSo] E. Sandier and M. Soret, S^1-valued harmonic maps with high topological degree: asymptotic behavior of the singular set, *Potential Anal. 13* (2000), no. 2, 169–184.

[S1] S. Serfaty, Local Minimizers for the Ginzburg-Landau Energy near Critical Magnetic Field, part I, *Comm. Contemp. Math., 1 , No. 2*, (1999), 213-254.

[S2] S. Serfaty, Local Minimizers for the Ginzburg-Landau Energy near Critical Magnetic Field, part II, *Comm. Contemp. Math., 1 , No. 2*, (1999), 295-333.

[S3] S. Serfaty, Stable Configurations in Superconductivity: Uniqueness, Multiplicity and Vortex-Nucleation, *Arch. for Rat. Mech. Anal., 149*, (1999), 329-365.

COURANT INSTITUTE OF MATHEMATICAL SCIENCES, 251 MERCER ST, NYC, NY 10012
E-mail address: serfaty@cims.nyu.edu

DÉPARTEMENT DE MATHÉMATIQUES, UNIVERSITÉ PARIS-12 VAL-DE-MARNE, 61 AVE DU GÉNÉRAL DE GAULLE, 94010 CRÉTEIL CEDEX, FRANCE
E-mail address: sandier@univ-paris12.fr

Contemporary Mathematics
Volume **350**, 2004

Some Regularity Problems of Stationary Harmonic Maps

Gang Tian

This paper is dedicated to H. Brezis and F. Browder

ABSTRACT. This paper gives a brief tour of some recent progress on the regularity theory of stationary harmonic maps. We will emphasize more on pseudo-holomorphic maps and quaternion maps which arise naturally from differential and symplectic geometry.

In this note, we discuss some recent results on regularity of stationary harmonic maps, particularly, special stationary harmonic maps, such as pseudo-holomorphic maps.

First let us recall some basic facts on harmonic maps. Let M be a submanifold of $\mathbb{R}^n$ with induced metric g and $\Omega \subset \mathbb{R}^m$ is a domain. The space $W^{1,2}(\Omega, M)$ consists of $W^{1,2}(\Omega, \mathbb{R}^n)$-maps $u : \Omega \mapsto \mathbb{R}^n$ such that $u(x) \in M$ for a.e. $x \in \Omega$. If we write such a u as $(u_1, \cdots, u_n)$, then

$$(0.1) \qquad ||u||_{1,2} = \sum_{i=1}^{n} \int_{\Omega} |u_i|^2 + |\nabla u_i|^2 < \infty.$$

We say u weakly harmonic if $u \in W^{1,2}(\Omega, M)$ and for any smooth $\xi : \Omega \mapsto \mathbb{R}^n$ with compact support

$$(0.2) \qquad \frac{d}{dt} \left(\int_{\Omega} |\nabla \pi_M (u + t\xi)|^2 \right) |_{t=0} = 0,$$

where π_M is an orthogonal projection from a neighborhood of M in $\mathbb{R}^n$ onto M. It was proved by F. Helein [**He**] that any weakly harmonic maps from 2-dimensional domains are smooth. However, in higher dimensions, weakly harmonic maps may not be regular in any open subset [**Ri**]. Therefore, we need to impose further constraints on weakly harmonic maps in order to a good regularity theory.

The notion of stationary harmonic maps was introduced by R. Schoen. A stationary harmonic map is a weak harmonic map u such that for any vector field

---

1991 *Mathematics Subject Classification.* Primary 58; Secondary 53.
Key words and phrases. Stationary, harmonic maps, regularity, monotonicity.
Supported partially by NSF grants and a Simons fund.

X with compact support in Ω,

$$(0.3) \qquad \frac{d}{dt}\left(\int_\Omega |\nabla u(x + tX(x))|^2 dx\right)\Big|_{t=0} = 0.$$

That is, it is stationary with respect to variation of the domain. A straightforward computation shows that (0.3) is the same as

$$(0.4) \qquad \int_\Omega |\nabla u|^2 \mathrm{div}X - 2\langle \frac{\partial u}{\partial x_j}, \frac{\partial u}{\partial x_l}\rangle \frac{\partial X_i}{\partial x_l} = 0.$$

Choosing X to be a cut-off of the scaling field, one can derive from this the monotonicity: $r^{2-m}\int_{B(x,r)} |\nabla u|^2$ is non-decreasing in r whenever the ball $B(x,r)$ lies in Ω.

It is clear that weakly harmonic maps are critical points of the energy functional

$$(0.5) \qquad E(u) = \int_\Omega |\nabla u|^2.$$

A map $u \in W^{1,2}(\Omega, M)$ is minimizing if $E(u) \leq E(v)$ for any $v \in W^{1,2}(\Omega, M)$ with $v = u$ on the boundary $\partial\Omega$. One can show that any minimizers are stationary harmonic maps.

In [**SU**], R. Schoen and K. Uhlenbeck proved that there is an $\epsilon = \epsilon(n, M) > 0$ such that for any minimizing harmonic map u on Ω and any ball $B(x,r) \subset \Omega$, whenever $r^{2-n}\int_{B(x,r)} |\nabla u|^2 < \epsilon$, we have $\sup_{B(x,r/2)} |\nabla u| \leq C/r$, where C is a uniform constant depending only n and M. The result of this sort is often called the ϵ-regularity and holds for many geometric partial differential equations. In [**Ev**], C. Evans extended this ϵ-regularity to stationary harmonic maps with $M = S^k$ as the target. The case for general targets was established later by F. Bethuel [**Be**].

The ϵ-regularity has an immediate corollary: If u is a stationary harmonic map and $\mathrm{Sing}(u)$ denotes its singular set, then $\mathcal{H}^{n-2}(\mathrm{Sing}(u)) = 0$, where $\mathcal{H}^k$ denotes the k-dimensional Hausdorff measure. A fundamental problem in studying harmonic maps is on the singular set of stationary harmonic maps. Given a stationary harmonic map, we can ask

(1) what is the dimension of $\mathrm{Sing}(u)$?

(2) What can we say about its structure?

(3) Can it be stratified by disjoint of topological submanifolds or even smooth submanifolds?

We also concern what are limits of a sequence of stationary harmonic maps. All these problems are very difficult, even for special classes of stationary harmonic maps which we will discuss in the following sections. If u is minimizing, a striking theorem of R. Schoen and K. Uhlenbeck states that $\dim(\mathrm{Sing}(u)) \leq n - 3$ [**SU**]. One of key arguments in their proof is to show that any minimizing harmonic map has tangents maps at any given point in the $W^{1,2}$-topology. This was needed to apply Federer's reduction technique and was achieved by making use of certain comparison functions. However, for general stationary harmonic maps, one may not have tangent maps in the $W^{1,2}$-topology. They exist only in a weaker topology. For example, consider the holomorphic map

$$u : (z_1, z_2) \in \mathbb{C}^2\backslash\{0\} \mapsto [z_1, z_2^2] \in \mathbb{C}P^1.$$

One can show that this map is stationary and harmonic. But if we scale $u_\epsilon(z_1, z_2) = [z_1, \epsilon z_2]$, u_ϵ converges weakly to a constant map, so u does not have a tangent map in the $W^{1,2}$-topology at 0. Therefore, one needs new techniques in the case of general stationary harmonic maps, even in the case of pseudo-holomorphic maps.

1. Tangent maps

Let $u : \Omega \mapsto M$ be a stationary harmonic map. For each $x \in \Omega$, scale $u_{x,\epsilon}(y) = u(x + \epsilon y)$, then for any $R > 0$,

$$\int_{B(0,R)} |\nabla u_{x,\epsilon}|^2 = \epsilon^{2-n} \int_{B(x,\epsilon R)} |\nabla u|^2.$$

Since $r^{2-m} \int_{B(x,r)} |\nabla u|^2$ is non-decreasing in r, we have a uniform bound on $\int_{B(0,R)} |\nabla u_{x,\epsilon}|^2$ for any fixed R and ϵ sufficiently small. It follows from the ϵ-regularity that any sequence $\{u_{x,\epsilon_i}\}$ with $\lim \epsilon_i = 0$ has a subsequence, say $u_{x,j}$, converging to a smooth harmonic map u_T on $\mathbb{R}^n \backslash \mathfrak{B}(u_T)$ in the open-compact topology, where $\mathfrak{B}(u_T)$ is a closed set such that its intersection with any compact set has finite $(n-2)$-dimensional Hausdorff measure. Such a map u_T may depend on the choice of the subsequence and clearly has bounded energy on any compact subset of $\mathbb{R}^n$. Moreover, the set $\mathfrak{B}(u_T)$ is called blow-up set and can be explicitly given by

$$\{y \in \mathbb{R}^n \mid \lim_{r \to 0} \overline{\lim}_{j \to \infty} r^{2-n} \int_{B(y,r)} |\nabla u_{x,j}|^2 \geq \epsilon\},$$

where ϵ is the constant given by the ϵ-regularity of Bethuel. By taking a subsequence again if necessary, we may also assume that $|\nabla u_{x,j}|^2 dv$ converge weakly to μ as measures. It was shown in [**Li**] that the blow-up set $\mathfrak{B}(u_T)$ is $(n-2)$-rectifiable and the limiting measure μ is of the form $|\nabla u|^2 dv + \theta(y)\mathcal{H}^{n-2} \llcorner \mathfrak{B}(u_T)$, where θ is called density function. The main reason for this is the monotonicity of $r^{2-n} \int_{B(x,r)} |\nabla u|^2$, because it implies that $r^{2-n}\mu(B(x,r))$ is non-decreasing, then it follows from a result of Prisse that $\mathfrak{B}(u_T)$ is rectifiable.

We do not know if u_T is stationary, and it is likely that it is not in general. However, we have the following blow-up formula first proved in [**LT**]: For any vector field X with compact support,

$$(1.1) \qquad -\int_{\mathfrak{B}(u_T)} \mathrm{div}_{\mathfrak{B}} X \theta d\mathcal{H}^{n-2} = \int_{\mathbb{R}^n} |\nabla u_T|^2 \mathrm{div} X - 2\langle \frac{\partial u_T}{\partial x_j}, \frac{\partial u_T}{\partial x_l} \rangle \frac{\partial X_i}{\partial x_l}.$$

This formula has a number of corollaries. First, it implies that u_T is stationary if and only if $\mathfrak{B}(u_T)$ together with the density θ is stationary, that is, $(\mathfrak{B}(u_T), \theta)$ has vanishing mean curvature and no boundary in a generalized sense. Secondly, we deduced in [**LT**]

THEOREM 1.1. *Let u be a stationary map and u_T be one of its limiting map together with the blow-up set $\mathfrak{B}(u_T)$ and the density θ as described above. Then $u_T(y) = u_T(y/|y|)$, $\theta(y) = \theta(y/|y|)$ and $\mathfrak{B}(u_T)$ is invariant under radial scaling. We call $(u_T, \mathfrak{B}(u_T), \theta)$ a (generalized) tangent map of u at $x \in \Omega$. Further, this tangent map is stationary in a common sense.*

If u is minimizing, Schoen and Uhlenbeck showed [**SU**] that $\mathcal{H}^{n-2}(\mathfrak{B}(u_T)) = 0$, so this (generalized) tangent map is just u_T. There may be some condition imposed on M such that we always have $\mathcal{H}^{n-2}(\mathfrak{B}(u_T)) = 0$ [**Li**]. Then one can apply Federer's reduction and show as Schoen and Uhlenbeck did in [**SU**] that the singular set of u is of at least codimension 3. For instance, if M has no harmonic spheres, then $\mathcal{H}^{n-2}(\mathfrak{B}(u_T)) = 0$ always holds (cf. [**Li**]).

2. Pseudo-holomorphic maps

In this section, we discuss a special class of stationary harmonic maps. Let (M, ω) be a symplectic manifold, where ω is a non-degenerate closed 2-form. Let J be any compatible almost complex structure, that is, $J : TM \mapsto TM$ is an endomorphism such that $J^2 = -Id$ and $\omega(Ju, Jv) = \omega(u, v)$ for any tangent vectors u and v. Then we have a compatible metric $g(u, v) = \omega(u, Jv)$. The simplest symplectic manifold is $\mathbb{R}^{2m}$ with the standard symplectic form $\omega_0 = \sum_{i=1}^{m} dx_i \wedge dx_{m+i}$. Let us fix an almost complex structure J_0 on $\mathbb{R}^{2m}$. A (weakly) pseudo-holomorphic map, with respect to J_0 on Ω and J on the target M, is a $W^{1,2}$-map $u : \Omega \subset \mathbb{C}^m \mapsto M$ whose restriction to any ball $B \subset \Omega$ is a limit of smooth maps from B into M in the $W^{1,2}$-topology and which satisfies

$$(2.1) \qquad\qquad du \cdot J_0 = J \cdot du.$$

If $u : \Omega \mapsto M$ is a limit of smooth (J_0, J)-holomorphic maps in the $W^{1,2}$-topology, then u is (weakly) pseudo-holomorphic. Note that such a map u is smooth outside a closed subset of Ω of codimension at least two.

LEMMA 2.1. *Any weakly pseudo-holomorphic maps are stationary harmonic maps.*[1]

PROOF. This lemma can be easily proved by using the fact: If u is a $W^{1,2}(\Omega, M)$-map, then

$$\int_{\Omega} |\nabla u|^2 \omega_0^n \geq 2n \int_{\Omega} u^* \omega \wedge \omega_0^{n-1},$$

where ω_0 is the standard symplectic form on Ω, and the equality holds if and only if u is weakly pseudo-holomorphic. $\qquad\square$

Of course, any pseudo-holomorphic maps from Riemann surfaces are smooth. In higher dimensions, pseudo-holomorphic maps may have singularities. If both J_0 and J are integrable, that is, Ω and M are complex manifolds with induced complex structures by J_0 and J respectively, then it was a classical result that any pseudo-holomorphic map u extends to a meromorphic map, in particular, there is an analytic subvariety $S \subset \Omega$ such that u is holomorphic on $\Omega \backslash S$. We conjecture that the same is true for weakly pseudo-holomorphic maps, that is, if u is any weakly pseudo-holomorphic map from Ω into M, then there is a locally closed set $S \subset \Omega$ such that u extends to a smooth map from $\Omega \backslash S$ and $S = \coprod_{i=0}^{m-2} S_i$ is a stratified set with each stratum S_i (possibly empty) being a J_0-submanifold of complex dimension i. If $m = 2$, then it says that u has only isolated singularities. This is a hard conjecture, nevertheless, Riviere and I was able to prove the following

THEOREM 2.2. [**RiT**] *Any (weakly) pseudo-holomorphic map from $\Omega \subset \mathbb{C}^2$ into an algebraic manifold M has only isolated singularities.*

This is sharp since there do exist examples of holomorphic maps with isolated singularities from $\mathbb{C}^2$. We do expect to remove the condition that the target manifold is algebraic. In fact, many arguments in the proof work for symplectic manifolds with an almost complex structure. It is also interesting to see if our arguments can be generalized to pseudo-holomorphic maps from higher dimensional domains.

---

[1]It is likely that any smooth pseudo-holomorphic map from $\Omega \backslash S$ with finite energy, where S is a closed subset of locally finite $(n-2)$-dimensional Hausdorff measure extends to a stationary harmonic map from Ω into M.

The proof of this theorem can be briefly described as follows: Let us assume that M is S^2 for simplicity. The general case can be reduced to this special case. Since u is stationary harmonic, it follows from the result of Bethuel [**Be**] that u is smooth outside a closed subset S defined by

$$S = \{x \in \Omega \mid \lim_{r \to 0} r^{2-2m} \int_{B(x,r)} |\nabla u|^2 \geq \epsilon\},$$

where ϵ is the constant given by the ϵ-regularity of Bethuel. One can show easily that S has vanishing 2-dimensional Hausdorff measure. For almost all $p \in S^2$, the level set $u^{-1}(p)$ is a J_0-holomorphic curve in $\Omega \backslash S$. Next we show that $u^{-1}(p)$ extends to an integral, rectifiable current whose almost all tangent spaces are J_0-invariant, that is, a $(1,1)$-current. One of key technical points is to show the regularity of $(1,1)$-currents in an almost complex 4-manifold. We proved that such $(1,1)$-currents have to be J_0-holomorphic curves. On the other hand, one can show that the singular set of u lies in the intersection of the closures of $\overline{u^{-1}(p)}$ and $\overline{u^{-1}(q)}$ for two generic distinct p and q. So the singular set of u is isolated. The actual proof is lengthy and we refer the readers to [**Ri**] for details.

3. Quaternion maps

Quaternion maps between hyperKähler manifolds arose from the study of higher dimensional gauge theory ([**FF**], [**Ch**]). Weakly quaternion maps and their regularity were first studied in [**LT**] and [**CL**]. They are stationary harmonic maps and minimize the energy among maps with fixed homology class as in the case of pseudo-holomorphic maps.

Recall that a hyperKähler manifold is a Riemannian manifold (M, g) with three parallel complex structures J_1, J_2 and J_3 compatible with the metric g such that

$$J_1^2 = J_2^2 = J_3^2 = J_1 J_2 J_3 = -Id.$$

A hyperKähler manifold has to be of dimension $4m$ and has an natural S^2-family of complex structures. The simplest hyperKähler manifold is the quaternion space $\mathbb{H}^m$. Now assume that Ω is a domain in $\mathbb{H}^m$ and J_{01}, J_{02}, J_{03} be the standard complex structures given by the quaternion structure.

A (weakly) quaternion map between two hyperKähler manifolds Ω and M is a $W^{1,2}$-map $u : \Omega \subset \mathfrak{H}^m \mapsto M$ whose restriction to any ball $B \subset \Omega$ is a limit of smooth maps from B into M in the $W^{1,2}$-topology and which satisfies

$$(3.1) \qquad du = A^{\alpha\beta} J_\alpha \cdot du \cdot J_{0\beta},$$

where $A = (A^{\alpha\beta})$ is an orthogonal matrix with determinant one. If $u : \Omega \mapsto M$ is a limit of smooth quaternion maps in the $W^{1,2}$-topology, then u is (weakly) quaternion. Note that such a map u is smooth outside a closed subset of Ω of codimension at least two.

LEMMA 3.1. *Any weakly quaternion maps are stationary harmonic maps.*[2]

---

[2]It is possible that any smooth quaternion map from $\Omega \backslash S$ with finite energy, where S is a closed subset of locally finite $(n-2)$-dimensional Hausdorff measure, extends to a stationary harmonic map from Ω into M. The reasons for this being true were given in [**LT**]. But a complete proof is needed. We will discuss it in a future paper.

We refer the readers to [**CL**] for its proof. The key point is the following identity: Define symplectic forms

$$\omega_{0\alpha}(v, v') = g_0(v, J_{0\alpha}v'), \quad \omega_\alpha(v, v') = g(v, J_\alpha v'),$$

where g_0 and g are the metrics on Ω and M, respectively. Then for any quaternion map u, we have

$$(3.2) \qquad E(u) = \frac{1}{2}\sum_{i=1}^{3}\int_\Omega A^{\alpha\beta}u^*\omega_\alpha \wedge \omega_{0\beta} + \frac{1}{8}\int_\Omega |du - A^{\alpha\beta}J_\alpha \cdot du \cdot J_{0\beta}|^2.$$

This implies that quaternion maps minimize the energy among maps with fixed homology class. If $\Omega \subset \mathbb{H}$ and $M = \mathbb{H}$, then u is a quaternion function and the quaternion map equation reduces to

$$\frac{\partial u}{\partial q} = 0,$$

where $\frac{\partial}{\partial q} = \frac{\partial}{\partial x_0} - i\frac{\partial}{\partial x_1} - j\frac{\partial}{\partial x_2} - k\frac{\partial}{\partial x_3}$, that is, $\bar{u}$ is quaternion holomorphic. Also, if u is anti-holomorphic with respect to some complex structures on Ω and M, then u is quaternion, but the converse is not true. A criterion was given in [**CL**].

An interesting question is to analyze structure of the singularity of a weakly quaternion map. This question has been explored and significant progress has been made in [**CL**]. By Bethuel's theorem, the singular set $S(u)$ of a weakly quaternion map has vanishing $(n-2)$-dimensional Hausdorff measure. In fact, we expect that $S(u)$ has Hausdorff codimension at least three or possibly four. If u is a tangent quaternion map with 1-dimensional singular set, then u is the pull-back of a map $\varphi : S^2 \mapsto M$ under $H \cdot \pi$, where $\pi : \mathbb{R}^4 = \mathbb{R} \times \mathbb{R}^3 \mapsto \mathbb{R}^3$ is the projection onto the second factor and $H : \mathbb{R}^3 \mapsto S^2$ is the Hopf map, furthermore, it follows from u being quaternion that φ satisfies the equation

$$(3.3) \qquad -\frac{1}{\sin\alpha}\frac{\partial\varphi}{\partial\theta} = A^{\beta\gamma}x_\beta J_\gamma\frac{\partial u}{\partial\alpha},$$

where θ and α are spherical coordinates of S^2 as the unit sphere in $\mathbb{R}^3$, and $x_1 = \sin\alpha\cos\theta$, $x_2 = \sin\alpha\sin\theta$, $x_3 = \cos\alpha$. This is a first order equation analogous to the Cauchy-Riemann equation. If this equation does not have non-trivial solution for hyperKähler manifolds, then the singular set of u should be of Hausdorff codimension at least 4. If the domain is 4-dimensional, then u has only isolated singularities. Even if (3.3) has a solution, the resulting u may not be stationary, so u may still have a chance to have codimension 4 singular set.

Finally, let me mention one geometric motivation for studying quaternion maps. When Ω is of real dimension 4, the quaternion equation is an elliptic system of first order, similar to the Cauchy-Riemann equation on Riemann surfaces. We hope to construct deformation invariants for hyperKähler manifolds by using the quaternion equation, which are analogous to the Gromov-Witten invariants for symplectic manifolds. More precisely, consider the moduli space of quaternion maps from either a K3 surface or an abelian surface into a hyperKähler manifold M with fixed homology class, we hope that after compactifying it appropriately, one can construct a fundamental class this moduli as one did in the Gromov-Witten theory (cf. [**RT**]). The key issue is how to compactify this space, analytically, it amounts to a comprehensive understanding of singularity of weakly quaternion maps.

Let u_i be a sequence of quaternion maps with uniformly bounded energy. By taking a subsequence if necessary, one can show that u_i converges to a limiting map u_∞ outside a closed subset $\mathfrak{B}(u_i)$ of finite $(4m-2)$-dimensional Hausdorff measure. We may also require $|\nabla u_i|^2 dv$ converges weakly to $|\nabla u_\infty|^2 dv + \mu$, where μ is a measure supported along $\mathfrak{B}(u_i)$. In fact,

$$\mu(S) = \int_{S \cap \mathfrak{B}(u_i)} \theta dH^{4m-2},$$

where θ is the density function given by

$$\theta(x) = \lim_{r \to 0} \lim_{i \to \infty} r^{2-4m} \int_{B(x,r)} |\nabla u_i|^2.$$

Note that $\theta \geq \epsilon$. One can show that θ is integer-valued modulo some universal constant. We may call the triple $(u_\infty, \mathfrak{B}(u_i), \theta)$ a generalized quaternion map. Note that the moduli space of generalized quaternion maps with fixed homology class is compact. The problem is about regularity of generalized quaternion maps:
1. Is the singular set of u_∞ of condimension at least 4? In particular, when $m = 1$, u_∞ may have only isolated singularities. If so, one can further classify them.
2. We expect that $(\mathfrak{B}(u_i), \theta)$ is an integral current calibrated by

$$(\sum_{i=1}^{3} \alpha_i \omega_i)^{2m-1}/(2m-1)!$$

for some α_i with $\alpha_1^2 + \alpha_2^2 + \alpha_3^2 = 1$. If so, by a result of J. King, one can deduce that $(\mathfrak{B}(u_i), \theta)$ is a holomorphic subvariety.

References

[Be]   F. Bethuel, *On the singular set of stationary harmonic maps*, Manuscripta Math., 78 (1993), no. 4, 417–443.

[Ch]   J.Y. Chen, *Complex anti-self-dual connections on product of Calabi-Yau surfaces and tri-holomorphic curves*, preprint, 1998.

[CL]   J.Y. Chen and Jiayu Li, *Quaternionic maps between hyperkhler manifolds*, J. Differential Geom., 55 (2000), no. 2, 355–384.

[Ev]   L.C. Evens, *Partial regularity for stationary harmonic maps into spheres*, Arch. Rational Mech. Anal., 116 (1991), no. 2, 101–113.

[FF]   J.M. Figueroa-O'Farrill, C. Khl and B. Spence, *Supersymmetric Yang-Mills, octonionic instantons and triholomorphic curves*, Nuclear Phys., B 521 (1998), no. 3, 419–443.

[He]   F. Helein, *Regularity of weakly harmonic maps from a surface into a manifold with symmetries*, Manuscripta Math., 70 (1991), no. 2, 203–218.

[LT]   Jiayu Li and G. Tian, *A blow-up formula for stationary harmonic maps*, Internat. Math. Res. Notices, 1998, no. 14, 735–755.

[Li]   F.H. Lin, *Gradient estimates and blow-up analysis for stationary harmonic maps*, Ann. of Math., 149 (1999), no. 3, 785–829.

[Ri]   T. Riviere, *Everywhere discontinuous harmonic maps into spheres*, Acta Math., 175 (1995), no. 2, 197–226.

[RT]   Y.B. Ruan and G. Tian, *A mathematical theory of quantum cohomology*, J. Differential Geom., 42 (1995), no. 2, 259–367.

[RiT]   T. Riviere and G. Tian, The singular set of J-holomorhic maps into algebraic manifolds, preprint, 2001.

[SU]   R. Schoen and K. Uhlenbeck, *A regularity theory for harmonic maps*, J. Differential Geom., 17 (1982), no. 2, 307–335.

DEPARTMENT OF MATHEMATICS, MASSACHUSETTS INSTITUTE OF TECHNOLOGY AND BEIJING UNIVERSITY

E-mail address: `tian@math.mit.edu`

Titles in This Series

For a complete list of titles in this series, visit the
AMS Bookstore at **www.ams.org/bookstore/**.